Methods in Molecular Biology

Series Editor
John M. Walker
School of Life Sciences
University of Hertfordshire
Hatfield, Hertfordshire, AL10 9AB, UK

For further volumes:
http://www.springer.com/series/7651

Metabolic Flux Analysis

Methods and Protocols

Edited by

Jens O. Krömer

Centre for Microbial Electrosynthesis (CEMES), Advanced Water Management Centre, The University of Queensland, Brisbane, QLD, Australia

Lars K. Nielsen

Australian Institute for Bioengineering and Nanotechnology (AIBN), The University of Queensland, Brisbane, QLD, Australia

Lars M. Blank

Institute of Applied Microbiology, RWTH Aachen University, Aachen, Germany

Humana Press

Editors
Jens O. Krömer
Centre for Microbial Electrosynthesis (CEMES),
Advanced Water Management Centre
The University of Queensland
Brisbane, QLD, Australia

Lars K. Nielsen
Australian Institute for Bioengineering
and Nanotechnology (AIBN)
The University of Queensland
Brisbane, QLD, Australia

Lars M. Blank
Institute of Applied Microbiology
RWTH Aachen University
Aachen, Germany

ISSN 1064-3745 ISSN 1940-6029 (electronic)
ISBN 978-1-4939-1169-1 ISBN 978-1-4939-1170-7 (eBook)
DOI 10.1007/978-1-4939-1170-7
Springer New York Heidelberg Dordrecht London

Library of Congress Control Number: 2014945361

Printed on acid-free paper

Humana Press is a brand of Springer
Springer is part of Springer Science+Business Media (www.springer.com)

Preface

Systems Biology has profoundly changed the way we approach the fundamental problem of understanding cellular processes. Besides this, it has transformed the way we design new strains for industrial applications in the field of metabolic engineering. Within this systems biology framework, we heavily rely on the so-called omics tools. These range from sequencing of an organism's genome (*genomics*), the measurement of gene transcription (*transcriptomics*) and protein abundance (*proteomics*) to the product and substrate concentrations in metabolism (*metabolomics*). Each of these tools delivers a dataset that represents a snapshot of information storage, information flow, available machinery, and building materials at a given time. However, they mostly fail to describe the current activities in the systems that ultimately represent the metabolic phenotype. In order for these tools to be quantitative, the absolute concentrations of enzymes and metabolites would be needed alongside the in vivo thermodynamic and kinetic parameters for all reactions in the network, including information about the regulatory status of each step along a pathway. The reason for this very complex puzzle is the fact that metabolic activity of (or flux through) a reaction is described as a function of *capacity-based* (enzyme abundance) and *kinetics-based* (enzyme activity) regulation:

$$v_i = f(\overbrace{E_i,}^{capacity} \underbrace{k_i, S_i, P_i, I_i}_{kinetics})$$

Where the enzyme concentration, E_i, regulates the flux on a capacity basis. The capacity is dictated by the regulation on transcriptional and translational level and also through the stability and activity of the protein (degradation, phosphorylation). E_i can be measured for many proteins with quantitative proteomics. The analysis of the kinetic parameters, however, is far more challenging. Current technology allows for the measurement of some of substrates, S_i, products, P_i, and effector molecules, I_i. But despite the technical progress in quantitative metabolite analysis not all concentrations are accessible with today's technology. More importantly, the in vivo (rather than *in-vitro*) kinetic parameters of enzymes, k_i, are only available for very few enzymes.

Because of this current limitation for the prediction of metabolic fluxes from a measured inventory of the cell, an alternative technology emerged and matured over the last decades: Metabolic Flux Analysis.

Metabolic fluxes are the end result of the interplay of gene expression, protein concentration, protein kinetics, regulation, and metabolite concentrations (thermodynamic driving forces)—basically the metabolic phenotype. A quantitative representation of this phenotype can be estimated with metabolic flux analysis, which was in analogy to other "omics-type" analyses, termed "fluxomics." The term fluxomics bundles a wide variety of tools that are at different stages of development. While some technology is well developed (steady-state flux analysis using flux balancing (FBA) or stable isotope metabolic flux analysis (^{13}C-MFA)) others are still specialized tools that only a few expert groups are currently able to apply (non-steady-state approaches).

$$\overbrace{\frac{dC_{MET_i}}{dt}}^{\textit{dynamic}} = \underbrace{S \times v \approx 0}_{\textit{stationary}}$$

This book is dedicated to Metabolic Flux Analysis in systems that can be studied under metabolic steady-state or pseudo-steady-state conditions, also called stationary approaches. Under these conditions the change of intracellular metabolite concentrations (C_{MET}) as a function of the stoichiometric matrix S and the flux vector v can be approximated to zero. This approximation converts the equations system from a differential to a linear one and greatly simplifies its solution.

We hope that this book will open the field of metabolic flux analysis to a wider scientific community by providing helpful tricks to those who want to start a new flux analysis project but are overwhelmed by the complexity of the approach. We tried to break it up into manageable bits of information that can be implemented step by step following the methods chapters provided. The overview chapters are there to explain the necessary basics to get started and also to review what has been achieved within a few model organisms.

The book is divided into several thematic parts. Part I focuses on the fundamental characteristics of the underlying networks and lays the basis for the new academic and student in the field to start developing their own metabolic networks for flux balance analysis. There has been significant progress in this area, and new algorithms (such as SEED) allow to automate the network building to a large extent. Part II then looks into the application of quantitative metabolite data and thermodynamic principles to constrain the solution space for flux balance analysis (FBA) in order to obtain more realistic results. The third and largest part of book (Part III) provides the experimental toolbox to conduct different types of flux analysis experiments. To start with, we highlight how to determine the biomass composition of a complex system, the marine sponge *Amphimedon queenslandica.* Determining the biomass composition is one of the most important steps in FBA, since many net-fluxes in a system get determined by growth rate and biomass composition. A prerequisite for thermodynamically constraining the solution space (Part II) is the conduction of quantitative metabolomics experiments. Special consideration is given to challenges that the experimenter faces in complex environments such as mammalian cell cultures or when trying to downsize flux analysis for high-throughput screening of *Saccharomyces cerevisiae* using Flux-P (a FIAT-FLUX automation). Finally three different ways to analyze labelling for ^{13}C-fluxomics are presented: Gas Chromatography-Mass Spectrometry (GC-MS), highlighted for *Corynebacterium glutamicum*, Nuclear Magnetic Resonance (NMR) and Membrane Inlet Mass Spectrometry (MI–MS) for in situ CO_2 labelling analysis for ^{13}C-fluxomics.

Part IV is dedicated to the processing of data from ^{13}C experiments in order to achieve flux distributions. Here it is important to correct for naturally occurring isotopes in the analytes. One state-of-the-art tool for the estimation of the fluxes (OpenFlux) is demonstrated in detail, while a new tool to quickly create flux maps for visualization rounds of this part.

Finally, Part V provides three overview chapters that summarize some key findings through FBA and ^{13}C-fluxomics in the important biotechnological organism *Escherichia coli* as well as recent progress with ^{13}C-fluxomics in the yeast *Pichia pastoris.*

The field is fast developing and we are sure while writing this, some other useful tools are appearing. Nevertheless, we are confident that this book will provide a solid basis to everybody interested in Metabolic Flux Analysis and provides protocols that cover a range of relevant organisms currently used in the field.

We thank all the contributors to this book and thank all our colleagues for their useful discussions and support during the journey of finalizing this book.

Brisbane, QLD, Australia — ***Jens O. Krömer***
Brisbane, QLD, Australia — ***Lars K. Nielsen***
Aachen, Germany — ***Lars M. Blank***

Contents

Contributors

JOAN ALBIOL • *Department of Chemical Engineering, Escola d'Enginyeria, Universitat Autònoma de Barcelona, Bellaterra, Catalonia, Spain*
JUDITH BECKER • *Institute of Systems Biotechnology, Saarland University, Saarbruecken, Germany*
LARS M. BLANK • *Institute of Applied Microbiology, RWTH Aachen University, Aachen, Germany*
RIC COLASANTI • *Computation Institute, University Chicago, Chicago, IL, USA*
JAN C. VAN DAM • *Department of Biotechnology, Delft University of Technology, Delft, Netherlands*
BERNARD DEGNAN • *School of Biological Sciences, The University of Queensland, Brisbane, QLD, Australia*
SANDIE DEGNAN • *School of Biological Sciences, The University of Queensland, Brisbane, QLD, Australia*
BIRGITTA E. EBERT • *Institute of Applied Microbiology, RWTH Aachen University, Aachen, Germany*
JANAKA N. EDIRISINGHE • *Computation Institute, University Chicago, Chicago, IL, USA; MCS Division, Argonne National Laboratory, Argonne, IL, USA*
JOSÉ P. FARIA • *Computation Institute, University Chicago, Chicago, IL, USA; Institute for Biotechnology and Bioengineering/Centre of Biological Engineering, University of Minho, Braga, Portugal*
PAU FERRER • *Escola d'Enginyeria, Edifici Q, Universitat Autònoma de Barcelona, Bellaterra, Catalonia, Spain; Department of Chemical Engineering, Escola d'Enginyeria, Universitat Autònoma de Barcelona, Bellaterra, Catalonia, Spain*
WALTER M. VAN GULIK • *Department of Biotechnology, Delft University of Technology, Delft, Netherlands*
VASSILY HATZIMANIKATIS • *Laboratory of Computational Systems Biotechnology, École Polytechnique Fédérale de Lausanne (EFPL), EPFL SB ISIC LCSB, Lausanne, Switzerland*
JOSEPH J. HEIJNEN • *Department of Biotechnology, Delft University of Technology, Delft, Netherlands*
CHRISTOPHER HENRY • *Computation Institute, University Chicago, Chicago, IL, USA; MCS Division, Argonne National Laboratory, Argonne, IL, USA*
PAULA JOUHTEN • *VTT Technical Research Centre of Finland, Helsinki, Finland*
ASTRID JUNKER • *Leibniz-Institut für Pflanzengenetik und Kulturpflanzenforschung (IPK), Gatersleben, Germany*
BJÖRN H. JUNKER • *Institute of Pharmacy, Martin-Luther-University, Halle/Saale, Germany*
TAHMINEH KHAZAEI • *Bioengineering, California Technology Institute, Pasadena, CA, USA*
CHRISTIAN KRACH • *Leibniz-Institut für Pflanzengenetik und Kulturpflanzenforschung (IPK), Gatersleben, Germany*
JENS O. KRÖMER • *Centre for Microbial Electrosynthesis (CEMES), Advanced Water Management Centre, The University of Queensland, Brisbane, QLD, Australia*

LARS KUEPFER • *Applied Microbiology, RWTH Aachen University, Aachen, Germany*
FABIEN LETISSE • *Université de Toulouse, INSA, UPS, INP, LISBP, Toulouse, France; INRA UMR792, Ingénierie des Systèmes Biologiques et des Procédés, Toulouse, France; CNRS, UMR5504, Toulouse, France*
HANNU MAAHEIMO • *VTT Technical Research Centre of Finland, Helsinki, Finland*
VERÓNICA SOFÍA MARTÍNEZ • *Australian Institute for Bioengineering and Nanotechnology (AIBN), The University of Queensland, Brisbane, QLD, Australia*
YU MATSUOKA • *Kyushu Institute of Technology, Iizuka, Fukuoka, Japan*
PIERRE MILLARD • *Université de Toulouse, INSA, UPS, INP, LISBP, Toulouse, France; INRA UMR792, Ingénierie des Systèmes Biologiques et des Procédés, Toulouse, France; CNRS, UMR5504, Toulouse, France*
LARS K. NIELSEN • *Australian Institute for Bioengineering and Nanotechnology (AIBN), The University of Queensland, Brisbane, QLD, Australia*
ANGELA TEN PIERICK • *Department of Biotechnology, Delft University of Technology, Delft, Netherlands*
JEAN-CHARLES PORTAIS • *Université de Toulouse, INSA, UPS, INP, LISBP, Toulouse, France; INRA UMR792, Ingénierie des Systèmes Biologiques et des Procédés, Toulouse, France; CNRS, UMR5504, Toulouse, France*
LAKE-EE QUEK • *Australian Institute for Bioengineering and Nanotechnology (AIBN), The University of Queensland, Brisbane, QLD, Australia*
COR RAS • *Department of Biotechnology, Delft University of Technology, Delft, Netherlands*
HENDRIK ROHN • *Leibniz-Institut für Pflanzengenetik und Kulturpflanzenforschung (IPK), Gatersleben, Germany*
FALK SCHREIBER • *Leibniz-Institut für Pflanzengenetik und Kulturpflanzenforschung (IPK), Gatersleben, Germany*
SAM SEAVER • *Computation Institute, University Chicago, Chicago, IL, USA; MCS Division, Argonne National Laboratory, Argonne, IL, USA*
REZA MALEKI SEIFAR • *Department of Biotechnology, Delft University of Technology, Delft, Netherlands*
KAZUYUKI SHIMIZU • *Kyushu Institute of Technology, Iizuka, Fukuoka, Japan; Institute of Advanced Biosciences, Keio University, Tsuruoka, Yamagata, Japan*
KENG CHER SOH • *École Polytechnique Fédérale de Lausanne (EFPL), Lausanne, Switzerland*
SERGUEI SOKOL • *Université de Toulouse, INSA, UPS, INP, LISBP, Toulouse, France; INRA UMR792, Ingénierie des Systèmes Biologiques et des Procédés, Toulouse, France; CNRS, UMR5504, Toulouse, France*
S. ALJOSCHA WAHL • *Department of Biotechnology, Delft University of Technology, Delft, Netherlands*
JABIN WATSON • *School of Biological Sciences, The University of Queensland, Brisbane, QLD, Australia*
CHRISTOPH WITTMANN • *Institute of Systems Biotechnology, Saarland University, Saarbruecken, Germany*
FANGFANG XIA • *Computation Institute, University Chicago, Chicago, IL, USA; MCS Division, Argonne National Laboratory, Argonne, IL, USA*
TAE HOON YANG • *Genomatica Inc., San Diego, CA, USA*

Part I

Models: Stoichiometry, Network Generation

Chapter 1

Stoichiometric Modelling of Microbial Metabolism

Lars Kuepfer

Abstract

Stoichiometric models describe cellular biochemistry with systems of linear equations. The models which are fundamentally based on the steady-state assumption are comparatively easy to construct and can be applied to networks up to genome scale. Fluxes are inherent variables in stoichiometric models and linear optimization can be used to identify intracellular flux distributions. Great caution, however, has to be paid to the selection of the specific objective function which inevitably implies the existence of a specific global cellular rationale. On the other hand, stoichiometric models provide an analytical platform for contextualization of experimental data. Equally important, the stoichiometric models can be used for structural analyses of metabolic networks as such supporting for example rational model-driven strategies in metabolic engineering.

Key words Stoichiometric model, Metabolic network, Metabolic engineering, Objective function, Data integration, Flux distribution, Fluxes, FBA

1 Introduction

1.1 Metabolic Network Models

Metabolic networks are relatively well understood when compared to other biological systems. This is because both the metabolites and the linking reactions converting the different constituents of metabolism are generally known. As such, the underlying structure of metabolic networks is well established and the connectivity, which basically represents fundamental cellular biochemistry, is largely understood. Metabolism plays a key role in homeostasis of living organisms since it continuously processes extracellular nutrients to form building blocks for cellular biomass and, equally important, to provide energetic equivalents for cellular fuelling. Metabolic networks vary in size from models of central carbon metabolism (~40 metabolites, 80 reactions) to genome-scale models of microbial metabolism (>500 metabolites, >1,000 reactions). Differences in the structure of metabolism between different microorganisms reflect for instance adaptation of organisms to specific environmental conditions and in particular nutrient availability.

Jens O. Krömer et al. (eds.), *Metabolic Flux Analysis: Methods and Protocols*, Methods in Molecular Biology, vol. 1191, DOI 10.1007/978-1-4939-1170-7_1, © Springer Science+Business Media New York 2014

In computational systems biology, stoichiometric models have for long been used to quantitatively analyze metabolic networks [1–4]. This is because stoichiometric models are on the one hand comparatively easy to construct, but on the other hand they represent cellular biochemistry in a mathematical formal way. In this chapter, we describe how stoichiometric models are constructed and what needs to be taken into account in order to perform simulations of actual biological relevance.

1.2 Model Construction

Generally speaking, there are three main approaches in microbial systems biology [5]:

1. *Interaction-based networks*, which focus on network connectivity but neglect network stoichiometry and assume stationarity of the system: Interaction-based networks are mainly used to describe propagation of information in large-scale systems such as gene transcription or proteomics.
2. *Dynamic models*, which are based on ordinary differential equations (ODEs) and which contain kinetic information about temporal state changes in a system: Dynamic models are for example used to quantitatively describe the dynamics of intracellular signalling cascades.
3. *Stoichiometric models* represent fundamental cellular biochemistry in metabolic networks; they quantitatively describe intracellular mass flow at steady state; that is, they assume stationary systems.

Notably the choice of a specific modelling approach has to be made with respect to the biological question which is under investigation and the experimental data available. While stoichiometric models clearly lack kinetic information when compared to dynamic models they nevertheless allow a quantitative representation of network structure up to genome scale. As such, stoichiometric models combine favorable properties of both interaction-based networks and dynamic models.

1.3 Construction of Stoichiometric Models

Consider a small network as the one shown in Fig. 1: This network considers five metabolites (A, B, C, D, E), six intracellular reaction rates (v_1: $A \rightarrow B$; v_2: $B \rightarrow C$; v_3: $B \leftrightarrow D$; v_4: $C \rightarrow D$; v_5: $C \rightarrow E$; v_6: $D \rightarrow E$), and four exchange reactions (v_{E1}: $\rightarrow A$; v_{E2}: $B \rightarrow$; v_{E3}: $C \rightarrow$; v_{E4}: $D \rightarrow$) which link the extracellular environment to the intracellular space. The first step in model building is to set up the system of equations describing the mass conservation of each of the metabolites. At this stage of the model-building process, we do not know much about the system and its behavior. We therefore have to assume that the concentrations, which represent intracellular pools of metabolites, change over time which in turn requires the use of ODEs to take into account system dynamics.

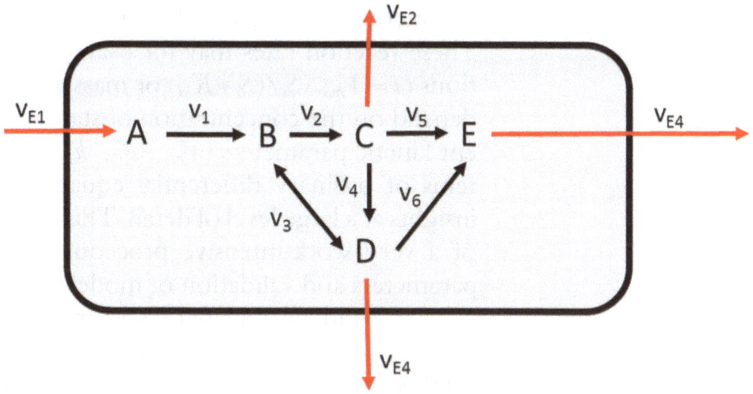

Fig. 1 Conversion of *A* to form various intracellular metabolites. The network considers five intracellular metabolites {*A*, *B*, *C*, *D*, *E*}, six intracellular reaction rates (v_1: $A \rightarrow B$; v_2: $B \rightarrow C$; v_3: $B \leftrightarrow D$; v_4: $C \rightarrow D$; v_5: $C \rightarrow E$; v_6: $D \rightarrow E$), and four exchange reactions (v_{E1}: $\rightarrow A$; v_{E2}: $B \rightarrow$; v_{E3}: $C \rightarrow$; v_{E4}: $D \rightarrow$)

By definition, producing reactions are counted as positive contributions to the respective conservation equation ("+") while consuming reactions are considered negative ("–"). Note that this could well be dealt with the other way around. Consistency in putting up the balance equations, however, is mandatory. All reactions in Fig. 1 are irreversible apart from equation v_3 which is reversible; that is, v_3 can produce *B* from *D* as well as *D* from *B*. To set up the system of equations we will assume that v_3 only produces *D* from *B*, such that v_3: $B \rightarrow D$. We will however keep in mind that the reaction directionality can also be other way around. If one now balances formation and consumption of each of the intracellular metabolites *A* to *E*, one obtains the following system of ordinary differential equations:

$$\frac{\mathrm{d}A}{\mathrm{d}t} = v_{\mathrm{E1}} - v_1.$$

$$\frac{\mathrm{d}B}{\mathrm{d}t} = v_1 - v_2 - v_3.$$

$$\frac{\mathrm{d}C}{\mathrm{d}t} = v_2 - v_4 - v_5 - v_{\mathrm{E2}}.$$

$$\frac{\mathrm{d}D}{\mathrm{d}t} = v_3 + v_4 - v_6 - v_{\mathrm{E3}}.$$

$$\frac{\mathrm{d}E}{\mathrm{d}t} = v_5 + v_6 - v_{\mathrm{E4}}.$$

Here, v_i is a placeholder for yet unknown reaction rate laws. These reaction rates may for example be Michaelis–Menten equations ($v_i = V_{max} \cdot S_i/(S_i + K_M)$ or mass action kinetics ($v_i = k_1 \cdot S_i$) which depend on the concentration of state variables (S_i) as well as different kinetic parameters (V_{max}, K_M, k_1). As discussed above, such systems of ordinary differential equations may provide mechanistic insights at a large level of detail. This comes, however, at the expense of a very work-intensive procedure needed for identification of parameters and validation of model structure, respectively [5].

To avoid dealing with systems of ODEs as shown above, a single assumption can be used to considerably simplify the above ODE system. During the exponential phase of microbial growth or during dormant states, it can be assumed that intracellular concentrations of metabolite A to E do not change. This implies on the one hand that the sum of all producing and the sum of all consuming reaction rate are equal and hence cancel out. With regard to above ODE system, this means in turn that there is no change over time since the system is in steady state. Therefore the derivatives on the left-hand side are all equal to zero. This state is referred to as the steady state and the so-called steady-state assumption is the fundamental assumption in stoichiometric modelling.

The original ODE system can thus be reduced to a linear system of equations, where each equation refers to the mass balance for one intracellular metabolite. Notably, metabolite concentrations are no longer considered in such systems. Since the rate laws have been significantly simplified due to the steady-state assumption they are no longer considered as functions of substrate conventions. Rather, these rates v_i represent the unknown variables of the system, for which the equations can be solved for. The rates v_i are usually referred to as fluxes and typically have the unit amount of moles per time per gram biomass (e.g., [mmol/h/g_{CDW}]). v_i are called intracellular fluxes while v_{Ei} are the so-called exchange fluxes, linking the intracellular space with the extracellular environment. The array $v = \{v_1, v_2, v_3, v_4, v_5, v_6, v_{E1}, v_{E2}, v_{E3}, v_{E4}\}$ of all fluxes is called flux distribution. Fluxes are inherent variables in metabolic network models for which the linear system of mass balance equations can be solved for. The identification of fluxes from a set of mass balance equations as such representing basic mass conservation laws is called flux balance analysis (FBA). FBA is a fundamental concept in metabolic modelling and allows to describe cellular biochemistry in a rigorous mathematical representation.

1.4 Matrix Representation of Stoichiometric Models

Using matrix notation the system of linear equations can be further summarized. In the so-called stoichiometric matrix S, the m rows represent the mass balances for each of the m intracellular metabolites $M = \{A, B, C, D, E\}$, while the columns represent the unknown fluxes. Notably, both intracellular fluxes (v_1 to v_6) and exchanges fluxes (v_{E1} to v_{E4}) need to be taken into account here, since otherwise mass conservation cannot be fulfilled. In our example the n columns represent the n rates $v = \{r_1, r_2, r_3, r_4, r_5, r_6, v_{E1}, v_{E2}, v_E, v_{E4}\}$. Note that the order of columns and rows is given by the index number of the corresponding identifier or metabolite, respectively. The order is hence arbitrary and no inherent property of the model structure itself:

$$S = \begin{pmatrix} -1 & 0 & 0 & 0 & 0 & 0 & 0 & 0 & 0 & 0 \\ 1 & -1 & -1 & 0 & 0 & 0 & 0 & 0 & 0 & 0 \\ 0 & 1 & 0 & -1 & -1 & 0 & 0 & -1 & 0 & 0 \\ 0 & 0 & 1 & 1 & 0 & -1 & 0 & 0 & -1 & 0 \\ 0 & 0 & 0 & 0 & 1 & 1 & 0 & 0 & 0 & -1 \end{pmatrix}.$$

When multiplying the matrix $S\,(m \times n)$ with the array of rates v $(n \times 1)$ one obtains an all-zero array $(m \times 1)$ which automatically satisfies the fundamental steady-state assumption:

$$S \cdot v = 0. \tag{1}$$

Equation 1 is the fundamental check of consistency for any stoichiometric model. Note that all entries of S are stoichiometric coefficients such that S is dimensionless.

1.5 Extracellular Metabolites

In contrast to the intracellular metabolites, the mass conservation for extracellular compounds is generally not fulfilled, since thus substances may be only consumed (substrates) or formed (products) in the extracellular space. In this case, the mass balance equations cannot be closed but in contrast need to remain open in order to allow production or consumption of intracellular metabolites. Consider the extension of our original model (Fig. 1), where extracellular (xt) compounds $\{A^{xt}, C^{xt}, D^{xt}, E^{xt}\}$ are now also considered (Fig. 2), while the set of rates $v = \{v_1, v_2, v_3, v_4, v_5, v_6, v_{E1}, v_{E2}, v_{E3}, v_{E4}\}$ remains the same. Setting up the mass balance equations for the intracellular metabolites leads to the same system as before.

However, the set of extracellular compounds now also needs to be taken into account:

$$\frac{dA^{xt}}{dt} = -v_{E1}.$$

$$\frac{dC^{xt}}{dt} = v_{E2}.$$

$$\frac{dD^{xt}}{dt} = v_{E3}.$$

$$\frac{dE^{xt}}{dt} = v_{E4}.$$

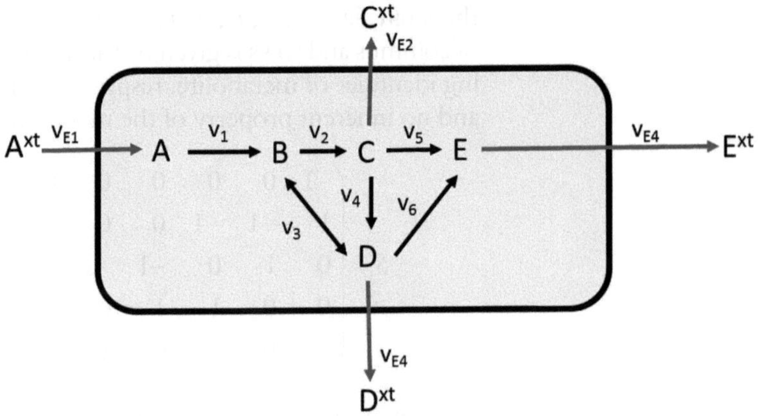

Fig. 2 An extended network which considers both intracellular {*A*, *B*, *C*, *D*, *E*} and extracellular metabolites (A^{xt}, C^{xt}, D^{xt}, E^{xt}). Notably, the inherent connectivity of the network as described by the intracellular fluxes and the exchange reactions remains the same as in Fig. 1. In contrast to the intracellular metabolites, the mass balances of the extracellular metabolites do not have to be closed

The corresponding, extended stoichiometric matrix S^* is

$$S^* = \begin{pmatrix} 0 & 0 & 0 & 0 & 0 & 0 & -1 & 0 & 0 & 0 \\ 0 & 0 & 0 & 0 & 0 & 0 & 0 & 1 & 0 & 0 \\ 0 & 0 & 0 & 0 & 0 & 0 & 0 & 0 & 1 & 0 \\ 0 & 0 & 0 & 0 & 0 & 0 & 0 & 0 & 0 & 1 \end{pmatrix}.$$

S^* has $m^*=4$ extracellular metabolites $M=\{A^{xt}, C^{xt}, D^{xt}, E^{xt}\}$, while the columns representing the n rates $v=\{v_1, v_2, v_3, v_4, v_5, v_6, v_{E1}, v_{E2}, v_{E3}, v_{E4}\}$ remain the same as in the former matrix S. Note that the mass balance equations for A^{xt}, C^{xt}, D^{xt}, and E^{xt} can never be closed since there is only a single entry for each of the rows of S^*such that the steady-state assumption cannot be fulfilled in this case, i.e.,

$$S^* \cdot v \neq 0. \tag{2}$$

This means that extracellular metabolites are in total either consumed or produced. In the example in Fig. 2 metabolite A^{xt} is a substrate in the surrounding medium, while C^{xt}, D^{xt}, and E^{xt} are products secreted from the intracellular space. It is rather a matter of model representation and visualization, whether external metabolites are considered in the network structure. Most importantly, the basic model reactions illustrated in Figs. 1 and 2 are the same; that is, S and S^* therefore have identical solutions.

1.6 Degree of Freedom

The intracellular flux distribution $v(n \times 1)$ represents the unknown variables for which the m linear mass balance equations can be solved for. An important property of the stoichiometric matrix is whether it allows determining a flux solution or not. There are stoichiometric models such as genome-scale networks with more than 1,000 reactions but only about 500 metabolites which obviously have much more unknown fluxes than balancing equations. Generally, the system of linear equations is called

1. Square, if $m = n$.
2. Underdetermined, if $m < n$.
3. Overdetermined, if $m > n$.

Note that in this regard m refers to the number of equations in the stoichiometric matrix S, which are linearly independent. The difference between the m linear independent set of metabolite mass balance equations and the n number of fluxes quantifies the degree of freedom of the stoichiometric model. Notably, the number of linear independent equations m must not be mixed with the number of rows of S. Consider for example the following reactions:

r_1: $A^{xt} + \text{ATP} \rightarrow A + \text{ADP}$.

r_2: $A \rightarrow B$.

r_3: $B + \text{ADP} \rightarrow B^{xt} + \text{ATP}$.

If one only considers the mass balances for ATP and ADP, one obtains the following stoichiometric matrix S^{ATP} in which the columns represent the array of fluxes $v = \{v_1, v_2, v_3\}$ while the rows represent the set $M = \{\text{ATP}, \text{ADP}\}$:

$$S^{\text{ATP}} = \begin{pmatrix} -1 & 0 & 1 \\ 1 & 0 & -1 \end{pmatrix}.$$

How many linear independent equations does S^{ATP} contain? Obviously, both rows of S^{ATP} can be converted into each other by multiplying either of them by −1. This means that they are linearly dependent and the second row does not contain any information in addition to equation one. Such linear dependencies are a frequent phenomenon in stoichiometric models, though they might not be as obvious as in the present example. They often occur for cofactor balances such as NADPH/NADP or ATP/ADP, but more complex structural dependencies might be difficult to identify due to the complexity of network interactions. Algorithms such as calculability analysis [6] allow the identification of dependencies in metabolic networks, but they become infeasible for larger networks due to the increasing number of combinatorial possibilities and numerical difficulties. The difference between the m linear independent set of metabolite mass balance equations and the n number of fluxes gives a rough estimate of the degree of freedom of the stoichiometric model.

1.7 Identifying Flux Distributions

In square stoichiometric models, the number of linear independent equations m is equal to the number of unknown fluxes n. In this case, an intracellular flux distribution can be obtained by Gaussian elimination. If S is square, one can also solve for the array of fluxes by matrix conversions. Consider the following example:

$$S \cdot v = b.$$

In this example, S may contain both intracellular and extracellular metabolites, such that the right-hand side b is not all zero for all components. If S is additionally a non-singular matrix so that the inverse matrix S^{-1} exists, then there is a unique solution given by

$$v = S^{-1} \cdot b.$$

A square stoichiometric matrix, however, is rather the exception in stoichiometric modelling. In most cases, stoichiometric models will be underdetermined which is obvious in genome-scale models with more than 1,000 biochemical reactions but a significantly lesser number of metabolite balance equations. Even if there are linear dependencies between some of the rows in the stoichiometric model, this will always be far from being enough to obtain a square matrix. To identify a particular flux distribution in such networks, linear optimization is needed which is discussed in more detail in the next paragraph.

Finally, overdetermined systems occur if there is too much and even contradicting information. Consider for example the case

$$\begin{pmatrix} 1 & 0 \\ 1 & -1 \\ -1 & 2 \end{pmatrix} \cdot \begin{pmatrix} v_1 \\ v_2 \end{pmatrix}^{\mathrm{T}} = \begin{pmatrix} 1 \\ 1 \\ 0 \end{pmatrix}.$$

Obviously, the rows of the matrix are neither linear dependent nor is the matrix square. Since there are less unknowns (v_1, v_2) than balance equations, there is just the trivial solution $v = (0\ 0)$. In FBA the case of overdetermined systems occurs sometimes when experimental data such as ^{13}C-labeling data or extracellular medium concentrations are considered. A possibility to still identify a flux solution in such a case is the usage of least square approaches, i.e., to minimize the violation of the constraints as imposed by the mass conservation laws.

1.8 Linear Optimizations in Underdetermined Systems

As discussed in the previous paragraph, the number of unknown variables will be significantly larger in genome-scale metabolic network models than the number of linear independent mass balance equations ($m < n$). The inherent degree of freedom of such stoichiometric models may be considerably large such that linear optimization needs to be used to identify an intracellular flux solution.

Linear optimization, also called linear programming (LP), is a relatively simple mathematical tool which is applied in various fields in natural sciences and engineering. The general structure of a linear optimization problem is as follows:

Max/min	$Z = obj \cdot v^{T}$
s.t.	$S \cdot v = 0$
	$S^{*} \cdot v \leq b$
	$lb_i \leq v_i \leq ub_i$

Here, Z is the scalar objective value (1×1) which is to be minimized or maximized. Mathematically, it is obtained by multiplying the objective function $obj(1 \times n)$ with the flux distribution $v^{T}(n \times 1)$. The objective function *obj* is often used to express a cellular rationale such as maximization of biomass production or minimization of glucose consumption. Various kinds of objective functions will be discussed in the next paragraph. Linear optimization problems additionally contain the so-called constraint section which is marked by "s.t." ("*subject to*"). In the case of FBA, the constraints section contains the following:

1. The steady-state assumption $(S \cdot v = 0)$ for intracellular metabolites.
2. Mass balances of extracellular metabolites which remain open $((S^{*} \cdot v \leq b))$; in this case the signs of the entries in the rows of S^{*} and b can be used to specify whether a compound is a substrate (net consumption) or a product (net production).
3. Upper (ub) and lower bounds (lb) for fluxes ($lb_i \leq v_i \leq ub_i$); this constraint provides a comfortable opportunity to specify the directionality of the fluxes in the network.

In our first example in Fig. 1, v_3 was reversible; that is, it could both as well produce B from D or D from B (Fig. 1). For the formulation of the stoichiometric matrix we had, however, neglected this specification and assumed v_3 to be irreversible. The reversibility of reactions can now easily be taken into account by setting the lower bound (lb_i) to an arbitrary low value below zero while the upper bound (ub_i) is set to an arbitrary high number greater than zero. In this case the corresponding flux v_i of a reversible reaction $((lb_i \leq v_i \leq ub_i)$ may go in one direction $(0 \leq v_i \leq ub_i)$ or the other $(lb_i \leq v_i \leq 0)$. In turn, the flux v_i of an irreversible reaction can thus be constrained to only go in one direction by setting the lower bound lb_i equal to zero $(0 \leq v_i \leq ub_i)$. Upper and lower bounds can also be used to impose experimental observations, such as limitations in enzymatic activity of the respiratory chain, which was used before to enforce acetate secretion in stoichiometric models of *E. coli* [4]. The consideration of stoichiometric models within a linear optimization problem is hence a direct extension of many of

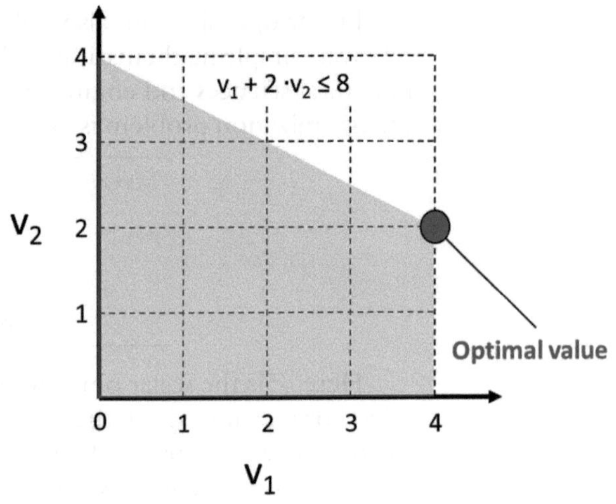

Fig. 3 An example for linear optimization. The feasible space (marked in *grey*) is bounded by the inequality constraint $v_1 + 2 \cdot v_2 \leq 8$ and upper and lower bounds, respectively, for both fluxes ($0 \leq v_1 \leq 4$, $0 \leq v_2 \leq 4$). The optimal value $Z^{opt} = 6$ is the maximal possible value of the objective function max $Z = (1\ 1) \cdot (v_1 v_2)^T$ which still satisfies the constraints

our earlier considerations (steady-state assumption, mass balance, flux directionality) which are all included in the mathematical formulation of the optimization problem.

Let us now take a brief look behind the general concept of linear optimization by considering the following example (Fig. 3):

$$\begin{aligned} &\max Z = (11) \cdot (v_1 v_2)^T \\ &\text{s.t.} \quad v_1 + 2 \cdot v_2 \leq 8 \\ &\qquad 0 \leq v_1 \leq 4 \\ &\qquad 0 \leq v_2 \leq 4. \end{aligned}$$

In this example, the objective function is obj = (1 1) which, when multiplied with the array of fluxes $v = (v_1\ v_2)$, gives the above equation for the calculation of the optimal value $Z = v_1 + v_2$. Constraint-wise, both fluxes in the system (v_1, v_2) are bounded between 0 and 4, such that all feasible combinations of fluxes lie in this square area. Another inequality constraint ($v_1 + 2 \cdot v_2 \leq 8$) cuts an additional slice from this region as such limiting the space of feasible flux solutions to the grayish area in Fig. 3. Without the inequality constraint, the optimal value would be on the bisecting line at $v_1 = 4$ and $v_2 = 4$, such that $Z = 8$. However, this point is infeasible due to the inequality constraint. The feasible optimal value, however, has to lie on the line given by $v_1 + 2 \cdot v_2 \leq 8$. Visual inspection shows that the intersection of the inequality constraint with the *y*-axis gives $Z = 4$, while the intersection with the upper bound for v_1 ($v_1 \leq 4$)

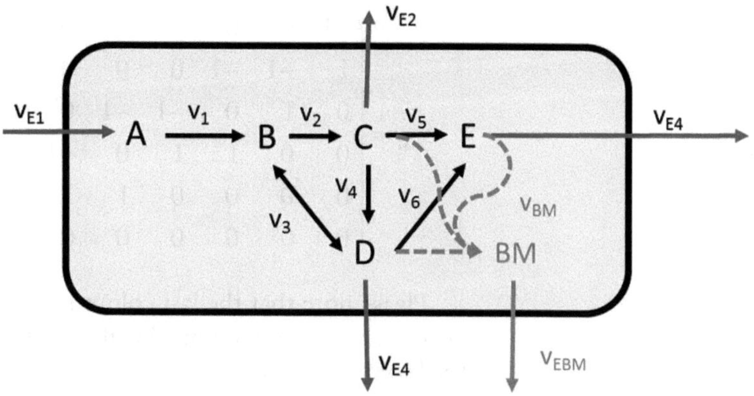

Fig. 4 Formation of biomass (BM) out of 1 mol *C*, 1 mol *D*, and 2 mol of *E* ($C+D+2E\rightarrow$ biomass)

gives an optimal value of $Z=6$. In between the former and the latter edge of the feasible space, Z increases linearly according to $v_1+2\cdot v_2$. The optimal value lies hence in the upper right corner of this region ($v_1=4$, $v_2=2$), giving an optimal value of $Z=6$. While the solution was found by visual inspection, mathematical approaches for linear optimization use computational algorithms such as the simplex algorithm to identify a solution for the linear optimization problem. Such algorithms are available in many software packages such as MATLAB®, CPLEX, or GLPK (GNU linear programming kit).

It should be noted that the optimal value Z will always represent the global optimum. This is due to the linearity of the underlying model which ensures mathematical convexity of the overall optimization problem. Uniqueness, of the flux distribution, however, cannot be guaranteed. In particular in larger networks with a lot of alternative yet equivalent pathway alternatives, different network routes cannot be discriminated with regard to their contributions to the objective function. It should hence be noted that such so-called flux variability [7] may always occur.

1.9 Maximization of Biomass Yield

The most widely used objective function is maximization of biomass formation [8]. Consider the following extension of the original model (Figs. 1 and 4). Here, it is assumed that in addition to the metabolites *C*, *D*, and *E* cellular biomass can also be formed in the reaction v_{BM} ($C+D+2E\rightarrow$ biomass). Notably, v_{BM} is rather a summation function than a physiological reaction and can be seen as a collector of all biomass constituents of the cell from all over the intracellular space. Basically, v_{BM} means that for 1 g_{CDw} of biomass (BM) a certain molar amount of *C*, *D*, and *E* are required. When constructing the stoichiometric matrix S_{BM} as in the example before (Fig. 1), there is hence an additional row and additional column needed in the stoichiometric matrix:

$$S_{BM} = \begin{pmatrix} -1 & 0 & 0 & 0 & 0 & 0 & 1 & 0 & 0 & 0 & 0 \\ 1 & -1 & -1 & 0 & 0 & 0 & 0 & 0 & 0 & 0 & 0 \\ 0 & 1 & 0 & -1 & -1 & 0 & 0 & -1 & 0 & 0 & -1 \\ 0 & 0 & 1 & 1 & 0 & -1 & 0 & 0 & -1 & 0 & -1 \\ 0 & 0 & 0 & 0 & 1 & 1 & -2 & 0 & 0 & 0 & -2 \\ 0 & 0 & 0 & 0 & 0 & 0 & 0 & 0 & 0 & 0 & 1 \end{pmatrix}.$$

Please note that the last column describing biomass formation is special in terms of its units. While all other columns in S_{BM} describe stoichiometric coefficients and are hence dimensionless, the new last column in S_{BM} has the unit mmol/g_{CDW} as such specifying the required molar amount of each metabolite to form 1 g_{CDW}. Remember that the remaining entries in S_{BM} are multiplied with fluxes having the unit mmol/h/g_{CDW}. To be consistent, v_{BM} has to have the unit 1/h; that is, it explicitly describes the biomass growth rate.

In the above example it is assumed that biomass composition is as follows: 1 mmol/g_{CDW} C, 1 mmol/g_{CDW} D, and 2 mmol/g_{CDW} E, respectively. Let us now consider the following flux distribution: $v_{E1} = 20$ mmol/h/g_{CDW}; $v_2 = 10$ mmol/h/g_{CDW}; $v_3 = 10$ mmol/h/g_{CDW}; $v_5 = 5$ mmol/h/g_{CDW}; and $v_6 = 5$ mmol/h/g_{CDW}, while v_4, v_6, v_{E2}, v_{E3}, and v_{E4} are equal to zero. Cross checking with the network structure in Fig. 4 shows that 5 mmol/h/g_{CDW} of C and D and 10 mmol/h/g_{CDW} of E, respectively, are used to form biomass such that $v_{BM} = 5$ 1/h.

Please note that maximization of $Z = \text{obj} \cdot v^T$ with the objective function

$$\text{obj} = \begin{pmatrix} 0 & 0 & 0 & 0 & 0 & 0 & 0 & 0 & 0 & 1 \end{pmatrix},$$

i.e., maximization of biomass formation generally does not mean that the fluxes through the stoichiometric network are maximized for growth rate μ. This is because the network will always produce as much biomass as provided by the substrate uptake rate $v_{\text{substrate}}$ such that the ratio between the uptake rate and growth rate is constant, which in turn is equal to biomass yield $Y^{\text{biomass}}_{\text{substrate}}$. Therefore, the equation

$$\mu = Y^{\text{biomass}}_{\text{substrate}} \cdot v_{\text{substrate}}$$

does always hold, such that the above function rather optimizes for the yield than for the growth rate [9, 10]. It is only in cases when additional constraints such as non-growth-associated maintenance are taken into account that maximal biomass is no longer a mere maximization of $Y^{\text{biomass}}_{\text{substrate}}$ but instead of growth rate μ [9]. This difference is important to make. Nevertheless, erroneous interpretations of optimization of biomass are fairly common, even in the literature.

1.10 The Choice of the Objective Function

When using linear optimization for the identification of flux distributions in stoichiometric modelling, the selection of the objective function has a huge impact in the final result. While the basic network reactions, medium composition, and reaction directionality are rigorously represented in the constraint section, the objective function is dependent on the choice of the modeller itself and presents as such a man-made interpretation of biological observations.

Objective functions which have been suggested and used before are for example

- Maximization of biomass yield, and maximization of growth rate [10–12].
- Minimization of glucose consumption [13].
- Maximization of ATP yield per flux [10, 14].

A comparison of computationally predicted flux distributions with experimentally measured ^{13}C-labeling data indicated that there is not a single, global objective function which is valid for all growth conditions, medium compositions, and even microorganisms [10]. In contrast, it could be shown that the choice of the objective function is heavily dependent on the environmental and experimental conditions. Maximization of biomass yield for example leads to flux distributions which were in agreement with fluxes measured in *E. coli* under nutrient scarcity in continuous cultures. In contrast, this objective function will never predict overflow metabolism in fast-growing batch cultures. It is well known from experimental practise that under these circumstances *E. coli* cells secrete large amounts of acetate instead of using the respiratory chain, which would be more efficient in terms of carbon usage. Such a fine-tuning and switching of metabolic modes of action can never be predicted with a single objective functions and show the obvious challenges when selecting a particular objective function for network analysis.

1.11 Genome-Scale Metabolic Network Reconstructions (GENREs)

The approaches introduced in this chapter are applicable to any kind of stoichiometric model no matter what size. This is one of the reasons why stoichiometric models are nowadays used for analyses in such diverse fields as microbiology, botany, and human physiology [15]. The relative easiness with which stoichiometric models can be constructed has also led to the establishment of the so-called genome-scale network reconstructions. These are metabolic network models that aim for an exhaustive representation of a specie's enzymatic setup based on annotated genome sequences. The number of genome-scale metabolic network reconstructions (GENREs) is constantly increasing since the first metabolic model on *Haemophilus influenzae* was published more than 10 years ago [16]. Most GENREs exist for bacteria where for example at least four reconstructions of *E. coli* exist up to now [17]. Among the eukaryotes, *S. cerevisiae* scores highest with at least five reconstructions [15] making it the organism with the most model curations so far. With regard to medical and pharmaceutical applications [18, 19], a growing number of human GENREs has been published, which comprise both generic models [20] as well as tissue-specific network reconstructions [21]. Also, an increasing number of stoichiometric models have been generated for plant metabolism [22] where GENREs will gain increasing importance in the future [23].

There are numerous reviews about the construction and validation of genome-scale models [15, 24]. In brief, an initial reconstruction is always developed in a first step based on pathway annotation data. The overall model structure then needs to be validated by detecting missing network connectivities in the stoichiometric model. In this so-called gap analysis, dead-end metabolites are identified which are either only produced and not consumed or vice versa only consumed but not produced. Most notably, such dead-end metabolites require inevitably that the linking reactions do not carry any flux due to the steady-state assumption. To obtain a network of biological relevance such gaps in the network structure hence need to be reconciled or removed. The final step in the construction of GENREs is the iterative comparison to experimental data to increase confidence in its predictive capabilities. In this regard, the most fundamental validation is to test whether essentiality of a gene deletion can be properly predicted, i.e., whether the lethality or the viability of a knockout mutation can be correctly described [25]. Further possibilities for validation are prediction of by-product secretion [4] or correct execution of specific metabolic functions [21]. In general, GENRE can be validated by comparison to multiple kinds of experimental data ranging from gene expression data [26] to fluxomics [10]. Examples for general fields of applications of GENREs are (1) contextualization of high-throughput data, (2) guidance for metabolic engineering, and (3) structural network analyses [15].

1.12 Analytical Algorithms

As discussed before, stoichiometric models describe cellular biochemistry as systems of linear equations. This linearity has significantly supported the development of specific computational algorithms for network analysis. The algorithms are mostly easy to implement, but allow nevertheless comprehensive insights into the basic network structure. Some examples for such algorithms for network analysis are as follows:

- Flux variability analysis [7], where the optimal value Z^{opt} is first determined in a preliminary step: Z^{opt} is then used as an additional constraint, while all fluxes v_i in the network are iteratively maximized and minimized to identify their possible range, i.e., their variability. Notably, global optimality of the flux solution is always guaranteed due to the additional constraint $Z^{obj} = obj \cdot v^T$.

- The flux coupling finder [27], which considers iteratively the dependencies between all pairs of fluxes v_i and $v_{j \neq i}$ in a network: Each flux v_i is considered once and set to 1, while all the remaining fluxes $v_{j \neq i}$ in the network are both minimized and maximized. The ratio v_j / v_i then tells which fluxes are inevitably equal (coupled) or otherwise dependent upon each other (e.g., directionality).
- Extreme pathways and elementary flux modes [28], which are matrix transformations of the null space of S: Both extreme pathways and elementary flux modes are sets of convex basis vectors that represent feasible flux distributions of a metabolic network.

Further examples of algorithms are OptKnock [29] for the identification of gene knockout strategies for microbial strain optimization or algorithms for the inclusion of thermodynamic information in stoichiometric network models [30]. Another important approach is the integration of omics data into network models data for data contextualization [26, 31].

1.13 Conclusions

Stoichiometric models are based on linear sets of equations to represent the biochemistry of cellular metabolism. Basically, the models describe the stoichiometry of metabolic networks at steady state. Fluxes are inherent variables in stoichiometric models and quantify intracellular carbon mass flow. Many stoichiometric models are underdetermined and linear programming is hence necessary to impose a specific cellular rationale in order to identify a flux distribution. Maximization of biomass yield is the most common objective function, but it has been shown that other objective functions are better suited for specific growth conditions. Genome-scale models of metabolism are increasingly gaining importance in microbiology, botany, and even medicine. Since stoichiometric models can well be used for integration and contextualization of omics data they will be of great use for the analysis of biological systems in the future.

References

1. Blank LM, Kuepfer L (2010) Metabolic flux distributions: genetic information, computational predictions, and experimental validation. Appl Microbiol Biotechnol 86(5):1243–1255
2. Domach MM, Shuler ML (1984) Testing of a potential mechanism for E. coli temporal cycle imprecision with a structural model. J Theor Biol 106(4):577–585
3. Lewis NE, Nagarajan H, Palsson BO (2012) Constraining the metabolic genotype-phenotype relationship using a phylogeny of in silico methods. Nat Rev Microbiol 10(4):291–305
4. Varma A, Palsson BO (1994) Stoichiometric flux balance models quantitatively predict growth and metabolic by-product secretion in wild-type Escherichia coli W3110. Appl Environ Microbiol 60(10):3724–3731
5. Stelling J (2004) Mathematical models in microbial systems biology. Curr Opin Microbiol 7(5):513–518

6. Klamt S, Schuster S, Gilles ED (2002) Calculability analysis in underdetermined metabolic networks illustrated by a model of the central metabolism in purple nonsulfur bacteria. Biotechnol Bioeng 77(7):734–751
7. Mahadevan R, Edwards JS, Doyle FJ 3rd (2002) Dynamic flux balance analysis of diauxic growth in Escherichia coli. Biophys J 83(3): 1331–1340
8. Schilling CH, Edwards JS, Letscher D, Palsson BO (2000) Combining pathway analysis with flux balance analysis for the comprehensive study of metabolic systems. Biotechnol Bioeng 71(4):286–306
9. Feist AM, Palsson BO (2010) The biomass objective function. Curr Opin Microbiol 13(3): 344–349
10. Schuetz R, Kuepfer L, Sauer U (2007) Systematic evaluation of objective functions for predicting intracellular fluxes in Escherichia coli. Mol Syst Biol 3:119
11. Edwards JS, Palsson BO (2000) The Escherichia coli MG1655 in silico metabolic genotype: its definition, characteristics, and capabilities. Proc Natl Acad Sci U S A 97(10): 5528–5533
12. Price ND, Reed JL, Palsson BO (2004) Genome-scale models of microbial cells: evaluating the consequences of constraints. Nat Rev Microbiol 2(11):886–897
13. Oliveira AP, Nielsen J, Forster J (2005) Modeling Lactococcus lactis using a genome-scale flux model. BMC Microbiol 5:39
14. Dauner M, Sauer U (2001) Stoichiometric growth model for riboflavin-producing Bacillus subtilis. Biotechnol Bioeng 76(2):132–143
15. Oberhardt MA, Palsson BO, Papin JA (2009) Applications of genome-scale metabolic reconstructions. Mol Syst Biol 5:320
16. Edwards JS, Palsson BO (1999) Systems properties of the Haemophilus influenzae Rd metabolic genotype. J Biol Chem 274(25): 17410–17416
17. Orth JD et al (2011) A comprehensive genome-scale reconstruction of Escherichia coli metabolism–2011. Mol Syst Biol 7:535
18. Krauss M et al (2012) Integrating cellular metabolism into a multiscale whole-body model. PLoS Comput Biol 8(10):e1002750
19. Kuepfer L (2010) Towards whole-body systems physiology. Mol Syst Biol 6:409
20. Duarte NC et al (2007) Global reconstruction of the human metabolic network based on genomic and bibliomic data. Proc Natl Acad Sci U S A 104(6):1777–1782
21. Gille C et al (2010) HepatoNet1: a comprehensive metabolic reconstruction of the human hepatocyte for the analysis of liver physiology. Mol Syst Biol 6:411
22. Dal'Molin CG, Quek LE, Palfreyman RW, Brumbley SM, Nielsen LK (2010) C4GEM, a genome-scale metabolic model to study C4 plant metabolism. Plant Physiol 154(4):1871–1885
23. de Oliveira Dal'molin CG, Nielsen LK (2012) Plant genome-scale metabolic reconstruction and modelling. Curr Opin Biotechnol 24(2): 271–277
24. Feist AM et al (2010) Model-driven evaluation of the production potential for growth-coupled products of Escherichia coli. Metab Eng 12(3): 173–186
25. Kuepfer L, Sauer U, Blank LM (2005) Metabolic functions of duplicate genes in Saccharomyces cerevisiae. Genome Res 15(10): 1421–1430
26. Shlomi T, Cabili MN, Herrgard MJ, Palsson BO, Ruppin E (2008) Network-based prediction of human tissue-specific metabolism. Nat Biotechnol 26(9):1003–1010
27. Burgard AP, Nikolaev EV, Schilling CH, Maranas CD (2004) Flux coupling analysis of genome-scale metabolic network reconstructions. Genome Res 14(2):301–312
28. Papin JA et al (2004) Comparison of network-based pathway analysis methods. Trends Biotechnol 22(8):400–405
29. Burgard AP, Pharkya P, Maranas CD (2003) Optknock: a bilevel programming framework for identifying gene knockout strategies for microbial strain optimization. Biotechnol Bioeng 84(6):647–657
30. Kummel A, Panke S, Heinemann M (2006) Putative regulatory sites unraveled by network-embedded thermodynamic analysis of metabolome data. Mol Syst Biol 2(2006):0034
31. Reed JL (2012) Shrinking the metabolic solution space using experimental datasets. PLoS Comput Biol 8(8):e1002662

Chapter 2

Tapping the Wealth of Microbial Data in High-Throughput Metabolic Model Reconstruction

Ric Colasanti, Janaka N. Edirisinghe, Tahmineh Khazaei, José P. Faria, Sam Seaver, Fangfang Xia, and Christopher Henry

Abstract

Genome-scale metabolic models are valuable tools in the metabolic engineering process, based on the ability of these models to integrate diverse sources of data to produce global predictions of organism behavior. At the most basic level, these models require only a genome sequence to construct, and once built, they may be used to predict essential genes, culture conditions, pathway utilization, and the modifications required to enhance a desired organism behavior. In this chapter, we address two key challenges associated with the reconstruction of metabolic models: (a) leveraging existing knowledge of microbiology, biochemistry, and available 'omics data to produce the best possible model; and (b) applying available tools and data to automate the reconstruction process. We consider these challenges as we progress through the model reconstruction process, beginning with genome assembly, and culminating in the integration of constraints to capture the impact of transcriptional regulation. We divide the reconstruction process into ten distinct steps: (1) genome assembly from sequenced reads; (2) automated structural and functional annotation; (3) phylogenetic tree-based curation of genome annotations; (4) assembly and standardization of biochemistry database; (5) genome-scale metabolic reconstruction; (6) generation of core metabolic model; (7) generation of biomass composition reaction; (8) completion of draft metabolic model; (9) curation of metabolic model; and (10) integration of regulatory constraints. Each of these ten steps is documented in detail.

Key words Genome assembly, Phylogenetic gene trees, Subsystems-based annotation, Metabolic reconstruction, Metabolic modeling, Flux balance analysis, RAST, Model SEED, Gap filling, Regulatory modeling

1 Introduction

Genome-scale metabolic models are becoming prevalent tools in the field of Metabolic Engineering, useful for their capacity to rapidly predict phenotype from genotype using few required parameters. These models integrate knowledge of genome annotations, metabolic pathways, biomass composition, and even transcriptional regulation to predict how an organism will respond to environmental and genetic perturbations. They have been applied extensively to

Jens O. Krömer et al. (eds.), *Metabolic Flux Analysis: Methods and Protocols*, Methods in Molecular Biology, vol. 1191, DOI 10.1007/978-1-4939-1170-7_2, © Springer Science+Business Media New York 2014

predict essential genes and culture conditions [1], design strategies to improve strain productivity [2–4], and identify gaps and errors in genome annotations [5, 6]. They even form a component of the first whole-cell dynamic model of bacteria [7].

A genome-scale metabolic model consists of three primary components: (1) a list of reactions that take part in the metabolic pathways of the organism including reaction stoichiometry and reversibility, (2) a set of gene-protein-reaction (GPR) associations that capture how gene activity is related to the activity of metabolic reactions, and (3) a biomass composition reaction that indicates which small molecules must be produced for an organism to grow and divide [8]. All of these components are used in a method called Flux Balance Analysis (FBA) to simulate microbial metabolism in a specified environmental condition [9–12].

One challenge that has hindered the widespread use of metabolic models is the time required to build a new model. Until recently, model reconstruction has been a laborious process, often requiring years of effort and expert knowledge to complete. More recently, the RAST [13] and Model SEED frameworks [14] were released (available at http://www.theseed.org/models/), providing a means to rapidly building new draft models from genome sequences in approximately 24 h. Today, the Model SEED has been applied to generate over 17,000 draft metabolic models, including a model of the over 3,500 complete prokaryotic genome sequences currently available in GenBank [15]. In a previous book chapter, we provide detailed instructions on the use of RAST and Model SEED to annotate a genome and produce a draft metabolic model [16].

In this chapter, we explore the automated model reconstruction process in greater detail, focusing on how data from microbiology, biochemistry, and 'omics studies are integrated in the reconstruction process to improve the models produced. We also discuss the reconstruction process starting with genome assembly, which is an earlier starting point than previous reviews. However, genome assembly is becoming an increasingly important step in the process as sequencing technology grows cheaper and sequencing machines become available to an increasing number of researchers. Finally, we complete this chapter with a discussion of the integration of constraints to represent transcriptional regulation in metabolic models. Like genome assembly, regulatory constraints are becoming increasingly important given the growth in gene expression datasets and our improved knowledge of regulatory networks and mechanisms. Overall, we divide our discussion of the model reconstruction process into ten distinct steps: (1) genome assembly from sequenced reads; (2) automated structural and functional annotation; (3) phylogenetic tree-based curation of genome annotations; (4) assembly and standardization of biochemistry database; (5) genome-scale metabolic reconstruction;

(6) generation of core metabolic model; (7) generation of biomass composition reaction; (8) completion of draft metabolic model; (9) curation of metabolic model; and (10) integration of regulatory constraints.

2 Materials

2.1 Genome Assembly from Sequenced Reads

1. A FASTA or FASTQ file containing all sequenced reads.
2. Appropriate selection of sequencing platform and library.
3. Sequence assembly software appropriate for selected sequencing platform (e.g., Newbler, Velvet, SOAPdenovo, MIRA, ALLPATHS-LG, ABySS, Kiki, Ray).
4. Assembly scaffold postprocessing software (e.g., BAMBUS, SSPACE, bwa, samtools).

2.2 Automated Structural and Functional Annotation

1. A curated, controlled vocabulary of functional roles that define the specific biological functions to be mapped onto genes during annotation.
2. An organization of related functional roles into curated subsystems.
3. A large number of diverse microbial genome sequences to serve as the references.
4. A database of protein families that represent isofunctional homologues.

2.3 Phylogenetic Tree-Based Curation of Genome Annotations

1. A large number of diverse microbial genome sequences to serve as the references.

2.4 Assembly and Standardization of Biochemistry Database

1. Compiled biochemistry from KEGG [17], SEED [18], and BioCyc [19].
2. Molecular structures for as many metabolites as possible.
3. Software for standardization and charging of molecular structures.

2.5 Genome-Scale Metabolic Reconstruction

1. A comprehensive database of the biochemical reactions.
2. An annotation ontology with a strict controlled vocabulary.
3. A curated mapping from the reactions to protein complexes and from protein complexes to functional roles in the annotation ontology.
4. A genome with genes consistently annotated with the annotation ontology.
5. A list of spontaneous and universal reactions that should be added to all models.

2.6 Generation of Core Metabolic Model

1. A database of central carbon and fermentation pathways.
2. A database of template electron transport chain pathways.
3. An annotation ontology with a strict controlled vocabulary.
4. A genome with genes consistently annotated with the annotation ontology.

2.7 Generation of Biomass Composition Reaction

1. An estimation of the fraction of biomass that consists of DNA, RNA, protein, lipids, cell wall, cofactors, and growth associated ATP consumption.
2. An approximate estimation of amino acid composition of protein, nucleotide composition of RNA, and GC content of the genome.
3. An annotation ontology with a strict controlled vocabulary and a database of genomes consistently annotated with the ontology.
4. A list of potential lipid, cell wall, and cofactor biomass components.

2.8 Completion of Draft Metabolic Model

1. A curated model-ready biochemistry database.
2. A media condition in which gap filling will be performed.
3. A mapping between compounds in the model and in the biochemistry database.
4. Software capable of solving large-scale mixed integer optimization problems.

2.9 Curation of Metabolic Model

1. Access to the Model SEED Web site (http://www.theseed.org/models/).
2. Cytoscape (http://www.cytoscape.org/) and CytoSEED plugin (http://sourceforge.net/projects/cytoseed/) for viewing metabolic models.
3. Software for running flux balance analysis on metabolic models using SBML files. For example, the COBRA toolbox (opencobra.sourceforge.net/) or OptFlux (www.optflux.org/).

2.10 Generation of Transcriptional Regulatory Constraints

1. Access to a high-quality template model of regulatory constraints for a closely related organism.
2. Access to gene expression data for modeled organism.

3 Methods

As described above, the automated genome annotation and metabolic model reconstruction process can be broken down into ten sequential steps, which we describe in detail here: (1) Genome assembly from sequenced reads; (2) automated structural and functional annotation of genome sequence; (3) phylogenetic tree-based curation of genome annotations; (4) curation and assembly of a biochemistry database; (5) genome-scale metabolic reconstruction, (6) generation of a core metabolic model, (7) generation of a biomass composition reaction, (8) completion of a genome-scale metabolic model, (9) curation of the metabolic model, and (10) integration of regulatory constraints into a model.

3.1 Genome Assembly from Sequenced Reads

When nearly all shotgun genome sequencing was conducted in a centralized manner at designated sequencing centers, few researchers were exposed to the genome assembly process. However, today shot-gun sequencing technology is becoming increasingly democratized, with individual labs often purchasing their own sequencer, and in this environment, it is no longer possible to delegate all genome assembly to a few designated experts. Accordingly, we have decided to include genome assembly as the first step in our discussion of the genome annotation and metabolic modeling process. We note, the process of obtaining a new genome sequence truly begins with the selection of a sequencing platform and library to be used for the sequencing itself. While this topic will not be discussed in detail, this is arguably the most important decision that affects the final assembly quality (*see* **Note 1**).

1. The first step of the assembly process is to determine which assembly approach to use. Currently, there are two approaches to genome assembly: reference mapping and *de novo* assembly. Reference mapping involves the alignment of sequenced reads on a reference genome. The use of the reference genome greatly simplifies the assembly process, and it reduces the sequencing depth required to obtain a complete genome. However, it requires that a sufficiently close reference genome exists, and the assumption that the sequenced genome is close to the reference genome can sometimes lead to errors and poor assumptions in the assembly process. Small variations in bacterial DNA can lead to significantly different phenotypes (e.g., virulence [20]), and the novel regions assembled often become the focus of ensuing annotation and modeling efforts. As the cost of sequencing has decreased, the reduced sequencing depth required by reference mapping no longer compensates sufficiently for the increased chance of error and loss of novel sequences when using this method. As such, de novo assembly is the recommended approach for microbial genomes, even when closely related genomes exist. The de novo assembly process can be broken down into three steps: preprocessing, assembly, and postprocessing. Depending on the platform and sequencing libraries, a wide array of computational tools can be used.

2. In the preprocessing step, raw reads that come in compressed form are converted into FASTQ or FASTA files. The reads can then be trimmed and/or filtered based on the quality score and whether ambiguous bases exist. Pattern matching is required to further remove contamination of adapter, barcode and eukaryotic sequences (*see* **Note 2**). It can be helpful to do more sophisticated preprocessing based on kmer abundance (e.g., error correction, coverage estimation), although some assemblers internally implement these analyses.
3. In the assembly step, preprocessed reads are connected into contigs that are as complete as possible. A variety of algorithms exist that automate this process, with the ideal algorithm depending on the sequencing platform being used. Here we introduce four options: (1) for 454 data, the Newbler assembler that comes with the platform is still the best; (2) for Illumina HiSeq data, Velvet [21] and SOAPdenovo [22] are commonly used (*see* **Note 3**); (3) for hybrid assembly with 454 and Illumina data, while relatively slow MIRA [23] implements many heuristics and gives very reliable output; and (4) for a very particular combination of 50× coverage overlapping short-insert paired reads, 50× coverage mate pair reads (2–10 kb) and 50× PacBio reads of insert size 1–3 kb, ALLPATHS-LG [24] can produce assemblies of finished quality for bacterial genomes [25].
4. In the postprocessing step, scaffolding data is used to assemble multiple partial contigs into longer more complete contigs. Although many assemblers integrate scaffolding data into their assemblies, stand-alone scaffolders such as SSPACE [26] and BAMBUS [27] often perform better. To inspect assembly quality in terms of base-level errors and chimeric assembly, a subsample of reads can be mapped to the contigs and analyzed using bwa [28] and samtools [29]. The metric for scoring assemblies is still being actively researched. The Mauve Assembly Metrics [30] used in the first Assemblathon [31] provides plots that assess accuracy, completeness and contiguity. For people who are not familiar with their sequencing protocol or target genomes, it is advisable to use the assembled contigs to reestimate parameters (insert size, sequencing error profile, genome size, etc.) and rerun assembly.

Here, we have focused on the assembly of single microbial genomes. Most of the aforementioned assemblers however fall short for metagenomic projects where terabases of sequencing data are being routinely generated. In these situations, distributed-memory parallel assemblers such as ABySS [32], Kiki [33], and Ray [34] are recommended.

3.2 Automated Structural and Functional Annotation

Once a complete genome sequence has been assembled from sequenced reads (or downloaded from available database, the next step in the model reconstruction process is the structural and functional annotation of the genome. This process has been completely automated in the RAST server (http://rast.nmpdr.org/), which enables users to upload genome sequence data in FASTA format and provides a complete structural and functional annotation approximately 24 h later. Detailed instructions on the use of RAST to perform annotation are provided in a separate chapter [16]. Here, we briefly describe the annotation process itself.

1. The first step in the genome annotation process is structural annotation, which involves the identification of all individual genes within the genome. In practice, this requires a systematic search for the start and stop codons for each gene. While stop codons are relatively simple to identify, start codons are notoriously difficult (varying with open reading frame). As such, in structural annotation of prokaryotes, it is the start sites of genes that are most likely to be inaccurate. Today, Glimmer [35, 36] is the most widely used algorithm for structural annotation of prokaryotic genomes. Structural annotation of eukaryotic genomes is far more challenging due to the existence of introns, exons, and splicing. As such, tools for eukaryotic gene calling are still evolving.
2. Once the structural annotation process is complete, an initial pass is made to assign functions to genes that can be reliably assigned functions based on the FIGfams/kmers [37]. This step often assigns functions to over 90 % of genes, unless the genome is distant from available references. Next, BLAST [38] is used to estimate similarity of genes in the new genome to genes in the reference genomes. Functions are then propagated from reference genomes based on similarity. In the SEED, most reliable and exact functional assignments are made in the first pass; the second pass typically assigns non-exact functions that are primarily useful as clues for the true function of the gene. It is of utmost importance that a controlled vocabulary be used and maintained when assigning gene function, as this facilitates the comparison of annotations and the use of annotations in downstream analysis.
3. In the SEED, annotated genes are gathered into subsystems, which are collections of closely related functions that work together to perform some biological task for the cell (e.g., glycolysis or flagella). Based on the genes represented in the subsystem, the organism is assigned a variant. In some cases, no subsystems variants are complete, and the subsystem is flagged for manual curation (*see* Subheading 3.3). Inevitably, there is an ongoing need to extend the controlled vocabulary of gene functions, as well as the organization of functions into

subsystems. This is achieved by the addition or extension of subsystems encoded in the annotation framework. Within the SEED Project, there is active encoding of new subsystems by experienced annotators, with support provided for user requests for new subsystems.

With the structural and functional annotation process complete, the next step is to curate the genome annotation to identify and correct any errors that arise from the automated annotation process.

3.3 Comparative Genomics and Tree-Based Curation of Genome Annotations

The genome annotation curation process primarily consists of a search for inconsistencies in the structural and functional annotations. For example, a gene that is too short or too long for its associated function is indicative of a truncation, a frameshift, or a misidentified gene fusion; a gene that is clustered with other genes with completely unrelated function may be indicative of a missannotation; and two genes with overlapping coordinates in a prokaryotic genome is indicative of an overzealous gene caller or a missed start site. Inconsistencies are most easily identified via comparative genomics approaches by comparing homologous genes across multiple genomes or by tree-based approaches where a gene is assessed in the context of a phylogenetic tree. Our curation protocol covers both approaches.

1. In the comparative genomics approach to annotation curation, one simply "walks the genome" gene by gene, comparing each gene with its homologs found in phylogenetically close neighbor genomes. Genome browsers that support this sort of gene-by-gene comparative-genomics view of a genome are a necessity for this task. For example, in the PubSEED (http://pubseed.theseed.org), the default view of a gene includes a display of the surrounding chromosomal region of the gene in the current genome, as well as the regions for the gene's homologs in several closely related genomes (Fig. 1).
2. When walking the genome in a comparative genomics viewer, one seeks to identify inconsistencies in the length, function,

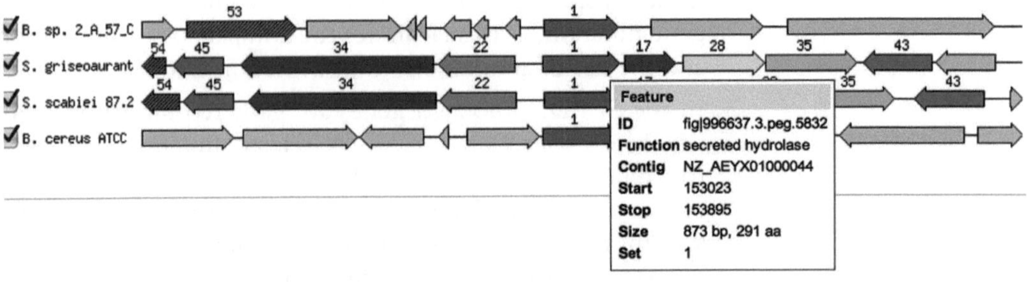

Fig. 1 Comparative genomics view of gene in PubSEED to support curation of genome annotations. Closely related genomes are aligned based on bidirectional best hits of the selected gene. Orthogous genes appearing in the aligned regions are identically colored, enabling the identification of gene clusters and inconsistent annotations at a glance

and context between the gene in the reference genome and the homologous genes in the nearby genomes. If the reference gene is too short, check for a truncation, and ensure that the gene actually has at least one complete functional domain. If the gene is too long, check of a fusion by testing to see if two functional domains are present. If the function of the gene is inconsistent with many of its homologues, check if the functions associated with the homologues are more consistent with the reference genome. Finally, if the gene is clustered with very different genes than its homologues, check if a missannotation has occurred. Chromosomal clustering is the strongest indicator of gene function. Once the gene-by-gene consistency analysis is complete, one can apply tree-based approaches to assess functional annotations.

3. Because trees capture evolutionary relationships much better than BLAST similarity scores, trees are routinely used to evaluate the consistency of functional annotations both manually curated and algorithmically propagated. The process of evaluating a particular gene annotation can be broken down into three steps: (1) tree selection, (2) annotation consistency inspection, and (3) annotation revision if needed.

4. In the tree selection process, one must first discover what trees include the gene of interest. In the PubSEED, this can be done by searching the gene's name or ID on the site homepage (http://pubseed.theseed.org). If the gene is associated with one or more trees, the gene page in the PubSEED will include the statement "*n* alignments and trees" in the left column of the gene information table. Clicking on the "tree" link will take the user to a list of trees that corresponds to an alignment containing at least part of the gene sequence. It is useful to look at the tree summaries on this page before you select the tree(s) you want to work with (*see* **Note 4**).

5. The next step is to inspect the annotation consistency of the selected gene tree. The tree page is rich with controls, coloring, symbols, and mouseovers. The user is advised to consult the "Alignment and Tree Help" button in the upper-left corner. Identical roles (components of the function) are presented on the same background color. A tree for a well annotated class of genes is expected to have visually compatible background colors. That is, genes of the same function should in general cluster into a subtree, and there should not be many more functional shifts along the internal branches of the tree than necessary. When evaluating a functional assignment, the following list of factors should be considered.

   (a) Whether a sequence is directly linked to literature (a dlit, displayed in bold).

   (b) Whether a sequence is in a manually curated subsystem [1] (⌘ sign)

(c) Whether a sequence shares chromosomal context with another sequence of high confidence (black bar) (*see* **Note 5**).

(d) Whether a sequence is truncated or a fused gene (*see* **Note 6**).

(e) How many copies of the gene (paralogs) exist (*see* **Note 7**).

(f) Subtrees connected by longer branches are more likely to have distinct functions.

6. The final step is to annotate genes within the selected tree. Based on the multiple sources of information displayed in a tree, the user can decide to assign a function to a previously hypothetical protein or correct an erroneous annotation with confidence. Once logged in, you should see a check box and a radio button next to each tip in the tree. The checkbox is for selecting genes to be assigned a new function. (All identical sequences collapsed at the tip will be assigned the same new function.) The radio button is for selecting a gene with the function to be applied to the checked sequences. Clicking on an internal node of a tree has the same effect of checking all the sequences in that subtree. This convenience function for batch assignment should however be exercised with extreme care.

Now that the genome annotation has been curated and corrected based on comparative genomics and phylogenetic gene tree analysis, we are ready to begin the process of constructing a genome-scale metabolic model based on the annotation. The first step in the model construction process is development and curation of a high-quality biochemistry database that is suitable for metabolic modeling.

3.4 Assembly and Standardization of Biochemistry Database

Before a model may be produced, a biochemistry database must be established containing all of the biochemical reactions that are required for the model to operate. This biochemistry database must also be prepared for modeling, a process that involves the removal of redundant copies of compounds and reactions in the database, the standardization of compound charges to a single pH (which matches the intracellular pH ideally), a prediction of reaction directionality based on thermodynamics, and the removal or correction of mass imbalanced reactions. We divide this process into four steps:

1. Create core biochemistry from multiple biochemistry databases. This ensures that the database used to construct the models is as complete as possible. Recommended databases include KEGG [39], MetaCyc [40], and other species-specific BioCyc databases [41] (*see* **Note 8**).
2. Match compounds within and between databases using chemical structure. Once compounds have been matched, the reactions they are involved in can be subsequently merged (*see* **Note 9**). Compounds with no known chemical structure can be matched using their names (*see* **Note 10**).

3. Balance every reaction in terms of atomic elements and charge. Unbalanced reactions should never be used within metabolic models (*see* **Note 11**). Due to the possible variation of protonation states, every compound for which structure data is available should be protonated at a pH of 7.
4. Calculate the Gibbs free energy change for every reaction using the group contribution method [42]. Gibbs energies, along with heuristics [6], can be used to determine the reversibility and directionality of each reaction (*see* **Note 12**).

Once a consistent biochemistry database has been established and prepared for modeling, the genome-annotations are used to identify which reactions in the database take part in the metabolic pathways of the organism of interest. This is the genome-scale metabolic reconstruction process.

3.5 Genome-Scale Metabolic Reconstruction

In the genome-scale metabolic reconstruction process, we identify the reactions known to take part in the metabolic pathways of our organism based on the annotated functions assigned to the genes in our genome. This is a two-step process, where reactions are first mapped to the biological functions used in our genome annotation, and then genes are associated with reactions in the form of rules that govern how gene activity relates to reaction activity.

1. A mapping is prepared between the genome annotation ontology and the reactions in the biochemistry database through protein complexes as an intermediate. This mapping is continuously maintained as the annotation ontology and reaction database evolve over time.
2. The annotation ontology map is used to translate gene annotations into a list of metabolic reactions with associated gene–protein–reaction (GPR) rules that govern how gene activity impacts reaction activity. A list of spontaneous reactions (e.g., $CO_2 + H_2O \Rightarrow HCO_3$) is added to every reconstruction, as these reactions occur regardless of the genes present in the genome. These reactions form the genome-scale metabolic reconstruction.

This completes the genome-scale metabolic reconstruction process. The next step in the model reconstruction process is to produce a high-quality core model, to ensure that central carbon pathways and electron transport chains are intact and functioning properly.

3.6 Generation of Core Metabolic Model

A core metabolic model represents the central carbon core metabolism and is designed to accurately simulate energy production of microbial organisms. Cellular energy generation in microbes is an essential part of metabolic modeling, and it depends on environmental factors such as carbon source/electron donor, fermentation capability of the organism, and presence of oxygen or anaerobic electron acceptors.

1. First, the biochemical reactions included in core model reconstruction pathways are carefully selected from the genome-scale metabolic reconstruction based on a well-studied, phylogenetically diverse set of model organisms such as *E. coli* and *B. subtilis*. The sugar degradation pathways (glycolysis, Entner–Doudoroff, citric acid cycle, pentose phosphate), fermentation pathways related to core metabolism and Electron Transport Chains (ETC) are considered the building blocks of the core model. Primarily microbes produce ATP through substrate level phosphorylation and/or oxidative phosphorylation. Electrons from primary electron donors are passed to electron carriers such as $NADH + H^+$ and $FADH_2$. Electron carriers are then oxidized by passing electrons to the ETC; electrons flow spontaneously down an energy gradient through a series of electron carriers; and at specific sites in the chain, the energy released is captured through the coupled pumping of protons out of the cell.
2. Reactions related to ETC are added to the core model to reflect the electron flow and the proton pumping reactions of the organism. Respiration is a complex process in microbes consisting of numerous proton-pumping reactions, oxidases, and reductases [43, 44]. Unlike mitochondria, bacterial ETCs are branched at the quinone or cytochrome and have the ability to alter the ETCs depending on the growth conditions. A facultative anaerobic organism like *E. coli* may use two routes during aerobic growth depending on the oxygen concentration (Fig. 2a) and also be able to utilize alternative terminal electron acceptors (such as fumarate, nitrate, DMSO, or TMAO) under anaerobic conditions. *Paracoccus denitrificans* contains a highly branched and diverse ETC where it grows aerobically and is able to reduce nitrate, nitrite, nitrous oxide, or nitric oxide anaerobically (Fig. 2b)[45–47]. In contrast to *E. coli* and *P. denitrifcians*, Clostridia does not use either aerobic or anaerobic respiration thus grows solely by fermentation. Microbes utilize fermentation processes to recycle electron carriers ($NADH + H^+$, $FADH_2$) and also produce ATP through substrate level phosphorylation.
3. Upon successfully building the core model, the Flux Balance Analysis (FBA) is performed under different media compositions to predict ATP yields, ETC pathways, and fermentation ability.

Once the core model is complete and displays accurate production of ATP via either ETC reactions or fermentation pathways, the next step in model reconstruction is the generation of a Biomass Composition Reaction.

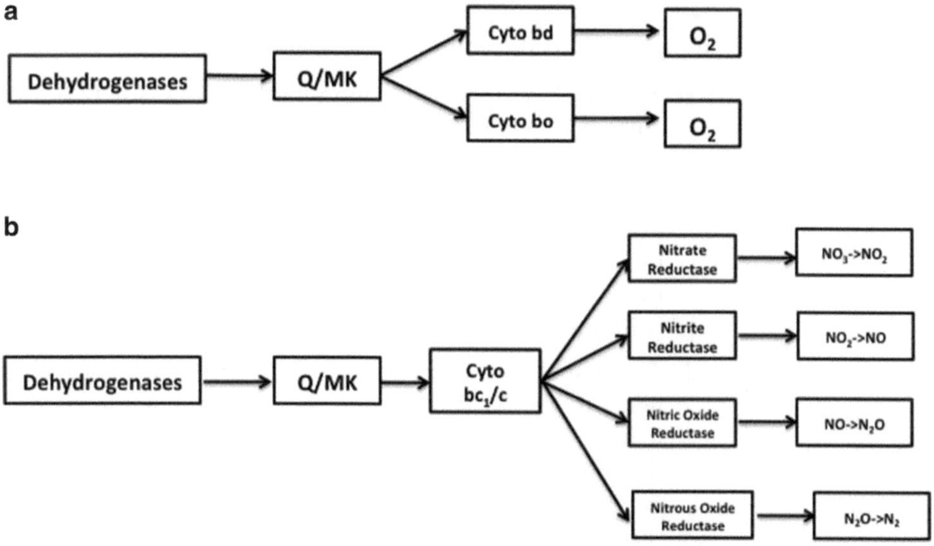

Fig. 2 ETC variations included in core model. Aerobic ETC variation of *E. coli* where cytochrome bd is synthesized under low oxygen conditions and cytochrome bo synthesized under high oxygen conditions (**a**). Anaerobic ETC variation of *P. denitrificans* where nitrate, nitrite, nitric oxide, or nitrous oxide may be used as the terminal electron acceptor (**b**)

3.7 Generation of Biomass Composition Reaction

A biomass composition reaction (BCR) is an abstract construct in metabolic models that describes all of the small molecules that a cell must produce in order to generate one gram of biomass. The metabolites in the BCR include the amino acid building blocks of proteins, the nucleotide building blocks of DNA and RNA, the lipid building blocks of cell walls, and even the trace cofactor compounds required to drive biochemical reactions. One of the most common objectives in metabolic modeling is the maximization of the flux through the BCR, representing the maximization of growth yield by the cell. As such, the content of the BCR has a great impact on the accuracy of the model's quantitative and qualitative predictions. The small molecules included in the BCR of a model depend upon the type of cell wall, the metabolic pathways present in the cell, and the electron transport chains utilized by the cell. It is important to note that an exact BCR cannot be automatically generated from genome sequence alone, because the relative quantity of biomass components cannot be predicted exactly from genome sequence. Thus, the coefficients for metabolites in automatically generated BCRs are only approximations, and experiments must be performed to obtain exact values for these coefficients. However, a qualitatively accurate BCR can be generated from a template based on the genome annotation. Previously in resources like the Model SEED, the template BCR was constructed manually based on literature review and functional curation, but here we will discuss an improved data-driven means of constructing a template BCR through machine learning.

1. The machine specific learning methodology for BCR design is a naïve Bayesian classifier (NBC). The NBC method relies on finding a statistical relationship between the occurrence of sets of functional roles of individual species, and their classification (e.g., as Gram positive or Gram negative, aerobic or anaerobic). A subset of the data, which has a known classification, is used to train the classifier, which can then be applied to classify subsequent unknown data in the same classifications.
2. The first step in creating an NBC for the inclusion of components in the BCR is to identify the readily available data that will be used as predictors in the BCR. In this case, we utilize the functional roles that are assigned to the genes in our genomes during the genome annotation process. This data is readily available even for new genomes, because it is automatically generated by our genome annotation process.
3. The next step is to select a training set of organisms with known BCRs and retrieve the functional roles for the genomes in this training set. Resources like the PubSEED can be readily used to retrieve genome annotations, and these annotations will be consistent with those generated by the RAST annotation pipeline. This consistency is critical for the success of the method. The training set data thus consists of a table of species, their known classification, and a list of the functional roles they exhibit. The selection of the training set requires some domain knowledge: (1) all desired classes must be represented in the set, and (2) the genomes selected for each class must be diverse. For example, if only *E. coli* genomes are selected as examples of Gram negative organisms, then the NBC produced will be a specific predictor of *E. coli* genomes rather than a general predictor of Gram negative genomes.
4. To train the classifier, the training set data must be split into three groups. The first group is used to derive the statistical relationships. The exact associative statistics will of course be totally dependent on the training data, and as such will differ in behavior and performance depending on which species were chosen for the training set. The optimum classifier will be the one derived from the optimum training set. As this cannot be known in advance, this has to be an iterative process of selecting a training set from a larger set. This larger set is the training set plus a second group of data, the validation set. The behavior and performance of a classifier is measured against this separate validation set. A note of the best performing classifier is maintained, and a new training set is selected from the combination of the previous training set and the previous validation set. The new training set is then used to produce a new classifier, which in turn is compared to the previous best

performing classifier. This process is repeated until either all combinations have been exhausted or until the optimum classifier has remained constant for a set number of iterations. This optimal classifier is then in turn tested against the third and final split in the sample data, the test set. Note that unlike the training set or the validation set, this data has not been used in the production of the classifier and is thus a more accurate measure of the true performance of the classifier against a fully unknown dataset. The use of this third dataset prevents the phenomenon of over fitting of the classifier to the sample dataset.

5. The NBC produced by this process then determines the classification of an unknown genome by looking at the functional roles of its genes. The NBC then uses the values derived from the training data to assign a probability that a genome is one class or another (e.g., Gram positive or Gram negative). The classification is chosen by the higher probability and the closer the higher value is to one, the more confidence the classifier places on its classification.
6. This classification approach can be used to determine which cofactors should be present, what kind of cell wall should be included, and approximate values for relative amounts of DNA, RNA, cell wall, and amino acids that should be included in the BCR. Additionally, all BCR include the ATP hydrolysis reaction, with ATP and H_2O serving as reactants and ADP and phosphate serving as products, to represent the aggregate energy costs associated with the synthesis and maintenance of 1 g of biomass. The coefficient on this ATP consumption term of the BCR has a large impact on predicted growth yields, and this value cannot be automatically predicted. It must be fit to experimental data. Typical values for a metabolic model fall between 30 and 100 mmol ATP per gram cell dry weight per hour [48].
7. DNA is the one category of BCR components for which stoichiometry can be calculated directly from the DNA sequence. Molar fractions of the deoxy nucleotides are set according to the GC content (e.g., deoxyguanine=deoxycytosine = GC and deoxyadenine and deoxythiamine = 1-GC).

At the end of this process, we have a draft BCR, which is the final piece required to have a complete genome-scale metabolic model, although curation of the metabolite coefficients and contents of the BCR is critical to model accuracy. The next step of the process is to assess the ability of the model to produce all BCR components and to gap-fill the model if necessary.

3.8 Completion of Draft Metabolic Model

The genome-scale metabolic reconstruction, core metabolic model and biomass objective function produced by the protocols described in Subheadings 3.5–3.7 contain all the data required to produce a functioning genome-scale metabolic model capable of predicting phenotypes with approaches like Flux Balance Analysis [9–12]. However, even when working with the most well known of microbial organisms (e.g., *E. coli*, *B. subtilis*), these models will be unable to produce some components of the biomass composition reaction due to gaps in the genome annotations and corresponding metabolic pathways. An optimization-based algorithm is commonly used to fill these gaps.

1. First, a biochemistry database in which all generic reactions, lumped reactions, and unbalanced reactions have been removed is merged with the reactions and compounds of the draft metabolic model. In this process, we ensure that identical compounds and reactions are unified to produce a single nonredundant biochemical network.
2. Next a flux balance analysis problem is formulated (described in detail elsewhere [6]) where the objective is to minimize the addition of reactions to the model while still enabling some flux through the biomass reaction of the model. This analysis may be run repeatedly to produce multiple alternative solutions to the gap-filling problem.
3. Each solution represents a set of reactions that must be either added or made reversible in order to enable the metabolic model to produce biomass in the media condition selected for autocompletion.

Once the autocompletion process is complete, the metabolic model will be capable of producing biomass in the media condition in which the autocompletion was performed (typically complete media). At this stage, flux balance analysis may be used to generate qualitative predictions of essential genes, growth conditions, growth phenotypes, and metabolic capabilities. But this metabolic model is still a draft model, and substantial curation must be performed before the model is capable of generating accurate quantitative predictions. In the next section, the tools available for viewing and curating metabolic models will be explored, with an emphasis on the Model SEED Web site.

3.9 Curation of Draft Metabolic Model

The automated model reconstruction process generates draft models capable of simulating growth from transportable nutrients. However, due to the current knowledge gaps in genome annotations and databases, generating high-quality predictive models requires further manual curation. Manual curation is an iterative process of network evaluation and verification, which is repeated until model predictions match the phenotypic characteristics of the organism of interest (Fig. 3). Here we describe the major steps involved in manual curation:

1. The greater the wealth of information available on the organism under study, the more extensively the model can be curated. Phenotypic data can be collected from the literature.

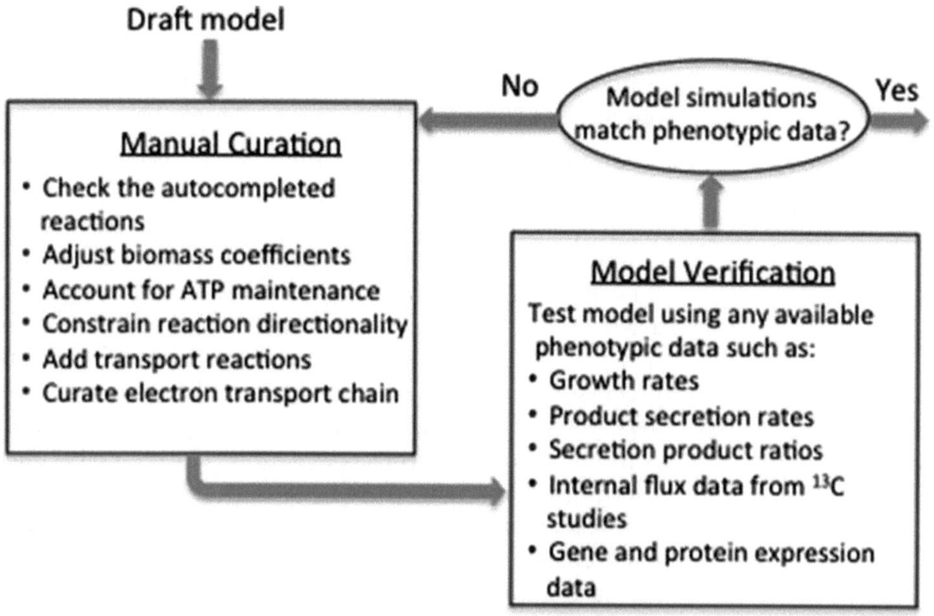

Fig. 3 Overview of the iterative manual curation process for the draft model developed by ModelSEED

Such information could include but are not limited to: (a) biochemical data that point to gene product function from gene expression studies, (b) genetic data that provide direct evidence for gene function, (c) physiological data that provide indirect evidence for biochemical reactions, such as substrate utilization, and product secretion from which information such as transport reactions and active metabolic pathways can be inferred. Further, physiological growth data available in the literature both with respect to what substrates the organism can grow on as well as "negative" data such as the substrates and conditions that the organism has been experimentally proven to be incapable of growing on are desirable. In the case that information on the organism under study is very limited, phylogenetically close organisms could be used as a starting point.

2. Sites like the ModelSEED (http://www.theseed.org/models) provide excellent interfaces for viewing and analyzing metabolic models once they are generated. The Model Viewer interface displays the model in a table, which contains tabs for reactions, compounds, biomass reactions, genes, media formulations, and maps that link to KEGG maps for the visualization of the pathways concerning the reaction of interest. Tools are available for performing analyses such as flux balance analysis, flux variability analysis, simulating knock outs of combinations of one or more genes, and creating and changing media conditions, to name a few. Furthermore, Model Viewer has

comparative tools allowing the user to compare the draft model with other previously published well-curated models. Model content can also be visualized through the Cytoscape SEED plugin for Cystoscape [1].

3. A template is used to determine which components of the biomass make up are necessary to simulate growth for each unique organism modeled. The coefficients for each of the components of the biomass equation require curation based on experimental data. In the case that no experimental data is available on the relative fractions (weight of biomass component/weight dry cell), approximate fractions of some biomass precursors can be determined from the genome (*see* **Note 13**). Of notable importance is the coefficient for the ATP component of the biomass equation. During cell replication, energy (in the form of ATP) is required for macromolecule synthesis. This energy is referred to as growth associated ATP maintenance. When the coefficients for the biomass precursors are adjusted, the growth associated ATP maintenance (GAP) can be determined by adjusting the coefficient to match experimentally determined growth yields when growth is computationally simulated.
4. The reactions used to generate the draft models are not organism specific, and are therefore mostly reversible. Biochemical data on the organism under study is necessary to constrain reactions to ensure that the flux through the metabolic pathway is physiologically relevant and reactions are thermodynamically feasible. Furthermore, if the system is under-constrained (i.e., has too many reversible reactions) futile cycles may appear in the model. Reaction directionality can be inferred from the Gibbs free energy of formation $\Delta_r G'^\circ$ of a reaction. Methods have been developed for the estimation of $\Delta_r G'^\circ$ for biochemical reactions [4].
5. The reactions added during the autocompletion process should always be checked for physiological relevancy. The autocompletion process uses an optimization algorithm to add the minimum number of reactions necessary to fill in the gaps. The added reactions are selected from a repository of mass and charge balanced reactions, and therefore could be completely unrelated to the organism of interest. It is especially important to check the reactions associated with the electron transport chain as autocompletion might result in the production of ATP in physiologically unreasonable ways.
6. Extracellular and intracellular transport systems are not well annotated, and for this reason, draft models generally lack transport reactions and manual curation is necessary. A general approach when experimental data is not available is to minimize

the number of transport reactions as to avoid bypass of essential intracellular pathways.

7. Aside from verifying the complete model at the genome scale, the verification step can also be approached by a pathway-by-pathway analysis, in that major pathways or functions that are known to exist in the organism under study are checked for functionality independently of the entire model. For such verification only the metabolic precursors for the individual pathway must be included in the medium and the metabolic output(s) of the pathway included as the only reactants in the biomass equation. This technique is especially useful when debugging potential problems in growth; each of the biomass substrates can be looked at individually to pinpoint which precursor is not being produced properly and thus hindering growth.

Comparison with existing published models of similar organisms is still one of the fastest ways to identify and correct errors that occur in the annotation and automated model reconstruction process. A detailed protocol [49] also exists for the manual model creation and curation process, which is one of the best available resources for the model curation and reconstruction process. We recommend reading this protocol in detail, comparing the protocol with the content of this book chapter, and using the protocol as a guide to the process of curating and revising your draft metabolic model.

3.10 Generation of Transcriptional Regulatory Constraints

While the reconstruction of metabolic pathways and metabolic models is starting to become routine for microbial organisms, we still lack a full understanding of regulatory mechanisms and in particular the interplay between metabolism and regulation [50]. Fully characterized regulatory networks are available for only a few well-known organisms such as *E. coli* [51] and *B. subtilis* [52]. However, data on transcriptional regulatory networks and gene expression profiles is becoming available for a growing number of organisms, inducing us to include the generation of transcriptional regulatory constraints in our discussion of the genome annotation and metabolic model reconstruction process. Different strategies have been employed to reconstruct transcriptional regulatory networks: (1) propagation of template networks to close genomes (*see* **Note 14**) [53], (2) prediction of *cis*-regulatory elements (*see* **Note 15**) [54] and (3) reverse engineering from expression data (*see* **Note 16**) [55]. These methods enable a characterization of regulatory networks, but an integration with metabolic models is required for the application of regulatory network knowledge to the prediction of phenotypes. Current integrated metabolic regulatory models [56] consist of the standard metabolic components augmented with a set of Boolean rules that connect gene activity to the presence or activity of external metabolites, stimuli or transcription factors (Fig. 4).

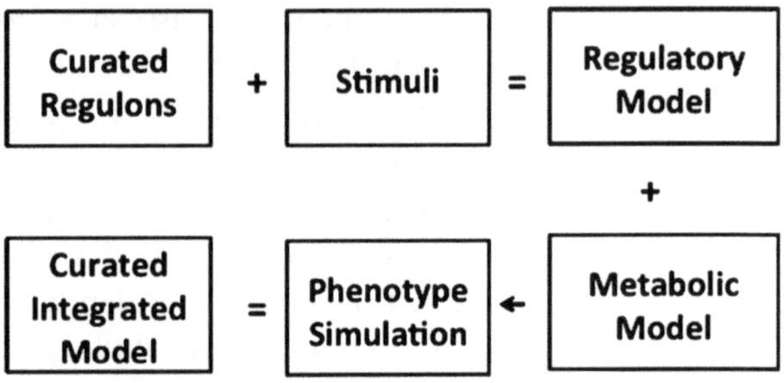

Fig. 4 Integrated metabolic regulatory model reconstruction. Starting with a set of curated regulons and the stimuli affecting the regulated genes, a simple model of *genes–effector response* can be built. Integration with the metabolic model for the same organism allows the prediction of growth phenotypes

Here, we outline a simple process applied to the reconstruction of regulatory constraints for a metabolic model:

1. The first step is to construct a high-quality template model for a model organism. This model contains all knowledge of the regulons (sets of co-regulated genes), transcription factors (which activate or inhibit regulons), and stimuli (which trigger the activity of transcription factors) that comprise the regulatory network of the model organism. This data can be collected from comprehensive databases, or computed from expression data, but we note that a template model can only be constructed for a small number of genomes (e.g., *E. coli*, *B. subtilis*, *S. cerevisiae*) due to the limited availability of the needed data.
2. Regulatory network data for the high-quality model must then be translated into Boolean rules governing the conditions when regulons are active or inactive. Stimuli and transcription factors become the triggers for the rules, while regulons are the entities triggered in each rule.
3. These regulatory constraints are then integrated with a metabolic model by mapping genes in the regulons to genes in the metabolic model. The two models are then simulated together using rFBA [56], enabling the integrated prediction of phenotypes. These phenotype predictions should be tested for quality to determine if some regulatory rules are overly constraining and must be relaxed. This process can unveil interesting biological questions and culminate with changes in the originally curated regulons, the metabolic model or fixing annotation errors.
4. Once complete, the template regulatory model may be propagated to close genomes by mapping genes in regulons to their orthologs in the close genomes. While the propagation of transcription factors in this manner is not recommended, regulons and their stimuli are often conserved across short phylogenetic distances.

5. Once the propagation of the template model regulons and stimuli is complete, the new propagated model may then be tested against all expression data available for the new genome. Regulons may be identified in the new genome based on expression data, and these expression-data-based regulons may be used to expand or split the regulons in the propagated model. In this way, we bring the propagated model into consistency with the expression data. The data necessary for this reconciliation is currently available for 24 genomes in the PubSEED (http://pubseed.theseed.org/seedviewer.cgi?page =AtomicRegulon&genome=all).
6. Finally, the propagated model constraints may be integrated with the metabolic constraints from the reconstructed genome-scale model to enable the prediction of phenotypes, flux profiles, and responses to genetic and environmental perturbations.

This completes the final step of a semi-automated reconstruction procedure for integrated genome-scale metabolic and regulatory models. We note that this process successfully integrates extensive biological data sources including biochemistry, genome sequences, subsystems and functional roles, phylogenetic gene trees, growth phenotypes, and gene expression data.

4 Notes

1. For a performance comparison of the leading next-generation sequencing platforms, *see* ref. [57] for benchmark results. For the sequencing protocol, it is almost always better to go for multiple paired-end libraries with a wide range of insert sizes than a single library because of the prevalence of repeat regions in DNA. There is a tendency for labs to use ultra-deep sequencing to compensate for the lack of multiple insert sizes because the former approach is less expensive, but it rarely pays off. In fact, high coverage reads create more burdens for preprocessing and can confuse de Bruijn graph based assemblers because of the sequencing errors they expose.
2. The most commonly used tools for general read conversion, trimming, and filtering are FASTX Toolkit (http://hannonlab.cshl.edu/fastx_toolkit/), cutadapt (http://code.google.com/p/cutadapt/), and seqtk (https://github.com/lh3/seqtk). We recommend seqtk for its superior efficiency when it is applicable.
3. Without long-insert libraries, most de novo assembly generates 100–200 contigs for a typical bacterial genome. Sequence coverage above 50× makes little difference. It is common practice to run Velvet or SOAPdenovo with a wide range of kmer sizes to find the optimal assembly or merge the contigs across runs.

4. The gene tree selection page lists the trees with their basic information: (1) the part of gene sequence covered, (2) the number of sequences, (3) a breakdown of the current function annotations. Because the SEED uses an iteratively profile trimming procedure to build the set of sequences nucleated by a gene family, each gene can be associated with a tree either as a trimmed domain or in its entirety. If you are interested in sorting out functions of very closely related sequences (e.g., enzyme specificity), you need to select a tree with near complete sequences. If you instead are looking to study a broad class of genes, you may prefer a large tree that shows how functions evolve.
5. The SEED computes function projection scores for bidirectional-best-hit genes (BBHs). The score takes into account how similar the DNA neighborhood of two genes is. For example, if a gene is surrounded by ten genes that are BBHs for the genes surrounding a dlit, this gene gets a full projection score (full black bar) and is strongly believed to share the same function with the dlit.
6. The background color of the check box and radio button reflects the amount of the given gene that is included in the alignment used to compute the tree. If the radio button next to a gene is too colorful, care should be exercised in assigning a function; it might be a multidomain (multifunctional) protein.
7. When multiple sequences from the same genome exist in a tree, it is often the case that these are paralogs that perform different functions. The locations of paralogs can also be good indicators about how far other sequence functions can be reliably propagated in the tree.
8. Most biochemistry databases are either free or offer a free academic license; KEGG however requires a paid academic license to access their FTP site.
9. For KEGG and MetaCyc, compound structures in the MOL format can be downloaded from their FTP site. For other species-specific BioCyc databases, the compound structures are available in InChI format within the flat files. The InChI format is by far the most appropriate format for comparing compound structures, as the algorithm used to calculate the connectivity and stereochemistry generates a string that can be considered canonical. Other chemical string formats such as SMILES cannot be used for this purpose. Furthermore, the InChI format separates the protonation state of a compound from the actual charge of the compound, allowing one to distinguish between multivalent inorganic elements, when using InChI format to compare compound structures, the “proton” or “p” layer must be removed. Most chemoinformatics software packages are able

to convert files in the MOL format into InChI format; examples of these packages include ChemAxon's MarvinBeans (commercial) and OpenBabel (open-source). Other databases have been formed specifically with the goal of merging biochemistry data for the purposes of metabolic modeling, notably BKM, MetRxn, and MetaNetX. At the time of writing, each of these had many incorrectly merged compounds, but may be updated in the near future and these databases are worth investigating. Manual review of a few simple compounds such as iron-containing compounds, or simple sugars, would indicate whether the problems still exist.

10. We have attempted to match names automatically by creating "searchname" strings, where every character is lowercase, and all punctuation is removed. We have found that outside of using compound structure, there are very few compounds in the biochemistry databases that can be merged on the basis of their names. However, this method has to be used when attempting to incorporate a metabolic model for which no compound structures have been generated. Manual review is needed to ensure that this process worked, especially if many of the names themselves do not contain stereochemical information. One key assumption that needs to be checked is that every compound in a metabolic model is unique. If any two or more compounds are matched to the same compound in the core biochemistry, then these need to be manually reviewed. Ensure that you are satisfied with the merging of compounds *before* merging reactions.

11. In order to balance every reaction at the correct pH, in terms of both atomic elements and charge, every compound must be charged at the neutral pH of 7. We use a plugin available through ChemAxon's MarvinBeans to do this, named "majorms." This plugin finds the dominant protonation state of a compound at a certain pH, it can be run using this command: "cxcalc–N hi majorms–H 7–f<format>:-a example>chargedExample," where the format can be either "inchi" or "mol". Several types of compounds cannot be charged in this manner: (1) Compounds with multiple non-covalently connected components. One unfortunate feature (it is not certain whether or not this is a bug) of the "majorms" plugin is that it is inconsistent as to whether it will charge every non-covalently bound component in a compound. As a result, the "majorms" plugin only outputs one component. Using the example of sodium hydroxide, the MOL file containing NaOH becomes a MOL file containing only Na; (2) Compounds with unknown atom groups (typically represented by a "R"); and (3) Polymeric compounds with an indeterminate number of repeats. For these compounds, we use the fully protonated

state. When balancing reactions including charged compounds in the InChI format, be aware that the InChI format includes the total possible number of protons in the formula, and uses the "proton" layer to indicate the protonation state. As a result, one must apply the output of the "proton" layer to the total number of protons for any given compound, before attempting to balance the reaction. If a reaction is imbalanced, it is not included in metabolic modeling, with the exception of reactions where the only reason for the imbalance are protons. It is acceptable to balance these reactions, manually, or automatically, by adding or deduction protons to/from the reaction order to balance the reactions.

12. Special care needs to be applied to transporters for which the Gibbs energy of transport also needs to be calculated, integrating the electrochemical gradient. For practical purposes, when using transporters to feed the in silico organism with nutrients or to allow undesired compounds to leave the in silico organism, the directionality of the transporters should be manually reviewed.
13. Fractional contributions of deoxynucleotide triphosphates (dATP, dGTAP, dCTP, and dTTP) can be readily calculated from the genome and although fractional contributions of amino acids as well as RNA nucleotides can also be estimated from the genome this is not recommended. This view would suggest that all genes and proteins in the cell are expressed in equal quantities at all times, which is a poor assumption. If experimental data is not available for determining the coefficients of the biomass precursors, it should be noted that ATP maintenance has the highest coefficient value and will exert the greatest influence on the yield computations. Obtaining experimental data on the growth of the organism, which is relatively easy to do will suffice for ATP coefficient estimations.
14. This particular methodology relies on previous observations of the conservation of prokaryotic gene networks [57]. A search for orthologous genes is conducted on the genome of interest. With a well-characterized network as starting point, the orthologous TFs and their targets noted on the target genome. In order to evaluate the significance of tendencies we observe in the well-characterized network and to assure that the observed trends are meaningful, the generation of random networks is the method to assess the statistical significance of the trend. After this analysis, the new interactions on the target genome are reconstructed.
15. To perform this task, the RegPredict Web resource [58] (http://regprecise.lbl.gov/RegPrecise/index.jsp) is the state-of-the-art tool for TRN reconstruction including Transcription

factor binding site data (TFBS data). RegPrecise contains both a repository of manually curated regulons and a large set of comparative genomics tools available in two reconstruction pipelines; the first reconstructs regulons for known position weight matrixes (PWMs) and the second performs ab initio regulon inference for unknown binding sites using analysis of regulon orthologues across closely related genomes or within the same pathway/SEED subsystem.

16. The universe of algorithms for reverse engineering from expression data is vast and different categories of algorithms have been proposed: (1) a mathematical modeling approach [59], (2) a module-based or direct inference methods [60], and (3) an unsupervised and (semi)-supervised methodologies [61].

Acknowledgements

We acknowledge the entire SEED, Model SEED, and CytoSEED teams at Argonne National Laboratory, Fellowship for Interpretation of Genomes, Hope College and University of Chicago for efforts on the frameworks described in this chapter. This work was supported by the US Department of Energy under contract DE-ACO2-06CH11357 (J.F., F.X., C.H.), and the National Science Foundation under grants PGRP-1025398 (S.S.), EFRI-1137089 (R.C., T.K.) and MCB-1153413 (J.E.). José P Faria acknowledges funding from his Ph.D. grant FRH/BD/70824/2010 of the FCT (Portuguese Foundation for Science and Technology).

References

1. Feist AM, Palsson BO (2008) The growing scope of applications of genome-scale metabolic reconstructions using Escherichia coli. Nat Biotechnol 26(6):659–667
2. Burgard AP, Pharkya P, Maranas CD (2003) OptKnock: a bilevel programming framework for identifying gene knockout strategies for microbial strain optimization. Biotechnol Bioeng 84(6):647–657
3. Pharkya P, Burgard AP, Maranas CD (2004) OptStrain: a computational framework for redesign of microbial production systems. Genome Res 14(11):2367–2376
4. Ranganathan S, Suthers PF, Maranas CD (2010) OptForce: an optimization procedure for identifying all genetic manipulations leading to targeted overproductions. PLoS Comput Biol 6(4):e1000744
5. Kumar VS, Maranas CD (2009) GrowMatch: an automated method for reconciling in silico/in vivo growth predictions. PLoS Comput Biol 5(3):e1000308
6. Henry CS et al (2009) iBsu1103: a new genome scale metabolic model of B. subtilis based on SEED annotations. Genome Biol 10(6):R69
7. Karr JR et al (2012) A whole-cell computational model predicts phenotype from genotype. Cell 150(2):389–401
8. Edwards JS, Palsson BO (2000) The Escherichia coli MG1655 in silico metabolic genotype: its definition, characteristics, and capabilities. Proc Natl Acad Sci U S A 97(10):5528–5533
9. Papoutsakis ET, Meyer CL (1985) Equations and calculations of product yields and preferred pathways for butanediol and mixed-acid fermentations. Biotechnol Bioeng 27(1):50–66
10. Jin YS, Jeffries TW (2004) Stoichiometric network constraints on xylose metabolism by

recombinant Saccharomyces cerevisiae. Metab Eng 6(3):229–238

11. Varma A, Palsson BO (1994) Stoichiometric flux balance models quantitatively predict growth and metabolic by-product secretion in wild-type Escherichia-coli W3110. Appl Environ Microbiol 60(10):3724–3731
12. Varma A, Palsson BO (1993) Metabolic capabilities of Escherichia-coli. 2. Optimal-growth patterns. J Theor Biol 165(4):503–522
13. Aziz RK et al (2008) The RAST Server: rapid annotations using subsystems technology. BMC Genomics 9:75
14. Henry CS et al (2010) High-throughput generation, optimization, and analysis of genome-scale metabolic models. Nat Biotechnol 28(9):977–982, Nbt.1672: p. 1-6
15. Schuler GD et al (1996) Entrez: molecular biology database and retrieval system. Methods Enzymol 266:141–162
16. Devoid S, et al. (2013) Automated genome annotation and metabolic model reconstruction in the SEED and Model SEED. Methods Mol Biol 985:17–45
17. Kanehisa M (2002) The KEGG database. Novartis Found Symp 247:91–101, discussion 101-3, 119-28, 244-52
18. Overbeek R, Disz T, Stevens R (2004) The SEED: a peer-to-peer environment for genome annotation. Commun ACM 47(11):46–51
19. Caspi R et al (2008) The MetaCyc database of metabolic pathways and enzymes and the BioCyc collection of Pathway/Genome databases. Nucleic Acids Res 36(Database issue): D623–D631
20. Mellmann A et al (2011) Prospective genomic characterization of the German enterohemorrhagic Escherichia coli O104:H4 outbreak by rapid next generation sequencing technology. PLoS One 6(7):e22751
21. Zerbino DR, Birney E (2008) Velvet: algorithms for de novo short read assembly using de Bruijn graphs. Genome Res 18(5):821–829
22. Li R et al (2010) De novo assembly of human genomes with massively parallel short read sequencing. Genome Res 20(2):265–272
23. Chevreux B et al (2004) Using the miraEST assembler for reliable and automated mRNA transcript assembly and SNP detection in sequenced ESTs. Genome Res 14(6): 1147–1159
24. Gnerre S et al (2011) High-quality draft assemblies of mammalian genomes from massively parallel sequence data. Proc Natl Acad Sci U S A 108(4):1513–1518
25. Ribeiro FJ et al (2012) Finished bacterial genomes from shotgun sequence data. Genome Res 22(11):2270–2277
26. Boetzer M et al (2011) Scaffolding pre-assembled contigs using SSPACE. Bioinformatics 27(4):578–579
27. Pop M, Kosack DS, Salzberg SL (2004) Hierarchical scaffolding with Bambus. Genome Res 14(1):149–159
28. Li H, Durbin R (2010) Fast and accurate long-read alignment with Burrows-Wheeler transform. Bioinformatics 26(5):589–595
29. Li H et al (2009) The Sequence Alignment/Map format and SAMtools. Bioinformatics 25(16):2078–2079
30. Darling AE et al (2011) Mauve assembly metrics. Bioinformatics 27(19):2756–2757
31. Earl D et al (2011) Assemblathon 1: a competitive assessment of de novo short read assembly methods. Genome Res 21(12):2224–2241
32. Simpson JT et al (2009) ABySS: a parallel assembler for short read sequence data. Genome Res 19(6):1117–1123
33. Xia F, Stevens RL Kiki: Terabase metagenome assembly with massively parallel computers. (Manuscript in preparaton)
34. Boisvert S, Laviolette F, Corbeil J (2010) Ray: simultaneous assembly of reads from a mix of high-throughput sequencing technologies. J Comput Biol 17(11):1519–1533
35. Delcher AL et al (1999) Improved microbial gene identification with GLIMMER. Nucleic Acids Res 27(23):4636–4641
36. Delcher AL et al (2007) Identifying bacterial genes and endosymbiont DNA with Glimmer. Bioinformatics 23(6):673–679
37. Meyer F, Overbeek R, Rodriguez A (2009) FIGfams: yet another set of protein families. Nucleic Acids Res 37(20):6643–6654
38. Altschul SF et al (1997) Gapped BLAST and PSI-BLAST: a new generation of protein database search programs. Nucleic Acids Res 25:3389–3402
39. Kanehisa M et al (2008) KEGG for linking genomes to life and the environment. Nucleic Acids Res 36(Database issue):D480–D484
40. Karp PD et al (2005) Expansion of the BioCyc collection of pathway/genome databases to 160 genomes. Nucleic Acids Res 33(19):6083–6089
41. Keseler IM et al (2011) EcoCyc: a comprehensive database of Escherichia coli biology. Nucleic Acids Res 39(Database issue): D583–D590
42. Henry CS, Broadbelt LJ, Hatzimanikatis V (2007) Thermodynamics-based metabolic flux analysis. Biophys J 92(7):1792–1805

43. Anraku Y (1988) Bacterial electron transport chains. Annu Rev Biochem 57:101–132
44. Unden G, Bongaerts J (1997) Alternative respiratory pathways of Escherichia coli: energetics and transcriptional regulation in response to electron acceptors. Biochim Biophys Acta 1320(3):217–234
45. Cox JC et al (1978) The variable cytochrome content of Paracoccus denitrificans grown aerobically under different conditions. FEBS Lett 93(2):261–265
46. Puustinen A et al (1989) Cytochrome o (bo) is a proton pump in Paracoccus denitrificans and Escherichia coli. FEBS Lett 249(2):163–167
47. de Gier JW et al (1994) The terminal oxidases of Paracoccus denitrificans. Mol Microbiol 13(2):183–196
48. Feist AM et al (2009) Reconstruction of biochemical networks in microorganisms. Nat Rev Microbiol 7(2):129–143
49. Thiele I, Palsson B (2010) A protocol for generating a high-quality genome-scale metabolic reconstruction. Nat Protoc 5:93–121
50. Schilling O et al (2007) Transcriptional and metabolic responses of Bacillus subtilis to the availability of organic acids: Transcription regulation is important but not sufficient to account for metabolic adaptation. Appl Environ Microbiol 73(2):499–507
51. Gama-Castro S et al (2011) RegulonDB version 7.0: transcriptional regulation of Escherichia coli K-12 integrated within genetic sensory response units (Gensor Units). Nucleic Acids Res 39(Database issue):D98–D105
52. Sierro N et al (2008) DBTBS: a database of transcriptional regulation in Bacillus subtilis containing upstream intergenic conservation information. Nucleic Acids Res 36(Database issue):D93–D96
53. Madan Babu M, Teichmann SA, Aravind L (2006) Evolutionary dynamics of prokaryotic transcriptional regulatory networks. J Mol Biol 358(2):614–633
54. Rodionov DA (2007) Comparative genomic reconstruction of transcriptional regulatory networks in bacteria. Chem Rev 107(8):3467–3497
55. Bansal M et al (2007) How to infer gene networks from expression profiles. Mol Syst Biol 3:78
56. Covert MW et al (2004) Integrating high-throughput and computational data elucidates bacterial networks. Nature 429(6987):92–96
57. Gelfand MS (2006) Evolution of transcriptional regulatory networks in microbial genomes. Curr Opin Struct Biol 16(3):420–429
58. Novichkov PS et al (2010) RegPredict: an integrated system for regulon inference in prokaryotes by comparative genomics approach. Nucleic Acids Res 38(Web Server issue): W299–W307
59. Hache H, Lehrach H, Herwig R (2009) Reverse engineering of gene regulatory networks: a comparative study. EURASIP J Bioinform Syst Biol 2009:617281
60. Michoel T et al (2009) Comparative analysis of module-based versus direct methods for reverse-engineering transcriptional regulatory networks. BMC Syst Biol 3:49
61. Elati M, Rouveirol C (2011) Unsupervised learning for gene regulation network inference from expression data: a review. Algorithms in Computational Molecular Biology. 955–978

Part II

Thermodynamics: A Step Beyond MFA

Chapter 3

Constraining the Flux Space Using Thermodynamics and Integration of Metabolomics Data

Keng Cher Soh and Vassily Hatzimanikatis

Abstract

Flux balance analysis of stoichiometric metabolic models has become one of the most common methods for estimating intracellular fluxes. However most of these networks are underdetermined and can have multiple alternate optimal flux distributions. Thermodynamic constraints can reduce the solution space significantly and at the same time provide a platform for the integration of metabolomics data. Here we go through the procedure to incorporate thermodynamic constraints and perform thermodynamic analysis of metabolic networks.

Key words Metabolic networks, Flux balance analysis, Thermodynamics-based flux balance analysis, Constraint-based analysis, Metabolomics, Fluxomics

1 Introduction

Metabolism is a complex network of intertwined biochemical reactions that transforms a set of available extracellular compounds into biomass and metabolic by-products. Mathematical modeling is an essential tool to help in the understanding of these complex biochemical networks and one of the most popular representation and form of analysis is through the use of stoichiometric models. Stoichiometric models have been traditionally built from a "bottom-up" approach whereby reactions were incrementally added to a set of existing known reactions for the organism. With the advent of full-genome sequencing and characterization, the size of metabolic networks experienced a significant leap, culminating in development of genome-scale metabolic reconstructions (GSMRs) of organisms. The first GSMR was completed in 1999 for *Haemophilus influenzae* [1], which comprised 343 metabolites and 488 reactions. Since then, the size of the GSMRs has increased even further with advances in sequencing and annotation methods, for example, the latest *Escherichia coli* reconstruction with 1,136 metabolites and 2,251 reactions [2] and the latest consensus *Saccharomyces*

Jens O. Krömer et al. (eds.), *Metabolic Flux Analysis: Methods and Protocols*, Methods in Molecular Biology, vol. 1191, DOI 10.1007/978-1-4939-1170-7_3,

cerevisiae model (Yeast6) consisting of 900 genes, 1,458 metabolites, and 1,888 reactions [3]. The development of each of these genome-scale reconstructions represents laborious efforts undertaken by many groups and the quality of these reconstructions is variable even though there are efforts to establish a standard protocol for their development [4].

These metabolic reconstructions represent "knowledge bases" for these organisms, capturing and organizing all the knowledge about the organism's metabolic capabilities in a format amenable to quantitative analysis. One of the primary usages of metabolic reconstructions or models is the determination of the intracellular fluxes through constraint-based methods such as flux balance analysis (FBA). One of the key reasons for the popularity of constraint-based methods is that they mainly require little input information besides stoichiometric information that is already encapsulated in the reconstructions and input-output flux measurements, which are highly tractable experimentally. Even though it has been shown that these stoichiometric models can "predict" the growth and other characteristics of the organisms accurately, these predictions pertain to the growth and the overall network behavior, and they cannot predict a unique intracellular flux distribution. Moreover, as the size of the metabolic models increases, the number of possibilities regarding the intracellular flux distribution increases "exponentially" [5]. The presence of alternate optimal flux distributions is a well-known challenge and the subject of many metabolic modeling studies. In order to be able to identify a unique flux distribution in silico that represents the intracellular flux distribution in vivo, additional constraints must be applied to the model to reduce the solution space [6].

Another key issue in the building and analysis of metabolic reconstructions is the assignment of reaction directionality. Reaction directionality remains a problematic issue in metabolic models and it is often based on the amount of information available on the specific enzyme of the organism at the point of curation and on information from labeling experiments. At this juncture, we would like to differentiate between "reaction reversibility" and "reaction directionality." *Reaction reversibility* is the catalytic property of the enzyme in carrying out (catalyzing) both the forward and reverse reaction; this information is usually scarce for most of the enzymes in the metabolic networks of an organism. If an enzyme is able to catalyze both the forward and reverse reactions, then the directionality of the reaction (*reaction directionality*) is dependent on the extent of displacement of the reaction from thermodynamic equilibrium. The Second Law of Thermodynamics states that a reaction can only proceed in a specific direction: the direction for which the Gibbs free energy of the reaction, $\Delta_r G$, is negative.

Thermodynamics-based flux balance analysis (TFBA) is an extension of FBA by incorporating thermodynamic constraints in the original FBA problem given by Eqs. 1 and 2. In TFBA and in the absence of specific information about the catalytic properties of the enzymes, we treat all reactions as catalytically reversible and we allow the thermodynamics of the network to determine the eventual directionality of the reaction in silico. Reversible reactions that are able to operate in both the forward and reverse directions given the network constraints (i.e., flux balance with thermodynamic constraints and other types of constraints) are termed as "bidirectional" whereas if the network constraints allow a reaction to operate only in one of the directions, we term this reaction as "unidirectional."

The utility and importance of network thermodynamics in metabolic network analysis have two aspects: (1) it reduces the allowable flux spaces by eliminating thermodynamically infeasible flux configurations and (2) it determines the thermodynamically feasible space of metabolite concentrations and reaction thermodynamic displacement [7, 8].

2 Thermodynamics of Biochemical Networks

2.1 Thermodynamics-Based Flux Balance Analysis Formulation

TFBA can be formulated as a mixed-integer linear programming (MILP) problem described in Table 1.

The TFBA problem incorporates thermodynamics-based constraints in the original FBA problem given by Eqs. 1 and 2. For biochemical reactions, the transformed Gibbs energy of reaction i (at a given pH and ionic strength), $\Delta_r G_i'$, which is a function of the chemical potentials of its reactants μ_j, determines the directionality of the reaction as shown in Eq. 3. The chemical potential of the reactants is a function of standard transformed Gibbs energy of formation of the compound $\Delta_f G_j'^{\circ}$ and the metabolite activity as shown in Eq. 4. The standard transformed Gibbs energy refers to the energies at specified pH and/or ionic strength conditions compared to standard conditions. Hence, we combine Eqs. 3 and 4 into the following equation:

Table 1
Thermodynamics-based flux balance analysis (TFBA) formulation

TFBA problem	FBA constraints	Mass balance	$S.v=0$	(1)
		Flux capacity	$v_{\min} \leq v_i \leq v_{\max}$	(2)
	Thermodynamics constraints	Gibbs energy of reaction	$\Delta_r G_i' = \Delta_{r,tpt} G_i' + \sum_{j=1}^{m} n_{i,j} \mu_j$	(3)
		Chemical potential	$\mu_j = \Delta_f G_j'^{\circ} + \Delta_{f,err} G_j'^{\circ} + RT \ln(x_j)$	(4)
		Thermodynamic feasibility	$\Delta_r G_i' - K + K z_i < 0$	(5)

Full Gibbs energy of reaction:

$$\Delta_r G_i' = \sum_{i=1}^{m} n_i \Delta_f G_i'^{\circ} + RT \ln\left(\prod_{j=1}^{m} x_j^{n_{i,j}}\right). \tag{6}$$

Thus, we obtain the common expression for the Gibbs free energy of the reaction which consists of two parts: (1) the standard transformed Gibbs free energy of the reaction, ΔG°, which is constant at specific pH and ionic strength, and (2) a part which is a function of the activities of the products and reactants. We will use here the common assumption that treats activities of the metabolites as being equal to their concentrations. In Eq. 4 we introduce the logarithm of the metabolite concentrations as new variables, and since the chemical potential depends linearly on these variables, the problem remains a linear programing problem.

In order for the Gibbs energy of reaction to determine the directionality of the reactions, we add binary "flux use" variables (z_i) and we add the constraint described by Eq. 5 that allows reactions to proceed only in the direction for which the $\Delta_r G_i'$ of the reaction is negative. The parameter K in Eq. 5 is a constant with an arbitrarily large value, and it is commonly used in this type of problem formulation for formulating on–off variables and conditions [9]. This last equation contains both continuous variables ($\Delta_r G_i'$) and discrete variables (z_i) and therefore the TFBA problem is a MILP problem.

In the following sections, we illustrate how to use TFBA to provide a platform for integration and analysis of metabolomics data and to reduce and characterize the flux space. We consider aerobic growth of *E. coli* and we use a small metabolic model of *E. coli* consisting of 72 metabolites and 95 reactions [10].

2.2 Thermodynamic Curation of Metabolic Model

In order to formulate the thermodynamic constraints, we need information regarding (1) the thermodynamic properties of the metabolites and the reactions, and (2) organism-specific properties. We call the assembly and organization of this information *the thermodynamic curation of the reconstruction* (Fig. 1).

2.2.1 Compound Properties

In order to include thermodynamic constraints into the FBA model, we first need the standard Gibbs free energy of formation of the compounds, which is defined as the change in the Gibbs free energy resulting from the formation of 1 mol of the compound in its standard state from its constituent elements in their standard states. These values can be obtained or estimated from:

1. Experimentally determined values, e.g., from NIST database [11] and National Bureau of Standards database [12].
2. Group contribution methods (GCM) which uses available experimental energies data to estimate the individual contributions of

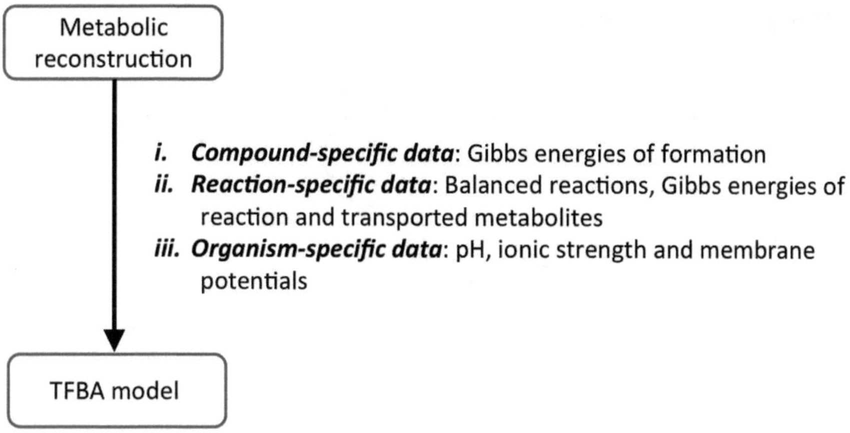

Fig. 1 Thermodynamic curation of metabolic reconstructions

different molecular groups to the overall compound formation energies [13–16].

3. Quantum mechanics/molecular mechanics (QM/MM) methods which are computationally intensive methods that estimate the energies from first principles [17–19].

Currently, the GCM provides the most comprehensive coverage of Gibbs energies of biological compounds within reasonable uncertainty (for example 84 % of the compounds and 96 % of the reactions in the genome-scale *E. coli* reconstruction [20], and similar coverage for all the genome-scale reconstructions that we have curated to date). Due to experimental uncertainty as well as uncertainty resulting from the fitting obtained by GCM methods, uncertainty in the estimated standard Gibbs free energy of formation of the compounds should be taken into consideration by adding a new variable term ($\Delta_{f,err}G_j'^{\circ}$) added to Eq. 4 (*see* **Note 1**). For compounds without estimated Gibbs energy of formation, we do not create the chemical potential constraint (Eq. 4).

2.2.2 Reaction Properties

The reactions in the model must be elementally and charge balanced (*see* **Note 2**) to ensure that their Gibbs free energies can be calculated correctly. The Gibbs free energy of reaction is calculated as shown in Eq. 3 as a function of the stoichiometry and chemical potentials of the reactants and products and there is also an additional component that takes into account a component for transport reactions.

Transport reactions are key reactions in cellular metabolism as nutrients have to be imported into the cell and metabolic byproducts have to be exported out of the cell. These reactions take place across membrane that are often charged and also between compartments with different physicochemical properties, e.g., different pH and ionic strength (*see* **Note 3**). Hence it is necessary and

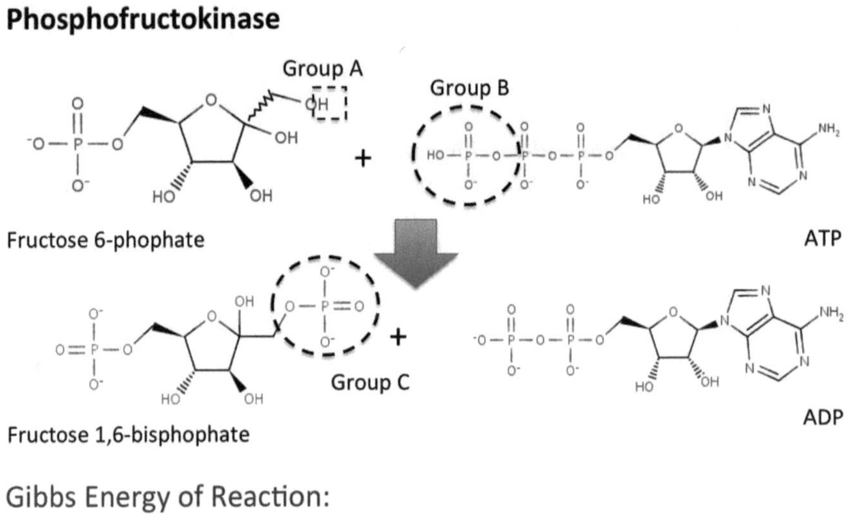

$$\Delta_r G = -\Delta G_{g,grpA} - \Delta G_{g,grpB} + \Delta G_{g,grpC}$$

Fig. 2 Group contribution method for estimation of Gibbs free energy of reactions. Example of phosphofructokinase reaction showing the three groups that are involved in the reaction

important that these effects are well accounted when performing TFBA. In TFBA, the thermodynamic calculations for transport reactions take into account all these considerations by adopting the formulation as described in Jol et al. [21]. The correction for transport reactions enters the TFBA problem formulation by including the transport term $(\Delta_{r,tpt}G_i')$ in Eq. 3, which can be calculated based on the Gibbs energy changes of the transported compounds (including protons) and this term is a constant under steady-state conditions. In the example *E. coli* model, there are 25 reactions that involve transport of metabolites and their calculated Gibbs energy of reaction are provided in the Supplementary Materials.

Therefore, the estimated Gibbs free energy depends only on the difference between the new groups produced and the groups that are transformed during the reaction. The rest of the groups and their associated error values cancel out without additional error propagation (*see* **Note 4**) (Fig. 2).

2.2.3 Organism-Specific Properties

The pH and ionic strength are important factors in biological systems as they affect the overall thermodynamic feasibility of reactions. Each compartment has its own specific pH and ionic strength, and we assume that pH and ionic strength are also subject to steady-state assumption of FBA. Under constant pH and ionic strength conditions, we adjust the standard Gibbs energy of reaction to *the transformed Gibbs free energy of reaction* using the methodologies introduced by Alberty [22]. It is then *the transformed Gibbs free energy of reaction* that is used to determine the directionality of the reaction.

In TFBA, we estimate *the transformed Gibbs free energy of reaction* using (1) the standard Gibbs free energy of formation of the compounds, (2) the specified pH and ionic strength, and (3) the *p*Ka values of the compounds [22] (*see* **Note 5**).

For *E. coli*, we assume optimal growth extracellular pH of 7.0 and the cytosolic pH to be 7.6 [25]. The membrane potential is assumed to be −180 mV [26], where the interior of the cell membrane is negatively charged. These properties can be very important for the study of the network thermodynamics, and there might be a great uncertainty in the reported values, which also depend on the experimental conditions. Moreover, for most of the organisms the pH, ionic strength, and membrane potential are unknown. Therefore, we recommend in all these studies to carry out sensitivity analysis in order to evaluate the potential effects of these parameters on the overall network.

In the absence of experimental information about the concentration of the metabolites, we assume minimum and maximum concentration bounds, e.g., between 0.01 μM and 0.1 M, which is wide enough to cover all the possible metabolite concentration ranges. If the concentration values of some metabolites are known, we set minimum and maximum concentration bounds for these metabolites determined by their corresponding reported mean value and standard deviation.

In organisms with multiple compartments, the metabolites are assigned to different compartments, and their transformed Gibbs free energy must be estimated according to the properties, i.e., pH and ionic strength, in these compartments.

3 Methods

After the thermodynamic curation, we can perform TFBA following a workflow that allows integration of available experimental information, or hypothesis testing by integrating *hypothetical* information along the various steps (Fig. 3).

3.1 TFBA Problem Formulation

First we convert the original FBA problem into a *bidirectional format*; that is, we split all catalytically reversible reactions into forward and reverse reactions whereby both the forward and reverse reaction variables represent positive variables. We then assign a binary "use variable" for each of the bidirectional flux variables that are linked to the transformed Gibbs free energy variable of the corresponding reaction as shown in Eq. 5; that is, there is a use variable assigned to each forward and reverse reaction flux variable. If the binary use variable is set to zero, the corresponding flux variable will also be set to zero. Hence, if the Gibbs energy of reaction is negative, only the forward use variable will be allowed to have a nonzero value and hence only the forward flux variable will be allowed to have a positive value as shown in Eq. 5 (*see* **Note 6**).

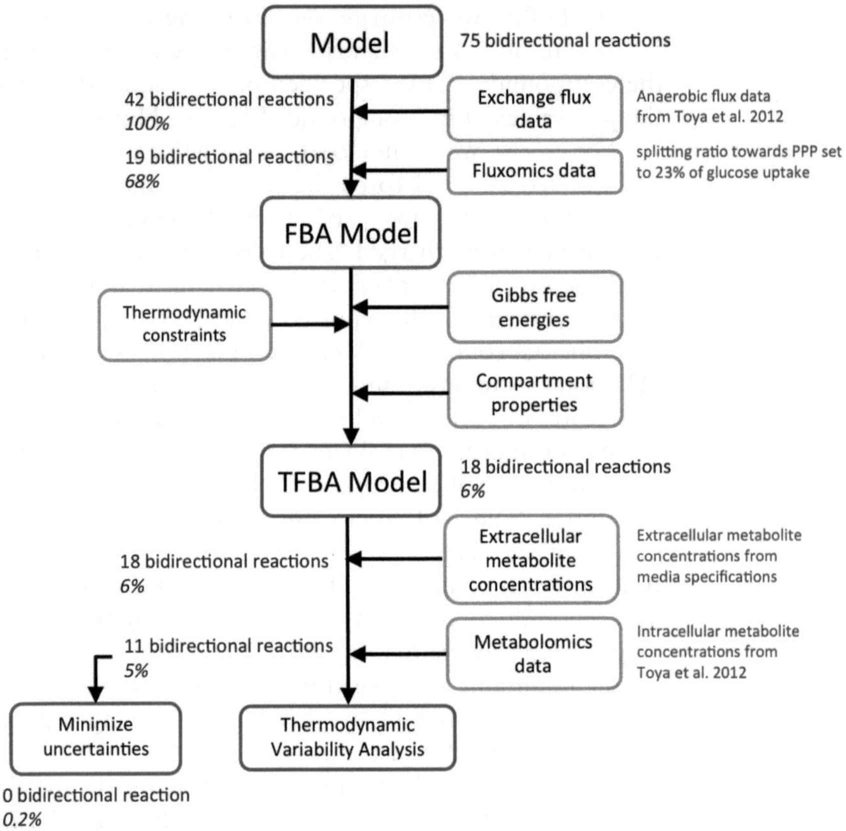

Fig. 3 Workflow for thermodynamics-based flux balance analysis (TFBA) and the reduction in the variability of the network at different stages based on the number of bidirectional reactions and the percentage of total variability in the example model

3.2 Reduction of the Flux Space Through the Integration of Flux and Concentration Information

In the following steps we reduce the flux space through the integration of experimental and/or hypothesis-based information. The basic procedure here is the addition of constraints to the TFBA problem and the performance of thermodynamics-based flux variability analysis (TFVA). In TFVA, we examine the minimum and maximum ranges of each net flux. Reactions that can operate in both directions are classified as bidirectional, and reactions that can operate only in one direction are classified as unidirectional. In some cases, some directions can carry flux under these constraints and they are classified as blocked reactions.

3.2.1 Integrating Flux Measurements

The original FBA model with all 75 reactions assumed to be bidirectional (*see* **Note 7**) is constrained based on *extracellular flux measurements*, i.e., on the specific uptake and product fluxes based on optimal growth data from Toya et al. [27], and we perform a TFVA. We find that 33 reactions become unidirectional, i.e., are allowed to operate only in one direction (Fig. 3). We can next integrate fluxomics information, which will constraint *splitting ratios in intracellular fluxes, flux ratios, or absolute fluxes* [28–30]. In this study we set only one such constraint: the pentose–phosphate pathway flux to 23 % of the glucose uptake rate as measured under the conditions considered here [27]. By setting only this additional constraint 23 more reactions become unidirectional (Fig. 3).

3.2.2 Adding Media Concentration Ranges

In these previous studies, the concentration bounds of extracellular and intracellular were set at the 1×10^{-5} M and 0.02 M [9, 31]. Next, we constrain the extracellular metabolite concentrations based on the media conditions specified. For this case, the common metabolites between the M9 glucose minimal media and extracellular metabolites in the model are only ammonium and phosphate and hence their concentrations were according to the M9 media concentrations. We set the concentration bounds of the measured metabolites to vary by 50% of their measured values were based on the starting concentrations in the media whereas the actual concentrations at the time point whereby the fluxes and metabolomics data were taken when the culture were at exponential growth conditions. For metabolites that were not detected in the media, we set them to the minimal value of 0.01 μmol and for metabolites that we do not have measurements, we leave their bounds as the default bounds. These media metabolite concentration constraints reduced the number of bidirectional reactions by one.

3.2.3 Integration of Metabolomics Data

We use the dataset of intracellular metabolite concentration from Toya et al. [27], where 20 of the metabolites in the model are part of the dataset. We set the bounds in the concentrations of these metabolites based on the reported mean values and taking into account the standard deviation in the measurements. After the integration of this metabolomics information, seven more reactions are constrained to be unidirectional. In the absence of absolute concentration values and large uncertainties in the metabolomics measurements, relative metabolite concentrations can be applied in the form of metabolite concentration ratios and if the uncertainty is quantified, it can be integrated in the form of bounds in the metabolite values and metabolite ratios (Figs. 4 and 5).

3.2.4 Characterization of the Thermodynamic Space

The thermodynamic space of the network comprises the space of *the thermodynamically allowable metabolite concentrations* and the space of *the reaction displacements from equilibrium*, which are connected through the chemical potentials of the metabolites. This thermodynamic space can be characterized and explored through the sampling of the concentrations of the metabolites and the corresponding Gibbs energies of reactions. Once the flux directionalities are specified the space is a convex linear space and we can use standard sampling algorithms [32, 33].

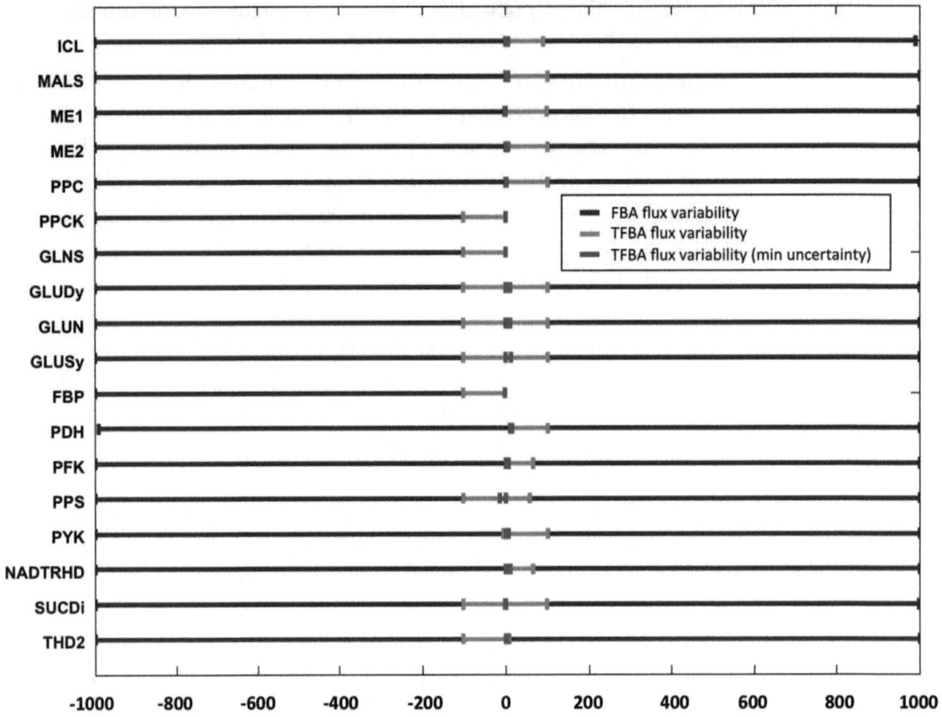

Fig. 4 Flux variability analysis (FVA) of selected reactions without and with thermodynamic constraints. (Color figure online)

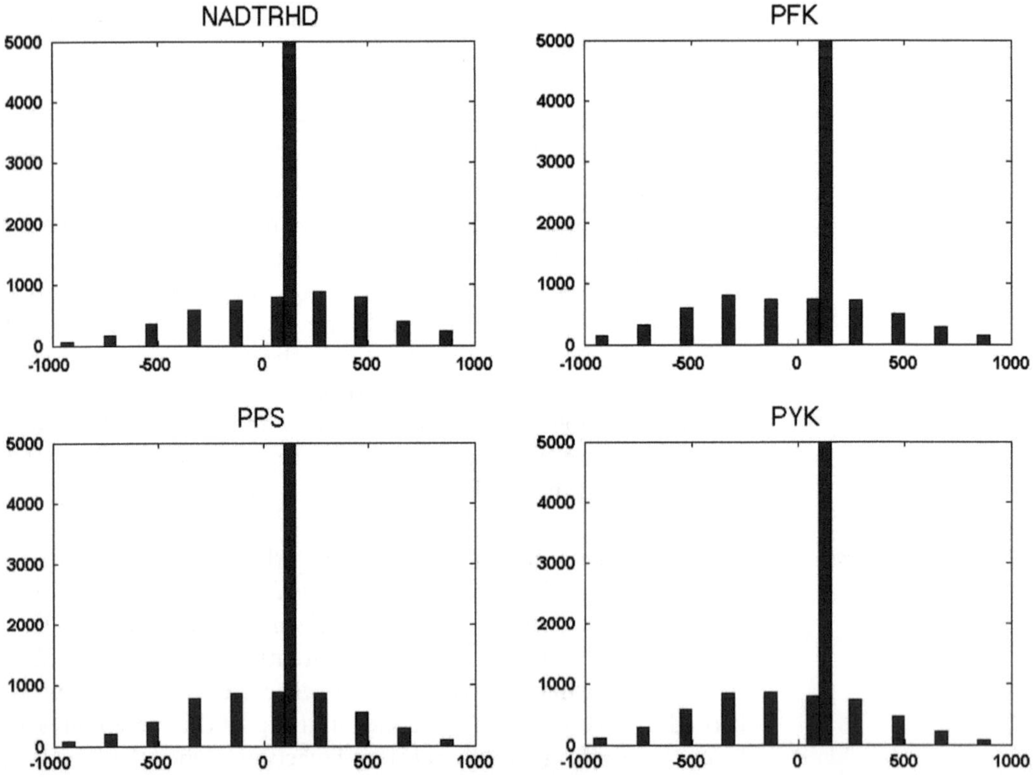

Fig. 5 Histograms of flux samples of selected reactions showing reduction in flux solution space without and with thermodynamic constraints. The *blue bars* represent flux values sampled from the FBA solution space and the *red points* represent flux values sampled from the TFBA solution space. (Color figure online)

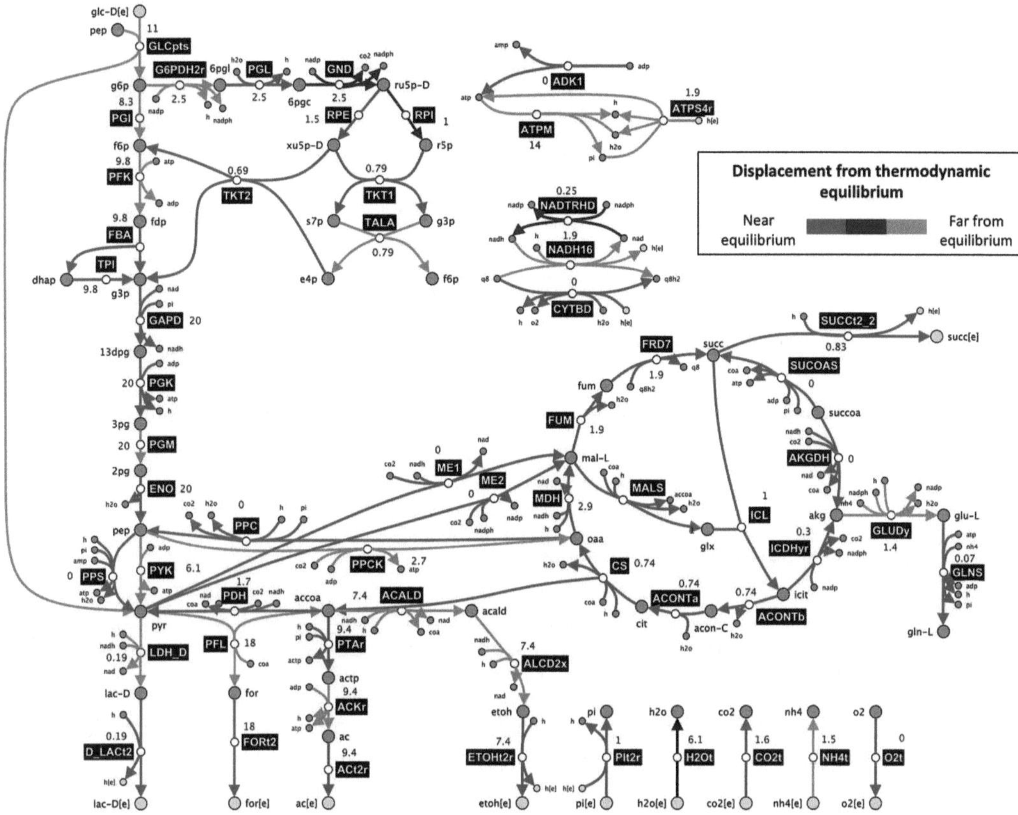

Fig. 6 Thermodynamic displacement of reactions in the network. (Color figure online)

By quantifying the extent of the displacement of the reactions from their thermodynamic equilibrium, we can identify the reactions with control over the network based on the principles from metabolic control analysis [34]. Metabolic control analysis theory states that reactions which are near equilibrium do not have a strong control on the fluxes. Figure 6 shows the thermodynamic displacement of the reactions in the network for one of the possible flux distributions of the network satisfying all the constraints.

3.2.5 Thermodynamic Variability Analysis

We can also perform thermodynamic variability analysis (TVA) to probe the thermodynamically feasible concentrations of the other intracellular metabolites that we did not have measurements for. As shown in Fig. 7, by minimizing the uncertainties in the Gibbs energy estimates, we can further reduce the possible thermodynamically feasible metabolite concentration ranges. In the set of intracellular metabolite measurements from Toya et al., many of the intracellular metabolite concentrations were not measured and hence we can estimate their thermodynamically feasible ranges using TVA as shown in Fig. 7. By minimizing the allowable ranges

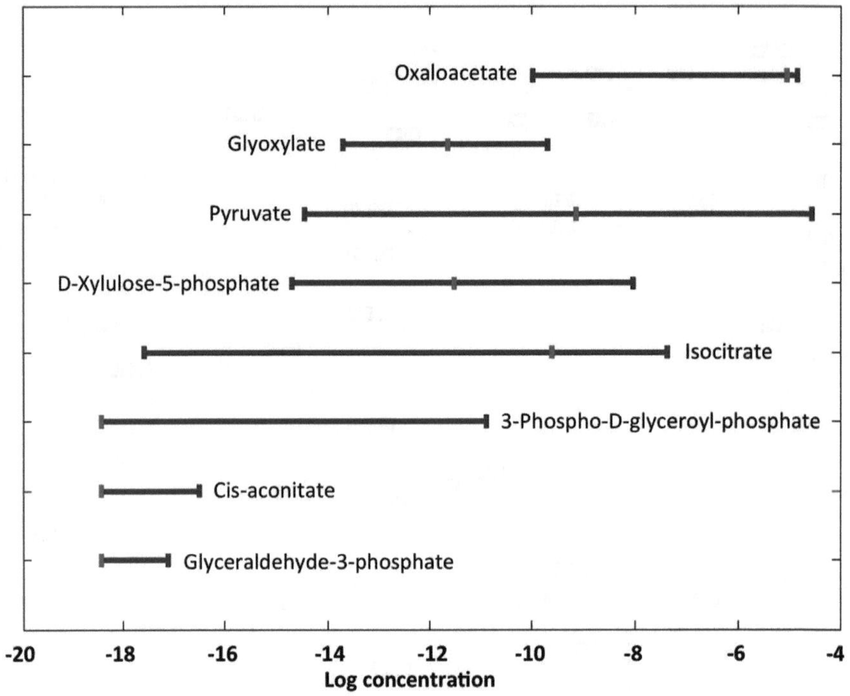

Fig. 7 Thermodynamic variability analysis of intracellular metabolite concentrations with uncertainty (*blue bars*) and after minimization of uncertainties (*red bars*). (Color figure online)

of the uncertainties in the Gibbs energies of formation and reaction, we can further reduce the feasible intracellular metabolite concentration ranges.

3.3 Further Analysis

By enumerating the different reaction directionality sets of the network, we can characterize the complete thermodynamic space of the network. As the directionality of the fluxes in the system represent the unique physiological state of the cell, hence each possible set of reaction directionalities of the system characterize the different possible physiological states given the data and constraints available. For each of these unique set of reaction directionalities, we can then apply kinetic modeling methods, such as ORACLE [35, 36], to further characterize their kinetic properties.

4 Notes

1. The uncertainty estimate is based on rigorous statistical analysis and it is the combination of the errors of the estimated group values. Therefore, it is different for different compounds.
2. In order to correctly calculate the Gibbs free energy of reaction, it is crucial that the reaction is both elementally and charge balanced. It has been shown that the balance of protons itself has significant implications on cellular growth and optimal flux distribution [37].

3. As the in vivo conditions are different from the standard conditions that most Gibbs free energies are based on, it is necessary to transform them to the relevant pH and ionic strength of the conditions that we are interested to study. The effects of pH and ionic strength on thermodynamic feasibilities of reactions are especially critical in determining the direction of metabolite transport. Alberty has developed a comprehensive treatment for the adjustment of the thermodynamic properties of biochemical reactions [22].
4. This error is significantly smaller than the error of the estimated Gibbs free energy of the reactants and products. This is because the Gibbs free energy of reaction is estimated from the sum of the total groups in the product compounds minus the sum of the total groups in the substrate compounds as shown in Fig. 2.
5. The pKa values of the compounds can be obtained from either experimental databases or chemoinformatics software packages such as ChemAxon [23], ACD/pKa [24], and SPARC (http://ibmlc2.chem.uga.edu/sparc/).
6. It is possible that the TFBA problem can become infeasible upon addition of the thermodynamic constraints. This could be due to two main reasons: (a) Uncertainty bounds for estimated Gibbs energies—Typically, we assume that both the estimated Gibbs energies of formation of compounds and Gibbs energies of reaction have one standard deviation of the uncertainty in their estimates. However, some of the bounds corresponding to these uncertainty estimates can be too tight and have to be relaxed to obtain a feasible solution. (b) Metabolite concentration bounds—The other possible reason for thermodynamic infeasibilities is that the metabolite concentration bounds are too constrained or measurement uncertainties are not taken into account. Hence in order to identify which bounds to relax, we can increase the bounds of the uncertainty variables first and minimize the sum of bounds (uncertainties) to identify which have to be increased to make the problem feasible.
7. Many reactions in the network should be bidirectional based on the range of their Gibbs energies in the network. Hence, it can lead to incorrect flux distributions if we constrain them in a specific direction a priori. Consider for example the anaplerotic reaction phosphoenoylpyruvate carboxykinase (PPCK):

   ATP + H + oxaloacetate <=> ADP + CO2 + phosphoenolpyruvate.

   PPCK is often constrained to be in the forward direction whereby it is ATP consuming but many experimental studies have found it to be capable of operating in the reverse direction [38, 39].

Acknowledgement

K.C.S. was supported by the Swiss National Science Foundation. V.H. is supported by funding from Ecole Polytechnique Fédérale de Lausanne (EPFL) and NEMO for Bioethanol, EU FP7 Programme. Support has also been provided from SystemsX.ch, The Swiss Initiative in Systems Biology, through the project MetaNetX.

References

1. Edwards JS, Palsson BO (1999) Systems properties of the Haemophilus influenzae Rd metabolic genotype. J Biol Chem 274(25): 17410–17416
2. Orth JD, Conrad TM, Na J, Lerman JA, Nam H, Feist AM, Palsson BO (2011) A comprehensive genome-scale reconstruction of Escherichia coli metabolism–2011. Mol Syst Biol 7:535
3. Heavner BD, Smallbone K, Price ND, Walker LP (2013) Version 6 of the consensus yeast metabolic network refines biochemical coverage and improves model performance. Database bat059
4. Thiele I, Palsson BO (2010) A protocol for generating a high-quality genome-scale metabolic reconstruction. Nat Protoc 5(1):93–121
5. Mahadevan R, Schilling CH (2003) The effects of alternate optimal solutions in constraint-based genome-scale metabolic models. Metabolic Engineering. p. 264–276
6. Price N, Reed J, Palsson B (2004) Genome-scale models of microbial cells: evaluating the consequences of constraints. Nat Rev Microbiol 2(11):886–897
7. Soh KC, Miskovic L, Hatzimanikatis V, Nielsen J (2011) From network models to network responses: integration of thermodynamic and kinetic properties of yeast genome-scale metabolic networks. FEMS Yeast Res 12(2):129–143
8. Soh KC, Hatzimanikatis V (2010) Network thermodynamics in the post-genomic era. Curr Opin Microbiol 13(3):350–357
9. Henry C, Broadbelt LJ, Hatzimanikatis V (2007) Thermodynamics-based metabolic flux analysis. Biophys J 92(5):1792–1805
10. Orth JD, Thiele I, Palsson BO (2010) What is flux balance analysis? Nat Biotechnol 28(3): 245–248
11. Goldberg RN, Tewari YB, Bhat TN (2004) Thermodynamics of enzyme-catalyzed reactions–a database for quantitative biochemistry. Bioinformatics 20(16):2874–2877
12. Wang PM, Neumann DB (1989) A database and retrieval-system for the Nbs tables of chemical thermodynamic properties. J Chem Inf Comput Sci 29(1):31–38
13. Jankowski MD, Henry CS, Broadbelt LJ, Hatzimanikatis V (2008) Group contribution method for thermodynamic analysis of complex metabolic networks. Biophys J 95(3): 1487–1499
14. Mavrovouniotis ML (1990) Group contributions for estimating standard Gibbs energies of formation of biochemical compounds in aqueous solution. Biotechnol Bioeng 36(10): 1070–1082
15. Mavrovouniotis ML (1991) Estimation of standard Gibbs energy changes of biotransformations. J Biol Chem 266(22):14440–14445
16. Noor E, Bar-Even A, Flamholz A, Lubling Y, Davidi D, Milo R (2012) An integrated open framework for thermodynamics of reactions that combines accuracy and coverage. Bioinformatics 28(15):2037–2044
17. Jorgensen WL (1989) Free-energy calculations—a breakthrough for modeling organic-chemistry in solution. Acc Chem Res 22(5): 184–189
18. Hu H, Yang WT (2008) Free energies of chemical reactions in solution and in enzymes with ab initio quantum mechanics/molecular mechanics methods. Annu Rev Phys Chem 59:573–601
19. Rizzo RC, Aynechi T, Case DA, Kuntz ID (2006) Estimation of absolute free energies of hydration using continuum methods: accuracy of partial, charge models and optimization of nonpolar contributions. J Chem Theory Comput 2(1):128–139
20. Feist A, Henry C, Reed J, Krummenacker M, Joyce A, Karp P, Broadbelt L, Hatzimanikatis V, Palsson B (2007) A genome-scale metabolic reconstruction for Escherichia coli K-12 MG1655 that accounts for 1260 ORFs and thermodynamic information. Mol Syst Biol 3:121
21. Jol SJ, Kummel A, Hatzimanikatis V, Beard DA, Heinemann M (2010) Thermodynamic calculations for biochemical transport and

reaction processes in metabolic networks. Biophys J 99(10):3139–3144
22. Alberty RA (2003) Thermodynamics of biochemical reactions. Wiley-Interscience, Hoboken, NJ, p ix, 397
23. Szegezdi F, Csizmadia, A (2007) A Method for Calculating the pK Values of Small and Large Molecules. 233rd ACS national meeting, CINF41 2007, Chicago, USA
24. ACD/pKa DB. A.C.D. Inc., Toronto, ON, Canada
25. Slonczewski JL, Rosen BP, Alger JR, Macnab RM (1981) pH homeostasis in Escherichia coli: measurement by 31P nuclear magnetic resonance of methylphosphonate and phosphate. Proc Natl Acad Sci U S A 78(10):6271–6275
26. Kaim G, Dimroth P (1998) ATP synthesis by the F1Fo ATP synthase of Escherichia coli is obligatorily dependent on the electric potential. FEBS Lett 434(1–2):57–60
27. Toya Y, Nakahigashi K, Tomita M, Shimizu K (2012) Metabolic regulation analysis of wild-type and arcA mutant Escherichia coli under nitrate conditions using different levels of omics data. Mol BioSyst 8(10):2593–2604
28. Zamboni N, Sauer U (2009) Novel biological insights through metabolomics and 13C-flux analysis. Curr Opin Microbiol 12(5):553–558
29. Sauer U (2006) Metabolic networks in motion: 13C-based flux analysis. Mol Syst Biol 2:62
30. Szyperski T (1995) Biosynthetically directed fractional 13C-labeling of proteinogenic amino acids. An efficient analytical tool to investigate intermediary metabolism. Eur J Biochem 232(2):433–448
31. Henry C, Jankowski MD, Broadbelt LJ, Hatzimanikatis V (2006) Genome-scale thermodynamic analysis of Escherichia coli metabolism. Biophys J 90(4):1453–1461
32. Schellenberger J, Palsson BO (2009) Use of randomized sampling for analysis of metabolic networks. J Biol Chem 284(9):5457–5461
33. Kaufman DE, Smith RL (1998) Direction choice for accelerated convergence in hit-and-run sampling. Oper Res 46(1):84–95
34. Heinrich R, Schuster S (1996) The regulation of cellular systems. Chapman & Hall, New York, NY
35. Miskovic L, Hatzimanikatis V (2011) Modeling of uncertainties in biochemical reactions. Biotechnol Bioeng 108(2):413–423
36. Miskovic L, Hatzimanikatis V (2010) Production of biofuels and biochemicals: in need of an ORACLE. Trends Biotechnol 28(8):391–397
37. Duarte NC, Herrgard MJ, Palsson BO (2004) Reconstruction and Validation of Saccharomyces cerevisiae iND750, a fully compartmentalized genome-scale metabolic model. Genome Res 14(7):1298–1309
38. Kwon YD, Lee SY, Kim P (2006) Influence of gluconeogenic phosphoenol pyruvate carboxykinase (PCK) expression on succinic acid fermentation in Escherichia coli under high bicarbonate condition. J Microbiol Biotechnol 16:1448–1452
39. Singh A, Soh KC, Hatzimanikatis V, Gill RT (2011) Manipulating redox and ATP balancing for improved production of succinate in E. coli. Metab Eng 13:76–81

Chapter 4

NExT: Integration of Thermodynamic Constraints and Metabolomics Data into a Metabolic Network

Verónica Sofía Martínez and Lars K. Nielsen

Abstract

Thermodynamic constraints are widely used in metabolic modelling such that calculated flux phenotypes are closer to real cell behavior. If metabolic data is also included in the analysis, a check of the thermodynamic consistency of the data can be realized and subsequently use the metabolic data to further constrain the solution space, giving a more specific representation of the cell metabolism under the studied conditions. Here *NExT*, a software based on network-embedded thermodynamic analysis, is presented, to integrate thermodynamics constraints and metabolomics data in the estimation of intracellular fluxes. New irreversible reactions can be inferred by calculating the thermodynamically feasible range of metabolite concentrations and Gibbs energy of reactions.

Key words Thermodynamic constraints, Metabolomics, NET analysis, *NExT*, Modelling, MATLAB

1 Introduction

Metabolic models are used to simulate metabolic phenotypes of cells, i.e., the quantification of metabolic fluxes through enzymatic reactions within the cells, by flux balance analysis (FBA) [1]. FBA relies on the use of constraints to bring calculated fluxes closer to physiological values [2]. Accurate formulation of constraints is therefore crucial in generating realistic fluxes.

The second law of thermodynamics enables determination of the direction of reactions on the basis of the standard Gibbs energy of formation ($\Delta_f G^o$) and metabolite concentrations. While enzymatic reactions by nature are reversible, it is common to define as irreversible those reactions that are restricted to operate in only one direction under all physiologically relevant conditions. If furthermore a subset of metabolite concentrations are measured, it is possible to determine the direction for a subset of reversible reactions under the given physiological condition, for which all flux directions are consistent with the sign of the respective Gibbs energy of reaction ($\Delta_r G$) [3–6]. The approach to check for

Jens O. Krömer et al. (eds.), *Metabolic Flux Analysis: Methods and Protocols*, Methods in Molecular Biology, vol. 1191, DOI 10.1007/978-1-4939-1170-7_4, © Springer Science+Business Media New York 2014

thermodynamic feasibility of pathways has evolved significantly over the years, from a component-based analysis of an individual metabolic pathway (e.g., glycolysis) [7] to network-based analysis using genome scale models (GeMs).

This chapter provides instructions on how to use *NExT* [8], a MATLAB software package based on network-embedded thermodynamic analysis (NET analysis) [6]. NET analysis is a network context nonlinear optimization approach to evaluate thermodynamic consistency of metabolomics data in a metabolic model, including compartmentalized models. NET analysis is used to estimate the feasible ranges of $\Delta_r G$ based on a given network and experimentally derived concentrations of intracellular metabolites. It can also be used to determine the feasible concentration ranges for the unmeasured metabolites, and to validate whether the experimentally derived metabolite concentrations are within thermodynamic feasible ranges. The directionality of reversible reactions for the given growth condition can be inferred from the calculated $\Delta_r G$ ranges.

2 Materials

1. Personal computer containing the following software: Microsoft Office Excel, MATLAB R2010a (or later versions), LINDO API or MATLAB Optimization Toolbox, *NExT* (*see* **Note 1**). *NExT* can be downloaded from http://web.aibn.uq.edu.au/cssb/Resources.html.
2. Biophysical compartment properties: pH, redox potential, ionic strength, and relative size (*see* **Note 2**).
3. Physical and chemical properties of metabolites: Standard Gibbs energy of formation ($\Delta_f G^o$) (in kJ/mol), number of hydrogen atoms, and charge (*see* **Note 3**).
4. Intracellular metabolite concentrations in mM (*see* **Note 4**).
5. A metabolic model (*see* **Notes 5** and **6**).

3 Methods

This section describes the required steps to check for thermodynamic feasibility of a metabolomic dataset in the context of a metabolic network. In addition, it explains how to use a metabolomics dataset, or specified physiological ranges of metabolite concentrations, to specify new thermodynamic constraints for a metabolic network (*see* **Note 4**). The section is divided into five subsections; first how to modify the model for thermodynamic calculations and readability by *NExT* is explained (Subheading 3.1). Then, how to upload the required data: the biophysical compartment properties (Subheading 3.2), the model's metabolites names link with the names specified in the metabolite properties database and additional metabolites to the database (Subheading 3.3) (*see* **Note 3**), and metabolite concentration (Subheading 3.4). Finally, how to run *NExT* in order to check for thermodynamic feasibility, to narrow the metabolite concentration range, and to identify reversible reactions behaving irreversibly under the specified conditions (Subheading 3.5). A yeast model is provided for the user as a working example (*see* **Note 5**).

3.1 Model Modifications

The following model modifications are required for the model to be suitable for thermodynamic calculations before importing to *NExT*:

1. Lump together identical reactions catalyzed by different enzymes in opposing directions as a reversible reaction (*see* **Note 7**).
2. Lump species HCO_3^-, CO_3^{2-}, CO_2, and H_2CO_3 as the reactant CO_{2tot}, since carbon dioxide equilibrates rapidly among these species in aqueous phase [9]. To balance the oxygen atoms, add H_2O to the product side of the reaction containing CO_{2tot} as a reactant, and vice versa.
3. Replace the oxidized and reduced FAD in mitochondria by the oxidized and reduced FAD_{enz} in order to represent the enzyme-bound FAD cofactor [9].

The following modifications are made to the model in order to generate the stoichiometric matrix from the list of metabolic reactions (*see* **Note 8**):

4. Always specify the metabolite's compartment. The compartment should be written between square brackets. A compartment can be specified by writing it after each metabolite or if the reaction takes place in just one compartment it may be specified at the beginning of the reaction, for example:

Original reaction	[Compartment] after each metabolite	[Compartment] at the beginning
_\2pg_c=h2o_c+pep_c	_\2pg[c]=h2o[c]+pep[c]	[c]_\2pg=h2o+pep
cit_m=icit_m	cit[m]=icit[m]	[m]cit=icit
pyr_c=pyr_m	pyr[c]=pyr[m]	Not possible

5. Write the stoichiometric coefficient in brackets. If coefficient is omitted, the default is 1.
6. Use equal (=) to separate reactant side and product side of a reaction equation.
7. Remove any of the following characters in the metabolite's names: (,), [,], =, and + (*see* **Note 9**).

	A	B	C	D
1	Abbreviation	reactions	reversible	
2	ENO	[c]_2pg=h2o+pep	1	
3	FBA	[c]fdp=dhap+g3p	1	
4	HEX1	[c]atp+glc_D=adp+g6p+h	0	
5	FBP	[c]fdp+h2o=f6p+pi	0	
6	GAPD	[c]g3p+nad+pi=_13dpg+h+nadh	1	
7	PYK	[c]adp+h+pep=atp+pyr	0	
8	PFK	[c]atp+f6p=adp+fdp+h	0	
9	MGSA	[c]dhap=mthgxl+pi	0	
10	PGI	[c]g6p=f6p	1	
11	PGM	[c]_2pg=_3pg	1	
12	PGK	[c]_3pg+atp=_13dpg+adp	1	
13	TPI	[c]dhap=g3p	1	
14	GND	[c]_6pgc+h2o+nadp=co2tot+nadph+ru5p_D	0	
15	PGL	[c]_6pgl+h2o=_6pgc+h	0	
16	RPI	[c]ru5p_D=r5p	0	
17	RPE	[c]ru5p_D=xu5p_D	1	
18	TALA	[c]g3p+s7p=e4p+f6p	1	
19	TKT1	[c]r5p+xu5p_D=g3p+s7p	1	
20	TKT2	[c]e4p+xu5p_D=f6p+g3p	1	
21	ACONT	[c]cit=icit	0	
22	ACS	[c]ac+atp+coa=accoa+amp+ppi	0	
23	FDH	[c]for+h2o+nad=co2tot+nadh	0	
24	LDH_D	[c]lac_D+nad=h+nadh+pyr	1	
25	LDH_L	[c]lac_L+nad=h+nadh+pyr	1	
26	PPA	[c]h2o+ppi=h+(2)pi	0	
27	ME2	[c]h2o+mal_L+nadp=co2tot+nadph+pyr	0	
28	TRDR	[c]h+nadph+trdox=nadp+trdrd	0	
29	GALK	[c]atp+gal=adp+gal1p+h	0	
30	GALT	[c]gal1p+h+utp=ppi+udpgal	1	
31	UDPG4E	[c]udpg=udpgal	1	

Model / Compartments / Names / Thermodynamic / Metabolites

Fig. 1 Example model to be loaded into *NExT*

8. Create a spreadsheet with the name "Model" in the provided Excel file (Fig. 1). Create the column headers: "abbreviation," "reactions," and "reversible" (*see* **Note 10**).
9. Copy the modified reaction equations under the heading "reactions."
10. Specify the reaction names in the first column for each reaction in the list.
11. In the last column specify reaction reversibility with "0" for irreversible and "1" for reversible.
12. Save the spreadsheet.

3.2 Saving the Biophysical Compartment Properties

1. In the same Excel file where the model was saved, create another spreadsheet named "Compartments" (Fig. 2).
2. In the first row create the column headers: "ID," "pH," "IS," "Potential mV," and "Volume" (*see* **Note 11**).
3. Under the header "ID" write the name of the compartment. The IDs must be identical to the ones previously provided in the square brackets for the reaction's equations (*see* Subheading 3.1, **step 5**).

	A	B	C	D	E	F	G	H
1	ID	pH	IS	Potential mV	Volume			
2	c	7.4	0.15	-315	0.9	cytosol		
3	e	7.1	0.15	0	0	extracellular		
4	m	8.0	0.15	-360	0.09	mitochondria		
5	i	6.7	0.15	-242.0	0.01	intermembrane mitochondrial		
6								

Model | Compartments | Names | Thermodynamic | Metabolites

Fig. 2 Example of compartment properties to be uploaded into *NExT*

4. Under the header "pH" write the pH of the corresponding cell compartment.
5. Under the header "IS" write the ionic strength of the corresponding cell compartment in molar (M).
6. Under the header "Potential mV" write the redox potential, in millivolts (mV), of the corresponding cell compartment (*see* **Note 12**).
7. Under the header "Volume" write the relative size of the corresponding cell compartment. If an external compartment was specified, the volume should be specified as zero. The sum of the column "Volume" has to be 1.
8. Save the spreadsheet.

3.3 Metabolite Names and Thermodynamic Properties

1. The thermodynamic database is pre-populated with physical and chemical information for 859 metabolites (*see* **Note 3**). Reactions containing metabolites not in the database will be omitted from the analysis. However, the user can add new metabolites to the database (see below).
2. The spreadsheet "Names" is used to connect metabolite names in the model with names in the database (Fig. 3). The first column is initially empty and the second contains the metabolite names linked to the thermodynamics database. Link metabolites in the model to metabolites in the database by adding the abbreviated metabolite names into the first column that match the metabolite in the adjacent cell (*see* **Note 13**).
3. Properties of new metabolites are specified in the "Thermodynamic" spreadsheet (Fig. 4). Leaving one space between the new metabolite and the previous one in the database, add the metabolite name as many times as the number of different species that can be found in aqueous solution (*see* **Notes 14** and **15**).
4. Add the species standard Gibbs energy of formation in kJ/mol in the second column and charge in the third column. Add the number of hydrogen atoms in the metabolite in the fourth column (*see* **Note 16**).

	A	B
199	ade	adenine
200	adn	adenosine
201	pap	Adenosine3-5-bisphosphate
202	aps	adenosinephosphosulfate
203	dcamp	adenylosuccinate
204	adp	ADP
205	adpglc	ADPglucose
206	adpman	ADPmannose
207	adprib	ADPribose
208	adprbp	ADPribose2-phosphate
209	adrnl	Adrenaline
210	aflatoxin	aflatoxinB1
211	agm	Agmatine
212	ala_D	alanine
213	ala_L	alanine
214	ala_B	alanine-B
215	aldstrn	Aldosterone
216		alpha-ketoglutaramate
217	akg	alpha-ketoglutarate
218	pheacgln	alpha-N-Phenylacetyl-L-glutamine
219	avite1	alpha-Tocopherol
220	avite2	alpha-Tocotrienol
221	aact	Aminoacetone
222	nh4	ammonia
223	amp	AMP
224	andrstandn	androstanedione
225	andrstndn	Androstenedione
226	andrstrn	Androsterone
227	andrstrnglc	androsteroneglucuronide
228	anth	anthranilate
229	arab_L	arabinose

Model / Compartments / Names / Thermodynamic / Metabolites

Fig. 3 Example of metabolite names into *NExT*

5. For metabolites with more than one specie specified, the pK_a between species can be specified in the fifth column if known. The pK_a between a pair of species is written in the row corresponding to the more negatively charged species.
6. Finally, the new names are connected to model metabolites on the "Names" spreadsheet and the spreadsheet is saved.

3.4 Metabolite Concentrations

1. The concentrations of intracellular metabolite can (optionally) be included in the analysis. These concentration ranges are specified in the spreadsheet "Metabolites." In the first row create the column headers: "Metabolite Name," "Lowest Concentration," and "Highest Concentration" (*see* **Note 17**).
2. Under the header "Metabolite Name" in the first column write the name of the measured metabolite; the software can handle sum and ratio of concentrations using the "+" and "/" operator, respectively (*see* Fig. 5 for an example of the spreadsheet "Metabolites") (*see* **Note 18**).

	A	B	C	D	E
377	2-Amino-4-hydroxy-6--erythro-1-2-3-trihydroxypropyldihydropteric	-2949.39	-3	13	
378					
379	5-Amino-1--5-Phospho-D-ribosylimidazole-4-carboxamide	-1409.32	-1	14	
380					
381	5-amino-1--5-phospho-D-ribosylimidazole	-1266.95	-1	13	
382					
383	alpha-ketoglutarate	-793.41	-2	4	
384					
385	alanine-B	-360.525	0	7	
386					
387	alanine	-314.663	-1	6	9.87
388	alanine	-371	0	7	2.35
389	alanine	-384.414	1	8	
390					
391	Aldosterone	-495.55	0	28	
392					
393	2-Aminomuconate6-semialdehyde	-308.023	0	7	
394					
395	S-Adenosyl-L-methionine	-278.966	1	23	
396					
397	S-Adenosylmethioninamine	54.47027	2	25	
398					
399	AMP	-1040.45	-2	12	
400	AMP	-1078.86	-1	13	
401	AMP	-1101.63	0	14	
402					
403	2-Aminomuconate	-493.959	-2	5	
404					
405	androstanedione	-151.813	0	28	
406					
407	Androstenedione	-77.1627	0	26	

Model / Compartments / Names / Thermodynamic / Metabolites

Fig. 4 Example of table with physical and chemical metabolite properties into *NExT*

	A	B	C	D
1	**Metabolite Name**	**Lowest Concentratio**	**Highest Concentration**	
2	o2[c]	0.001	0.1	
3	o2[m]	0.001	0.1	
4	ppi[c]	1	100	
5	ppi[m]	1	100	
6	AEC	0.0001	10	
7	nadh/nad	0.0001	10	
8	nadph/nadp	0.0001	10	
9	lac_L+lac_D	0.0001	10	
10	icit+cit	0.0001	10	
11	pro_D+pro_L	0.0001	10	
12	atp	0.0001	10	
13	accoa	0.0001	10	
14	gly	0.0001	10	
15	f6p	0.0001	10	
16				

Model / Compartments / Names / Thermodynamic / Metabolites

Fig. 5 Example of table with metabolites' concentrations into *NExT*

3. Specify the lower and upper values of metabolite concentrations in the second and third column, respectively.
4. Save the spreadsheet.

3.5 Running NExT in MATLAB

1. Run MATLAB. Change the current folder to *NExT* directory.
2. In the MATLAB command line, type "NExT" and enter. The program user interface will appear (*see* Fig. 6).
3. Click the blue x button to select the Excel file which contains the input data (metabolic model, biophysical compartment properties, physical and chemical metabolite properties and internal metabolites' concentrations). Click the "Open" button once the Excel file is selected.
4. In the text box to the right side of "Min Conc [mM]" and "Max Conc [mM]" set the minimum and maximum concentrations in mM. These boundary values will be used for the metabolites that were not specified in the Excel file in the spreadsheet "Metabolites." The default values are 0.0001 mM and 10 mM for the minimum and maximum concentrations, respectively.
5. Choose the nonlinear solver to be used by *NExT*. Click the drop-down list to select between LINDO and fmincon. By default LINDO will be displayed (*see* **Note 19**).
6. Save and close the Excel file containing the metabolic model, compartment properties, and metabolite concentrations and properties (*see* **Note 20**).
7. Click the button "Start Optimization" to start running *NExT* (*see* **Note 21**).
8. When the optimization is completed a text similar to "Finished running, check results in Excel file modelYeastTotal.xls" will

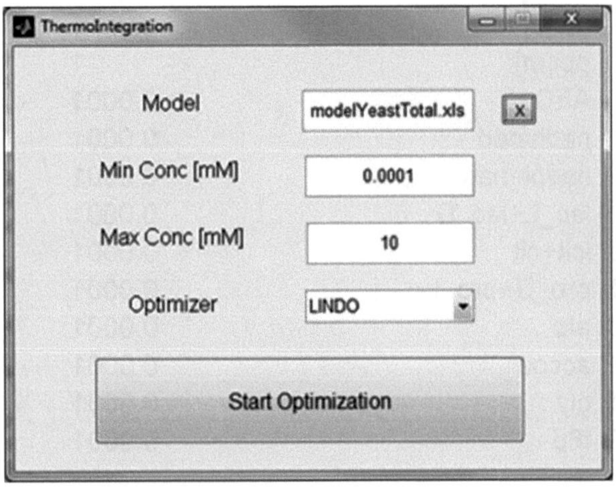

Fig. 6 *NExT* graphical interface

be displayed in the MATLAB command window. The results will be written in the same Excel file provided with the model, in the spreadsheets: "metabolites_results" and "DrG_results."

9. If the constraints of metabolite concentrations and irreversibilities given are thermodynamically infeasible, the message "Model Infeasible" will appear instead. In this case *NExT* will suggest a list of reactions so that, if the reaction direction is modified to reversible, the model will become thermodynamically feasible. The list will be shown in the spreadsheet "Wrong_Directions" of the Excel file.

10. In the spreadsheet "metabolites_results" (Fig. 7) the list of compartment-specific metabolites is shown. In the second and third columns are the minimum and maximum metabolite concentrations specified as input to the software, and in the fourth and fifth columns the reduced concentration range obtained after adding thermodynamic constraints (*see* **Note 22**).

	A	B	C	D	E	F
1	metabolite name	min[mM]	max[mM]	Opt min[m	Opt max[mM]	
2	nadh/nad	0.0001	10	0.0001	10.00005	
3	nadph/nadp	0.0001	10	9.99E-05	10	
4	AEC	0.0001	10	0.046783	0.999985	
5	ACP[c]	0.0001	10	1E-04	10	
6	ACP[m]	0.0001	10	1E-04	10	
7	L2aadp6sa[c]	0.0001	10	1E-04	10	
8	L2aadp6sa[m]	0.0001	10	1E-04	10	
9	L2aadp[c]	0.0001	10	1E-04	10	
10	L2aadp[m]	0.0001	10	1E-04	10	
11	L_dpchrm[c]	0.0001	10	1E-04	10	
12	Lcyst[c]	0.0001	10	1E-04	10	
13	Lcyst[m]	0.0001	10	1E-04	10	
14	Lcystin[c]	0.0001	10	1E-04	10	
15	Lfmkynr[c]	0.0001	10	1E-04	10	
16	Lkynr[c]	0.0001	10	1E-04	10	
17	Lpipecol[c]	0.0001	10	1E-04	0.003327	
18	Nacasp[c]	0.0001	10	1E-04	10	
19	Nacasp[m]	0.0001	10	1E-04	10	
20	Nacsertn[c]	0.0001	10	1E-04	10	
21	Sfglutth[c]	0.0001	10	1E-04	10	
22	Ssq23epx[c]	0.0001	10	1E-04	10	
23	T4hcinnm[c]	0.0001	10	1E-04	10	
24	T4hcinnm[m]	0.0001	10	1E-04	10	
25	_10fthf[c]	0.0001	10	1E-04	10	
26	_10fthf[m]	0.0001	10	1E-04	10	
27	_11docrtsl[c]	0.0001	10	1E-04	10	
28	_11docrtsl[m]	0.0001	10	1E-04	10	
29	_11docrtstrn[c]	0.0001	10	1E-04	10	
30	_11docrtstrn[m]	0.0001	10	1E-04	10	
31	_12RHPET[c]	0.0001	10	1E-04	10	

Model / Compartments / Names / Thermodynamic / Metabolites / Sheet1 / metabolite_results / DrG_results

Fig. 7 The spreadsheet "metabolites_results" resulting from *NExT*. The list contains all metabolites in their respective compartments with the minimum and maximum concentrations specified as inputs to the software and the resulting narrowed concentration range

	A	B	C	D	E	F	G
1	ReactionID	Extended rxn	Model irreversible rxn	tDrGo	min DrG	max DrG	New irreversible rxn
2	ENO	[c]_2pg=h2o+pep	0	-3.94506	-32.4853	24.59515	0
3	FBA	[c]fdp=dhap+g3p	0	22.13034	-46.3662	39.25447	0
4	HEX1	[c]atp+glc_D=adp+g6p+h	1	-26.8762	-66.7682	0	1
5	FBP	[c]fdp+h2o=f6p+pi	1	-13.2266	-52.6784	0	1
6	GAPD	[c]g3p+nad+pi=_13dpg+h+nadh	0	-0.28803	-30.3349	40.8252	0
7	PYK	[c]adp+h+pep=atp+pyr	1	-26.9024	-62.3855	0	1
8	PFK	[c]atp+f6p=adp+fdp+h	1	-24.9164	-64.9577	0	1
9	MGSA	[c]dhap=mthgxl+pi	1	-51.4699	-97.1342	-45.8264	1
10	PGI	[c]g6p=f6p	0	3.163821	-19.1639	31.70403	0
11	PGM	[c]_2pg=_3pg	0	-6.02075	-16.3408	22.51946	0
12	PGK	[c]_3pg+atp=_13dpg+adp	0	8.239714	-31.8017	25.50276	0
13	TPI	[c]dhap=g3p	0	7.66	-20.8802	36.20021	0
14	GND	[c]_6pgc+h2o+nadp=co2tot+nadph+ru5p_D	1	3.410842	-54.0985	0	1
15	PGL	[c]_6pgl+h2o=_6pgc+h	1	-33.2361	-61.7763	-4.69588	1
16	RPI	[c]ru5p_D=r5p	1	-0.46	-29.0002	0	1
17	RPE	[c]ru5p_D=xu5p_D	0	-1.58124	-30.1215	26.95897	0
18	TALA	[c]g3p+s7p=e4p+f6p	0	2.527143	-48.3408	59.60756	0
19	TKT1	[c]r5p+xu5p_D=g3p+s7p	0	-2.02423	-59.1047	55.05619	0
20	TKT2	[c]e4p+xu5p_D=f6p+g3p	0	-20.7995	-71.6674	36.28092	0
21	ACONT	[c]cit=icit	1	6.649282	-22.1521	0	1
22	ACS	[c]ac+atp+coa=accoa+amp+ppi	1	-14.1286	-44.4826	0	1
23	FDH	[c]for+h2o+nad=co2tot+nadh	1	-21.679	-40.6261	0	1
24	LDH_D	[c]lac_D+nad=h+nadh+pyr	0	21.48695	-20.1864	45.47605	0
25	LDH_L	[c]lac_L+nad=h+nadh+pyr	0	21.48695	-20.5476	21.63516	0
26	PPA	[c]h2o+ppi=h+(2)pi	1	-24.0658	-48.6931	-30.4064	1
27	ME2	[c]h2o+mal_L+nadp=co2tot+nadph+pyr	1	-2.28319	-59.7925	0	1
28	TRDR	[c]h+nadph+trdox=nadp+trdrd	1	-1.96138	-59.0414	0	1
29	GALK	[c]atp+gal=adp+gal1p+h	1	-26.5576	-66.599	0	1
30	GALT	[c]gal1p+h+utp=ppi+udpgal	0	-2.96095	-37.2092	55.25263	0
31	UDPG4E	[c]udpg=udpgal	0	0	-28.5402	28.54021	0

metabolite_results | DrG_results

Fig. 8 The spreadsheet "DrG_results" resulting from *NExT*. The list contains the model reactions, the specified direction, the estimated range of $\Delta_r G$, and the new direction due to the thermodynamic constraints

11. In the spreadsheet "DrG_results" (Fig. 8) a list of the metabolic reactions is shown in the first and second columns, followed by the model-specified direction in the third column (with "1" for irreversible reactions and "0" for reversible). The transformed standard $\Delta_r G$ (*see* **Note 23**) and the estimated minimum and maximum $\Delta_r G$ are presented in columns 4, 5, and 6. Reactions that have "1" in the seventh column are reversible reactions that were constrained to irreversible under the specific condition, and reactions that were initially specified irreversible in the model (*see* **Note 22**).

4 Notes

1. The software package *NExT* provides the option to choose between the MATLAB Optimization Toolbox with the function fmincon or the optimization software LINDO API to solve the nonlinear optimization problem. The results obtained with either solver are comparable, but LINDO API is generally faster.
2. Compartment-specific biophysical properties can be found in the literature. These properties are essential for the calculation of Gibbs energies of reactions ($\Delta_r G$) in order to represent real

physiological conditions. The relative compartment size is only needed when working with compartmentalized models. The value is used to estimate compartment-specific concentrations based on average cell concentrations.

3. The software includes a database with physical and chemical information for 859 metabolites. The $\Delta_f G^o$ of 177 metabolites were estimated from published data [10], while the remaining were estimated from group contribution method [4, 11], which are highlighted in light blue in the Excel database. If a studied reaction contains a metabolite that is not available in the provided database, the metabolite properties are to be manually included in order to estimate the Gibbs energy of the reaction of interest.
4. The software handles metabolite concentration ranges, sums of metabolite concentrations, and ratios of metabolite concentrations. If a concentration is unknown, a broad range of concentrations can be specified. It is suggested to work with concentrations ranges between 0.0001 and 10 mM to conservatively cover the possible physiological concentration [6]. Exceptions should be made for oxygen, with a maximum concentration of 0.1 mM, and for carbon dioxide and inorganic phosphate, which should be modified to range between 1 and 100 mM. The software will recognize the following metabolite names: "o2," "co2tot," "pi," "phosphate," and "ppi," and will automatically fix their concentration ranges to the previous specified values if no concentration range was specified by the user. Some of the ratios that *NExT* can deal with are the adenylate energy charge, which is recognized in the software by the name "AEC," and ratios for the redox cofactors.
5. A modified version of the consensus yeast GeM [12] that contains only two compartments (mitochondria and cytoplasm) is provided (modelYeast5Full.xls). The reduction was made in order to work with the main cellular compartments for which biophysical properties are available in the literature.
6. In order to calculate accurate $\Delta_r G$ of the reactions in the network, the reaction equations should be balanced and should contain all metabolites and cofactors involved. It is also important that protons keep the name "h" or "H" and water the name "h2o," "H2O," or "water"; the software is case sensitive, and thus the names should be exactly matching the ones provided here.
7. The reactions are lumped together to avoid a thermodynamic equilibrium state where a zero $\Delta_r G$ is considered for each of them.
8. A MATLAB program that generates the Model in an Excel file from an SBML file is provided. The program is called "SBML_Excel" and can be run by writing SBML_Excel in the

MATLAB Command Window. The result is an Excel file with a spreadsheet named "Model," which contains the metabolic model in the format required for running *NExT*.

9. *NExT* recognizes the text between square brackets as cellular compartments, the text between brackets as stoichiometric coefficients, and the text between +/= as reactants. Therefore, if any of the previous characters is used in the name of a metabolite the network will be erroneously recognized by the software, resulting in a wrong stoichiometric matrix and erroneous metabolite specifications. The spaces are not used by *NExT* to divide the different reactants; these are ignored by the software and only +/= are considered as separators between reactants.
10. The specific names of headers for the Model spreadsheet are not essential; they can be renamed but the order is crucial: in the first column a reaction name, in the second column the reaction equation, and in the last column the reaction reversibility.
11. The header names are irrelevant; however it is important to keep the compartments' biophysical properties in the same order.
12. If the redox potential of any compartment is specified as zero all reactions that take place in that compartment will not be taken into account for any calculation.
13. The metabolite names specified in the first column should be the same that were named in the reaction equations. The metabolite names should be written without the specific compartment. Each metabolite name has to be listed once only; otherwise the software will generate an error message. If two metabolites are linked to the same specified name in the second column a new row can be created repeating the name in the second column and adding the other name in the first column. An example are optical isomers, which by group contribution will have the same $\Delta_f G^o$, same number of hydrogen atoms in the structure, and the same charge. For example, alanine will appear as:

ala_D	alanine
ala_L	alanine

14. Enzyme-catalyzed reactions are written in terms of reactants (e.g., ATP) that are made up of a sum of species (e.g., ATP^{4-}, $HATP^{3-}$, and $MgATP^{2-}$). In order to get an accurate estimation of the Gibbs energy of a reactant the physical and chemical properties of all the possible species in aqueous solution are necessary.

15. It is essential for the software to retain the structure of the database when new metabolites are added. Otherwise wrong Gibbs energy estimations will result, and hence errors in the estimation of reaction direction and thermodynamic feasibility check. It is recommended that any new metabolite to be included is appended to the end of the database list, since the current database has been checked carefully.
16. If more than one species are specified for an added metabolite, columns 2 (standard Gibbs energy of formation), 3 (charge), and 4 (number of hydrogen atoms) should be filled for all the species. The species must be ordered by charge, with the most negatively charged species listed first.
17. The header names are irrelevant; however it is important to keep the inclusion of metabolite concentrations in the same order.
18. *NExT* is able to handle the input of metabolite concentration, sum of concentrations, ratio of concentrations, and the adenylate energy charge (AEC) ratio. For the software to be able to distinguish between the different inputs of concentration data, it is essential that the name is written in the first column of the spreadsheet "Metabolites." If a metabolite concentration was measured by compartment, the metabolite name should include the metabolite compartment between square brackets (e.g., 2pg[c]); if the concentration represents a cell average concentration, the metabolite name should not include any compartments (e.g., 2pg); if the concentration represents the sum of two metabolite concentrations, the names of both metabolites need to be included with a plus (+) sign in between (e.g., 2pg + 3pg); if a ratio of metabolites' concentrations is specified, the character "/" should be written between the metabolite names (e.g., nadh/nad); if the adenylate energy charge ratio is specified, it should be written with the name AEC (with capital letters).
19. The selected nonlinear solver should be properly installed on your computer. If fmincon is selected the MATLAB Optimization Toolbox needs to be installed. If LINDO is selected the LINDO API software needs to be properly installed on the computer and linked with the MATLAB version being used by the proper startup m-file indicating the location of the license file.
20. The Excel file containing all the input data for the software should be closed before pressing the button "Start Optimization" because the final results will be saved by *NExT* in the same Excel file in a new spreadsheet. MATLAB cannot write in an opened file and an error message will be displayed by MATLAB even if *NExT* has completed the optimization.

21. While the program is running, text will be written in the MATLAB Command Window to inform what task the software is performing. Text similar to the following will appear as the program progresses:

    Loading data

    Calculating stoichiometric matrix

    Calculating thermodynamic constraints

    Loading concentration constraints

    Targets to optimize 1110: 10.20.30.40.50.60.70.>>

22. The list of metabolites in the spreadsheet "metabolites_results" contains all model metabolites that are linked to their physical and chemical properties in the spreadsheet "Thermodynamic." Metabolites with no properties provided to *NExT* are not considered in the thermodynamic feasibility analysis. The same is true for the spreadsheet "DrG_results," which only reports reactions for which all reactants are specified in the spreadsheet "Thermodynamic."
23. The transformed standard $\Delta_r G$ takes into account the cell physiological conditions, which differ from the standard conditions in pH and ionic strength.

References

1. Varma A, Palsson BO (1994) Metabolic flux balancing—basic concepts, scientific and practical use. Bio/Technol 12:994–998
2. Thiele I, Palsson BO (2010) A protocol for generating a high-quality genome-scale metabolic reconstruction. Nat Protoc 5: 93–121
3. Fleming RM, Thiele I, Nasheuer HP (2009) Quantitative assignment of reaction directionality in constraint-based models of metabolism: application to Escherichia coli. Biophys Chem 145:47–56
4. Henry CS, Broadbelt LJ, Hatzimanikatis V (2007) Thermodynamics-based metabolic flux analysis. Biophys J 92:1792–1805
5. Hoppe A, Hoffmann S, Holzhutter HG (2007) Including metabolite concentrations into flux balance analysis: Thermodynamic realizability as a constraint on flux distributions in metabolic networks. BMC Syst Biol 1:23
6. Kummel A, Panke S, Heinemann M (2006) Putative regulatory sites unraveled by network-embedded thermodynamic analysis of metabolome data. Mol Syst Biol 2:2006.0034
7. Mavrovouniotis ML (1993) Identification of localized and distributed bottlenecks in metabolic pathways. Proc Int Conf Intell Syst Mol Biol 1:275–283
8. Martínez VS, Queek L, Nielsen LK (2014) Network Thermodynamic Curation of Human and Yeast Genome-Scale Metabolic Models. Biophys J 107
9. Alberty RA (2003) Thermodynamics of biochemical reactions. Wiley, Hoboken, NJ
10. Alberty RA (2006) Biochemical thermodynamics: applications of Mathematica. Methods Biochem Anal 48:1–458
11. Jankowski MD et al (2008) Group contribution method for thermodynamic analysis of complex metabolic networks. Biophys J 95: 1487–1499
12. Herrgard MJ et al (2008) A consensus yeast metabolic network reconstruction obtained from a community approach to systems biology. Nat Biotechnol 26:1155–1160

Part III

Experiments: Design Considerations, Fermentation, Analytics

Chapter 5

Customization of ^{13}C-MFA Strategy According to Cell Culture System

Lake-Ee Quek and Lars K. Nielsen

Abstract

^{13}C-MFA is far from being a simple assay for quantifying metabolic activity. It requires considerable up-front experimental planning and familiarity with the cell culture system in question, as well as optimized analytics and adequate computation frameworks. The success of a ^{13}C-MFA experiment is ultimately rated by the ability to accurately quantify the flux of one or more reactions of interest. In this chapter, we describe the different ^{13}C-MFA strategies that have been developed for the various fermentation or cell culture systems, as well as the limitations of the respective strategies. The strategies are affected by many factors and the ^{13}C-MFA modeling and experimental strategy must be tailored to conditions. The prevailing philosophy in the computation process is that any metabolic processes that produce significant systematic bias in the labeling pattern of the metabolites being measured must be described in the model. It is equally important to plan a labeling strategy by analytical screening or by heuristics.

Key words ^{13}C-MFA, Steady-state, Non-stationary, Computation, Labeled substrates

1 Introduction

^{13}C tracer-based metabolic flux analysis (^{13}C-MFA) is a powerful approach for quantifying intracellular metabolic fluxes using ^{13}C enrichment data obtained from metabolites. Any differences between ^{13}C enrichment signatures are the direct result of having different internal metabolic flows or configurations [1–4]. These enrichments are then used to back-calculate the metabolic flows at a metabolic steady state. Oxidative pentose-phosphate pathway, glycolysis by Entner-Doudoroff pathway, gluconeogenesis, anaplerosis, and glyoxylate shunt, as well as the reversible exchange or conversion between adjoining metabolite pools, are metabolic reactions or pathways that can be resolved by ^{13}C-MFA but not by conventional MFA [5]. ^{13}C-MFA is therefore applied on top of MFA to quantify the activity of these alternative pathways under different growth conditions.

Jens O. Krömer et al. (eds.), *Metabolic Flux Analysis: Methods and Protocols*, Methods in Molecular Biology, vol. 1191, DOI 10.1007/978-1-4939-1170-7_5, © Springer Science+Business Media New York 2014

A successful ^{13}C-MFA experiment is the combined outcome of (1) a good a priori understanding of the cells' metabolic behavior, (2) a well-designed cell culture system, (3) efficient and sufficient sampling, (4) a proper sample preparation, (5) a quality and sensitivity analytics, and (6) an adequate computational framework [1, 5–9]. While well-designed ^{13}C-MFA experiments covered by conventional computational frameworks are fairly simple to solve, it only takes a few seemingly harmless modifications to the experiment to make analysis very demanding. Full dynamic ^{13}C-MFA, for example, is only performed by a handful of specialist groups worldwide [1, 6, 9, 10]. The challenge is reflected in the fact that many key studies still are as much about the new ^{13}C-MFA methodology used to generate those findings as the metabolic findings themselves [6, 11–14].

Packaged ^{13}C-MFA software have been made available, such as 13C-FLUX [15], FiatFlux [16], METRAN [17], and OpenFLUX [18], to streamline the workflow for flux estimation using enrichment data in standard cell culture system. These packages have distinct as well as overlapping capabilities. Atypical problems, however, will require the development of a new strategy, or the customization of existing ones. Whether a problem is standard or atypical, a priori knowledge of the metabolic behavior of the system under investigation is critical to guide the decision-making process. ^{13}C-MFA will typically return a solution; the question is whether this solution is accurate, or even valid. The quality and confidence of flux results are improved by adopting the right modeling assumptions and computational framework, both of which require users to be diligent and be aware of the scope and limitations of various approaches.

While ^{13}C-MFA is a significant step change in terms of modeling complexity when compared to conventional MFA, ^{13}C-MFA is a necessary step forward in the effort to quantify metabolic phenotype at a higher resolution. This chapter describes three major ^{13}C-MFA strategies, their underlying assumptions about metabolic processes, and how these affect the appropriateness for a given cell culture system.

2 Isotopic Stationary ^{13}C-MFA Using Proteinogenic Amino Acids

The most widely employed ^{13}C-MFA strategy is the use of ^{13}C enrichment of proteinogenic amino acids [9]. The enrichment of proteinogenic amino acids is measured by GC-MS [1, 19] or—less frequently—NMR [4, 20]. The ^{13}C enrichment of 16 out of 20 amino acids (cysteine, glutamine, asparagine, and tryptophan excluded) can be determined. These amino acids together provide isotopic labeling information of eight precursor metabolites located across central carbon metabolism (glycolysis, pentose-phosphate pathway, and TCA cycle): pyruvate, acetyl-CoA, 3-phosphoglycerate, phosphoenolpyruvate, erythrose-4-phosphate, oxaloacetate, 2-oxo-glutarate, and ribose-5-phosphate [21]. Note that the MS fragments of some amino acids should not be used for ^{13}C-MFA [22].

Proteinogenic amino acids are very abundant, typically 2–3 orders of magnitude more abundant than free amino acids and other major metabolites. The analyte signal is strong and stable since it is, in effect, the integration precursor metabolite signals for several cell generation times. This dramatically reduces background noise and enables ^{13}C-MFA on small samples, e.g., high-throughput screening of microbial strain libraries. On the flipside, the dynamic response is very slow and this approach is unsuited for capturing transient processes.

Conventional steady-state ^{13}C-MFA assumes that cells are at a metabolic and isotopic pseudo-steady state [15, 16], which typically requires operations for 3–5 doubling times at constant conditions. Accordingly, the application of proteinogenic amino acids for steady-state ^{13}C-MFA is largely limited to batch (exponential growth phase) or steady-state continuous cultures. It is important to minimize the size of the inoculums used and any external disturbances (e.g., pH, temperature). Using rich media, sustaining steady-state growth may not be adequate as the exchange of one or more intracellular metabolites with a large, external pool of metabolites can introduce non-isotopic effects despite cells [13]; this effect can potentially be minimized by carefully modifying the media [23]. Despite having a narrow scope, the majority of metabolic characterizations are based on the use of proteinogenic amino acids [12, 19, 24].

3 Isotopic Stationary ^{13}C-MFA Using Free Metabolites

The enrichment of free intracellular metabolites (or amino acids) can be used for steady-state ^{13}C-MFA [25, 26], and is typically the second choice when proteinogenic amino acids cannot be used for ^{13}C-MFA. The enrichments of free intracellular metabolites are significantly more responsive and can quickly equilibrate to changes in metabolic activity or to the switching of input substrates. The observed turnover rates (i.e., time constants) of most intracellular metabolites are in the order of seconds to minutes, though can be hours in a few cases [25, 27, 28]. Standard steady-state ^{13}C-MFA can be applied as long as the free intracellular metabolites have reached an isotopic steady state. It is important to characterize the trajectory of enrichment of metabolites such that isotopic steady state is confirmed, and to only perform ^{13}C-MFA on such data [28].

The rate of turnover is largely influenced by metabolite pool sizes and proximity to the entry point of labeled input substrates. For example, when labeled glucose is applied, TCA cycle intermediates often take longer to reach isotopic steady state compared to glycolytic intermediates because the former are present in greater concentrations and are downstream of glycolysis [27]. Additionally, the dynamics of labeling of a free metabolite can be affected by tributary metabolites. If protein turnover is significant, unlabeled carbon backbones can enter the TCA via transamination reactions given the impression that the size of the keto-acid (e.g., pyruvate, oxo-glutarate, and oxaloacetate) pools are significantly larger than real [29, 30]. Even if some metabolite pools have not reached isotopic steady state, ^{13}C-MFA can still be performed using only the subset of enrichment data that has reached isotopic steady state.

The analytical capabilities to measure concentration and enrichment of intracellular metabolites have advanced significantly in recent years [26, 31]. Intracellular metabolites, however, still come with significantly larger measurement errors, which can be up to 24 % by LC-MS/MS [30] compared to the <1 % error for proteinogenic amino acids (quantified by GC-MS). The larger errors on metabolite measurements propagate making the flux determination less precise. Compartmentalization of metabolism in eukaryotic cells adds a layer of complexity in ^{13}C-MFA because the metabolites measured are derived from whole-cell extracts [7]. The model must therefore account for the fact that the overall enrichment of a metabolite can be a mixture of two or more distinct enrichments. Additional simplification can be achieved by assuming that the transfer of metabolites between compartments are absent, unidirectional, or fully reversible. These assumptions are more often validated by ad hoc model fitting, and less by direct biological evidence.

Another issue is metabolic channeling [14, 32], which is characterized by downstream metabolites having a greater enrichment than intermediate(s) in a linear reaction pathway. By incorporating a "dilution parameter" in the fitting process, 10 % of the total intracellular concentrations of metabolites involved in the Calvin-Benson-Bassham cycle (fructose-6-phosphate, glyceraldehydes-3-phosphate, dihydroxyacetone phosphate and ribose-5-phosphate) for a cyanobacterium *Synechocystis* were found to be unlabeled [14]. In CHO cells consuming [1,2-$^{13}C_2$]-glucose, the fraction of total enrichment of malate was found to be higher than pyruvate during non-growth phase despite the only bypass reaction phosphoenolpyruvate carboxykinase being absent [32]. Metabolic channeling violates the assumption in ^{13}C-MFA that metabolites within a compartment are well mixed, and it is important to manually inspect the enrichment data to identify possible deviations from this assumption.

4 Isotopic Non-stationary ^{13}C-MFA

Isotopic non-stationary ^{13}C-MFA does represent a major step-up both in terms of experimental and computational complexity [6]. This approach is therefore reserved for scenarios when metabolites do not reach isotopic steady state due to experimental or cost limitations, or when fluxes can only be time resolved (e.g., a linear series of reversible reactions).

Successfully implemented isotopic non-stationary ^{13}C-MFA has the potential to deliver significant benefits [6]. The scope of application is wider than stationary ^{13}C-MFA because the non-stationary computation framework can utilize both non-stationary and stationary datasets, and arguably every metabolite undergoes a transient phase before reaching isotopic stationary. Even though non-stationary ^{13}C-MFA involves the estimation of both fluxes and metabolite pool sizes, flux resolution tends to be better than stationary ^{13}C-MFA because non-stationary approach utilizes a much larger set of redundant data [28, 30]. The ability to indirectly estimate metabolite concentrations, particularly in specific compartments, is another significant benefit. This ability, however, is somewhat opportunistic since it is not obvious up front—before computation—which metabolite concentrations can be determined.

A significant lead-up in design effort is required before the actual experiment, particularly for atypical systems [33]. Flux estimation still operates on the assumption that metabolism is at steady state; that is, fluxes and metabolite pool sizes are constant. Labeling experiments often are initiated when cells are at steady state, and any disturbance introduced during the switch to labeled substrates should be minimized [27]. In fed-batch, the labeled substrate mixture used in the feed is often customized such that the relative proportions of isotopic substrates in the culture medium are maintained; otherwise any changes in the enrichment of the substrates must be accounted for separately [34]. Fluxes tend to be better resolved if the enrichment of metabolites directly downstream of the fluxes is measured [7].

While the duration of the labeling experiment is flexible and significant cost saving can be achieved by conducting a shorter experiment, frequent sampling is needed during the transient phase. A sampling frequency of one order of magnitude faster than the turnover time of metabolites was suggested [6], for example, to take 20 samples within the first 40 s of switching to labeled substrates, followed by exponentially increasing sampling intervals [27]. It may be important to pace the sampling frequency according to the turnover rates of metabolites that belong to pathways prioritized to be resolved. For example, in some cases glycolysis and oxidative pentose-phosphate pathways can be resolved by steady-state ^{13}C-MFA due to the fast turnover of intermediates [32, 35], while flux estimates for TCA cycle are deferred to the non-stationary approach.

Computationally, a system of ordinary differential equations (ODE) is used to simultaneously estimate steady-state fluxes and metabolite pool sizes based on the time profiles of ^{13}C enrichment of metabolites. A typical metabolic problem involves in excess of 500 ODEs, and a single simulation can take seconds to minutes to complete even using high-performance workstations or computing clusters [14, 30]. This is in contrast to the millisecond time scale for an isotopic steady-state approach that involves only algebraic equations.

Non-stationary ^{13}C-MFA can proceed without metabolite concentrations, but metabolite concentrations should be provided if they can be obtained with reasonable effort. The importance of using experimentally derived metabolite concentrations to improve flux estimates is still being debated [6, 14]. The dynamics of metabolite enrichments mainly place upper constraints on pool sizes, whereas the bulk of flux estimates come from differences in the enrichment patterns [14]. Arguably, however, reversibility and pool size are two parameters poorly estimated simultaneously, and therefore reversible fluxes, as well as the overall fluxes, are better estimated by providing pool size constraints [36]. More experiments are required before the value of measured metabolite concentrations as well as the effect of network topology can be properly accessed.

5 Choice of Labeled Input Substrates

The effectiveness of a labeled substrate in probing the activity of specific reactions depends on the cell's actual metabolic processes, i.e., network topology and flux magnitude, and on the enrichments that can be measured. A cost-effective strategy should use the least expensive substrates that bring about high levels of relevant flux information in the measured enrichments. Sometimes this means ignoring the inexpensive substrates (e.g., [U-^{13}C]-glucose and [1-^{13}C]-glucose) and using expensive, but informative substrates. This is exemplified by the success of using expensive labeled glutamine to elucidate the role of glutamine in lipogenesis in mammalian cells [37].

There is a complex three-way relationship between labeled substrates, metabolic processes, and measured enrichments that dictates the effectiveness of a labeled substrate in resolving a flux of interest. Significant effort has been invested to develop rational methods to distinguish fluxes that are identifiable given a mixture of labeled substrates [38–44]. These approaches, based on structural identifiability, can reveal fluxes that can or cannot be resolved depending on the labeled substrates used. The same approaches can also be used to identify metabolites that carry significant flux information and subsequently to prioritize analytical efforts to quantify these metabolites.

The complexity of developing an optimal design is nontrivial when multiple factors are considered: (1) the cost and availability of labeled substrates, (2) the combinatorial effects of using different types and quantities of substrates, (3) enrichments that can be quantified, (4) data that are noisy and non-redundant, (5) the fluxes that are to be resolved, (6) the dependency of flux parameter sensitivity on network topology and actual flux distribution, (7) the opportunity to perform multiple labeling experiments, and (8) computation framework (stationary vs. non-stationary, global vs. flux ratio, qualitative vs. quantitative). Given the complexity of the design problem, the task may be more achievable by designing an experiment that is sufficient, rather than emphasizing on a thorough and formal optimization. A more tractable approach may therefore involve generating a shortlist of labeled substrates by preliminary screening—eliminating substrates that will perform poorly—based on fluxes that are to be resolved. For example, in mammalian cell culture, only a few types of labeled glucose are effective at elucidating TCA cycle due to a high rate of glutaminolysis [45–47].

Despite the advancement in design strategies, the types of labeled glucose used have not changed over decades. Early ^{13}C-MFA experiments in yeast saw the paired use of a 1:4 mixture of [U-^{13}C]-glucose and natural glucose, and of 100 % [1-^{13}C]-glucose labeling setups for flux estimation [24]. The same experimental setups are still being used for ^{13}C-MFA on a large array of *E. coli* mutants [48], different yeast strains [12], and yeast growing on various carbon substrates [49]. The continued practice suggests that the conventional labeling strategy is so far effective for a variety of metabolic problems. This presents an interesting contrast to the aforementioned rational approaches, in which finding the right labeled substrates to apply is largely empirical and not necessarily an elaborate design process.

6 Conclusion

Flux estimation by ^{13}C-MFA is ultimately a retrospective investigation of what happened to the labeled substrates when these substrates were consumed by the cells. An estimated set of fluxes is the result of having simulated ^{13}C enrichments closely reproducing measured values. The fluxes generated are therefore closer to physiological values from an experimental standpoint than conventional flux simulation by optimization.

The computational complexity appears to scale with metabolic complexity. The choice of computation framework is tiered according to the cell culture system, with stationary ^{13}C-MFA using proteinogenic amino acids as the primary choice, otherwise with stationary ^{13}C-MFA using free intracellular metabolites, or ultimately with non-stationary ^{13}C-MFA. It is important to provide enough modeling parameters and constraints to describe any metabolic processes that have significant involvement in the redistribution of ^{13}C label among metabolites. It is desirable to strive to be comprehensive, as indicated by the developments in large-scale ^{13}C-MFA [50]. This is counterbalanced by the need to be parsimonious, since an excessively comprehensive or heavily parameterized model will simply be over-fitted and slow to compute, as well as generate findings that have weaker significance [51, 52].

There is a variety of commercially available labeled substrates, and each has potential in resolving different fluxes. It is possible to bypass any design effort and straightaway adopt conventional labeling strategies, which are likely to be effective as well since central metabolism is largely invariant for most organisms. More exotic systems will require further planning, which can be facilitated by analytical tools already available. It is more sensible to work towards multiple ^{13}C-MFA experiments that each has been designed to resolve a few key fluxes at high precision, rather than resolving all fluxes at moderate precision in a single experiment. ^{13}C-MFA is ultimately an iterative learning process, in which the modeling and experimental strategy improves concomitantly with our understanding of the cell's metabolic phenotype.

References

1. Wittmann C (2007) Fluxome analysis using GC-MS. Microb Cell Fact 6:6
2. Szyperski T (1998) ^{13}C-NMR, MS and metabolic flux balancing in biotechnology research. Q Rev Biophys 31:41–106
3. Wiechert W, de Graaf AA (1996) In vivo stationary flux analysis by ^{13}C labeling experiments. Adv Biochem Eng Biotechnol 54:109–154
4. Massou S, Nicolas C, Letisse F, Portais JC (2007) NMR-based fluxomics: quantitative 2D NMR methods for isotopomers analysis. Phytochemistry 68:2330–2340
5. Sauer U (2006) Metabolic networks in motion: ^{13}C-based flux analysis. Mol Syst Biol 2:62
6. Noh K, Wiechert W (2011) The benefits of being transient: isotope-based metabolic flux analysis at the short time scale. Appl Microbiol Biotechnol 91:1247–1265
7. Zamboni N (2011) 13C metabolic flux analysis in complex systems. Curr Opin Biotechnol 22:103–108
8. Tang YJ, Martin HG, Myers S, Rodriguez S, Baidoo EE, Keasling JD (2009) Advances in analysis of microbial metabolic fluxes via (13)C isotopic labeling. Mass Spectrom Rev 28:362–375
9. Zamboni N, Fendt SM, Ruhl M, Sauer U (2009) (13)C-based metabolic flux analysis. Nat Protoc 4:878–892
10. Antoniewicz MR, Kelleher JK, Stephanopoulos G (2007) Elementary metabolite units (EMU): a novel framework for modeling isotopic distributions. Metab Eng 9:68–86
11. Ahn WS, Antoniewicz MR (2012) Towards dynamic metabolic flux analysis in CHO cell cultures. Biotechnol J 7:61–74
12. Christen S, Sauer U (2011) Intracellular characterization of aerobic glucose metabolism in seven yeast species by 13C flux analysis and metabolomics. FEMS Yeast Res 11:263–272
13. Niklas J, Sandig V, Heinzle E (2011) Metabolite channeling and compartmentation in the human cell line AGE1.HN determined by 13C labeling experiments and 13C metabolic flux analysis. J Biosci Bioeng 112: 616–623
14. Young JD, Shastri AA, Stephanopoulos G, Morgan JA (2011) Mapping photoautotrophic metabolism with isotopically nonstationary (13)C flux analysis. Metab Eng 13:656–665
15. Wiechert W, Mollney M, Petersen S, de Graaf AA (2001) A universal framework for ^{13}C metabolic flux analysis. Metab Eng 3:265–283
16. Zamboni N, Fischer E, Sauer U (2005) FiatFlux—a software for metabolic flux analysis from ^{13}C-glucose experiments. BMC Bioinformatics 6:209

17. Yoo H, Antoniewicz MR, Stephanopoulos G, Kelleher JK (2008) Quantifying reductive carboxylation flux of glutamine to lipid in a brown adipocyte cell line. J Biol Chem 283: 20621–20627
18. Quek LE, Wittmann C, Nielsen LK, Kromer JO (2009) OpenFLUX: efficient modelling software for 13C-based metabolic flux analysis. Microb Cell Fact 8:25
19. Becker J, Klopprogge C, Zelder O, Heinzle E, Wittmann C (2005) Amplified expression of fructose 1,6-bisphosphatase in *Corynebacterium glutamicum* increases in vivo flux through the pentose phosphate pathway and lysine production on different carbon sources. Appl Environ Microbiol 71:8587–8596
20. Schmidt K, Nielsen J, Villadsen J (1999) Quantitative analysis of metabolic fluxes in *Escherichia coli*, using two-dimensional NMR spectroscopy and complete isotopomer models. J Biotechnol 71:175
21. Dauner M, Sauer U (2000) GC-MS analysis of amino acids rapidly provides rich information for isotopomer balancing. Biotechnol Prog 16:642–649
22. Antoniewicz MR, Kelleher JK, Stephanopoulos G (2007) Accurate assessment of amino acid mass isotopomer distributions for metabolic flux analysis. Anal Chem 79:7554–7559
23. Deshpande R, Yang TH, Heinzle E (2009) Towards a metabolic and isotopic steady state in CHO batch cultures for reliable isotope-based metabolic profiling. Biotechnol J 4:247–263
24. Blank LM, Kuepfer L, Sauer U (2005) Large-scale ^{13}C-flux analysis reveals mechanistic principles of metabolic network robustness to null mutations in yeast. Genome Biol 6:R49
25. Kromer JO, Sorgenfrei O, Klopprogge K, Heinzle E, Wittmann C (2004) In-depth profiling of lysine-producing *Corynebacterium glutamicum* by combined analysis of the transcriptome, metabolome, and fluxome. J Bacteriol 186:1769–1784
26. van Winden WA, van Dam JC, Ras C, Kleijn RJ, Vinke JL, van Gulik WM, Heijnen JJ (2005) Metabolic-flux analysis of Saccharomyces cerevisiae CEN.PK113-7D based on mass isotopomer measurements of (13)C-labeled primary metabolites. FEMS Yeast Res 5:559–568
27. Noack S, Noh K, Moch M, Oldiges M, Wiechert W (2011) Stationary versus non-stationary (13)C-MFA: a comparison using a consistent dataset. J Biotechnol 154:179–190
28. Schaub J, Mauch K, Reuss M (2008) Metabolic flux analysis in Escherichia coli by integrating isotopic dynamic and isotopic stationary 13C labeling data. Biotechnol Bioeng 99:1170–1185
29. Grotkjaer T, Akesson M, Christensen B, Gombert AK, Nielsen J (2004) Impact of transamination reactions and protein turnover on labeling dynamics in (13)C-labeling experiments. Biotechnol Bioeng 86:209–216
30. Noh K, Gronke K, Luo B, Takors R, Oldiges M, Wiechert W (2007) Metabolic flux analysis at ultra short time scale: isotopically non-stationary ^{13}C labeling experiments. J Biotechnol 129:249–267
31. Bennett BD, Kimball EH, Gao M, Osterhout R, Van Dien SJ, Rabinowitz JD (2009) Absolute metabolite concentrations and implied enzyme active site occupancy in Escherichia coli. Nat Chem Biol 5:593–599
32. Ahn WS, Antoniewicz MR (2011) Metabolic flux analysis of CHO cells at growth and non-growth phases using isotopic tracers and mass spectrometry. Metab Eng 13:598–609
33. Shastri AA, Morgan JA (2007) A transient isotopic labeling methodology for 13C metabolic flux analysis of photoautotrophic microorganisms. Phytochemistry 68:2302–2312
34. Antoniewicz MR, Kraynie DF, Laffend LA, Gonzalez-Lergier J, Kelleher JK, Stephanopoulos G (2007) Metabolic flux analysis in a nonstationary system: fed-batch fermentation of a high yielding strain of *E. coli* producing 1,3-propanediol. Metab Eng 9: 277–292
35. Sengupta N, Rose ST, Morgan JA (2011) Metabolic flux analysis of CHO cell metabolism in the late non-growth phase. Biotechnol Bioeng 108:82–92
36. Noh K, Wahl A, Wiechert W (2006) Computational tools for isotopically instationary 13C labeling experiments under metabolic steady state conditions. Metab Eng 8:554–577
37. Metallo CM, Gameiro PA, Bell EL, Mattaini KR, Yang J, Hiller K, Jewell CM, Johnson ZR, Irvine DJ, Guarente L, Kelleher JK, Vander Heiden MG, Iliopoulos O, Stephanopoulos G (2012) Reductive glutamine metabolism by IDH1 mediates lipogenesis under hypoxia. Nature 481:380–384
38. van Winden WA, Heijnen JJ, Verheijen PJ, Grievink J (2001) A priori analysis of metabolic flux identifiability from (13)C-labeling data. Biotechnol Bioeng 74:505–516
39. Isermann N, Wiechert W (2003) Metabolic isotopomer labeling systems. Part II. Structural flux identifiability analysis. Math Biosci 183: 175–214
40. Chang Y, Suthers PF, Maranas CD (2008) Identification of optimal measurement sets for complete flux elucidation in metabolic flux analysis experiments. Biotechnol Bioeng 100: 1039–1049

41. Rantanen A, Mielikainen T, Rousu J, Maaheimo H, Ukkonen E (2006) Planning optimal measurements of isotopomer distributions for estimation of metabolic fluxes. Bioinformatics 22:1198–1206
42. Mollney M, Wiechert W, Kownatzki D, de Graaf AA (1999) Bidirectional reaction steps in metabolic networks: IV. Optimal design of isotopomer labeling experiments. Biotechnol Bioeng 66:86–103
43. Pey J, Rubio A, Theodoropoulos C, Cascante M, Planes FJ (2012) Integrating tracer-based metabolomics data and metabolic fluxes in a linear fashion via Elementary Carbon Modes. Metab Eng 14:344–353
44. Crown SB, Antoniewicz MR (2012) Selection of tracers for 13C-metabolic flux analysis using elementary metabolite units (EMU) basis vector methodology. Metab Eng 14:150–161
45. Walther JL, Metallo CM, Zhang J, Stephanopoulos G (2012) Optimization of 13C isotopic tracers for metabolic flux analysis in mammalian cells. Metab Eng 14:162–171
46. Crown SB, Ahn WS, Antoniewicz MR (2012) Rational design of 13C-labeling experiments for metabolic flux analysis in mammalian cells. BMC Syst Biol 6:43
47. Metallo CM, Walther JL, Stephanopoulos G (2009) Evaluation of 13C isotopic tracers for metabolic flux analysis in mammalian cells. J Biotechnol 144:167–174
48. Haverkorn van Rijsewijk BR, Nanchen A, Nallet S, Kleijn RJ, Sauer U (2011) Large-scale 13C-flux analysis reveals distinct transcriptional control of respiratory and fermentative metabolism in Escherichia coli. Mol Syst Biol 7:477
49. Fendt SM, Sauer U (2010) Transcriptional regulation of respiration in yeast metabolizing differently repressive carbon substrates. BMC Syst Biol 4:12
50. Ravikirthi P, Suthers PF, Maranas CD (2011) Construction of an E. coli genome-scale atom mapping model for MFA calculations. Biotechnol Bioeng 108:1372–1382
51. Leighty RW, Antoniewicz MR (2011) Dynamic metabolic flux analysis (DMFA): a framework for determining fluxes at metabolic non-steady state. Metab Eng 13:745–755
52. Suthers PF, Burgard AP, Dasika MS, Nowroozi F, Van Dien S, Keasling JD, Maranas CD (2007) Metabolic flux elucidation for large-scale models using ^{13}C labeled isotopes. Metab Eng 9:387–405

Chapter 6

Quantitative Metabolomics Using ID-MS

S. Aljoscha Wahl, Reza Maleki Seifar, Angela ten Pierick, Cor Ras, Jan C. van Dam, Joseph J. Heijnen, and Walter M. van Gulik

Abstract

Quantitative intracellular metabolite measurements are essential for systems biology and modeling of cellular metabolism. The MS-based quantification is error prone because (1) several sampling processing steps have to be performed, (2) the sample contains a complex mixture of partly compounds with the same mass and similar retention time, and (3) especially salts influence the ionization efficiency. Therefore internal standards are required, best for each measured compound. The use of labeled biomass, ^{13}C extract, is a valuable tool, reducing the standard deviations of intracellular concentration measurements significantly (especially regarding technical reproducibility). Using different platforms, i.e., LC-MS and GC-MS, a large number of different metabolites can be quantified (currently about 110).

Key words Quantitative metabolite analysis, Isotopomer dilution, Mass spectrometry, Extraction, LC-MS, GC-MS

1 Introduction

Cellular metabolism is organized in complex biochemical reaction networks and regulated by various mechanisms. Unraveling the cellular pathways and metabolic regulation requires quantitative metabolite measurements and modeling. In the following we discuss protocols for the rapid quenching of metabolism, metabolite extraction, and subsequent quantification based on isotope dilution mass spectrometry (ID-MS).

Quenching of metabolism is mostly performed by fast cooling of fermentation broth. de Koning et al. [1] introduced the use of a methanol/water mixture (60:40 v/v) at low temperature (−40 °C) that leads to a mixing temperature of approximately −20 °C. In later years, a variety of different compositions have been proposed [2] that stop metabolism and prevent the loss of intracellular metabolites into the extracellular space (leakage). The lesson learned so far is that there is no method that performs well with all organisms under all conditions. In turn, this requires a thorough

Jens O. Krömer et al. (eds.), *Metabolic Flux Analysis: Methods and Protocols*, Methods in Molecular Biology, vol. 1191, DOI 10.1007/978-1-4939-1170-7_6,

evaluation of the protocol—Canelas et al. [3, 4] have proposed a method for the validation of quenching and extraction procedures based on rigorous balancing of all involved solutions and spiking at different steps in the procedure. In this publication, we focus on protocols that were validated for *Saccharomyces cerevisiae*.

After quenching, a separation of intra- and extracellular space has to be achieved. Most popular is centrifugation at low temperature with one to three washing steps [3, 5] or rapid cold filtration which at very high extracellular concentrations seems to perform more reliably [5, 6] (Figs. 1 and 2).

For the extraction of intracellular metabolites from the cell pellet, mostly ethanol boiling is used. Alternatives are boiling water [7, 8], cold methanol/chloroform mixtures [1], freeze/thaw

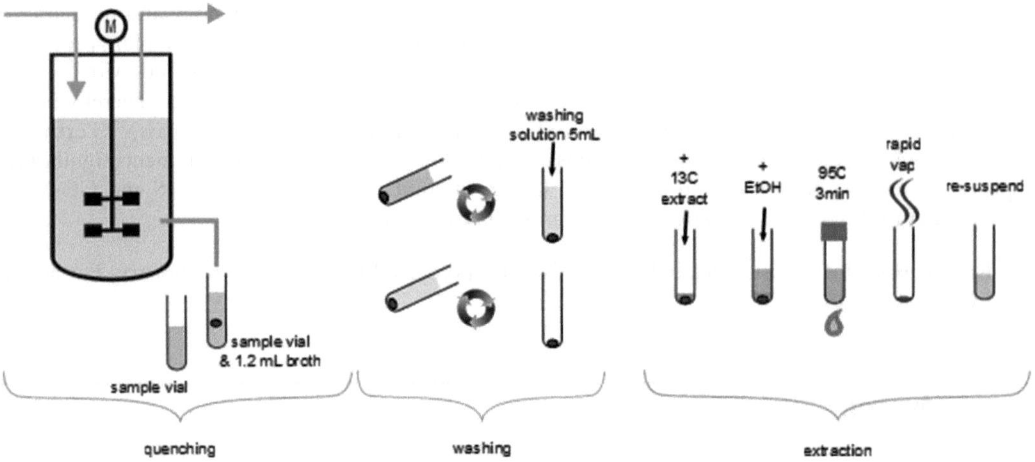

Fig. 1 Steps for sample processing: (1) quenching, (2) washing, and (3) extraction

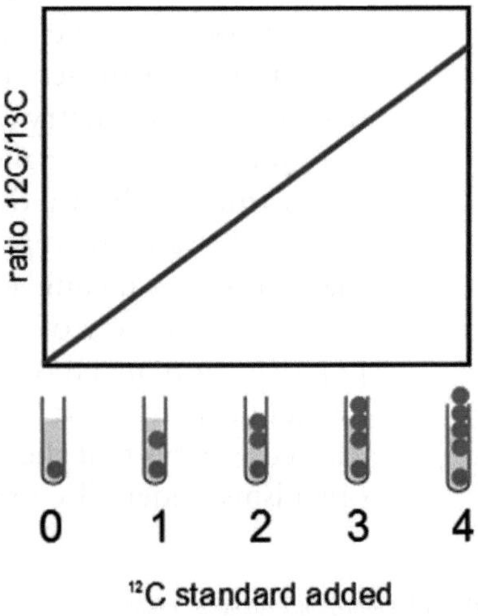

Fig. 2 Standard calibration with internal ^{13}C standard at constant concentration (*green*). The calibration line is measured as $^{12}C/^{13}C$ ratio. (Color figure online)

cycles [9], or cold acidic acetonitrile/methanol [10]. Compared to acid or alkali extraction, like perchloric acid or potassium hydrochloride extraction [11], the salt content is not increased which is advantageous for most analysis methods.

An internal isotopic standard containing a labeled isotopologue for each analyzed compound is produced from a fully labeled culture. The method of using U-$^{13}C_6$-labeled glucose as substrate to produce labeled cell extract as internal standard has been introduced by Mashego et al. [12] and Wu et al. [2]. With this internal standard, the majority of intracellular metabolites of central carbon metabolism—glycolysis, pentose phosphate pathway, TCA, and most intracellular amino acids—can be measured. The approach has been applied for metabolite measurements for the model organisms *S. cerevisiae* [12, 13], *Penicillium chrysogenum* [14], *Escherichia coli* [15, 16], and further organisms like *Corynebacterium glutamicum*. Lafaye [17] used a ^{15}N-labeled nitrogen source to label N-containing metabolites and obtained absolute concentrations for intermediates of the glutathione biosynthesis.

Summarizing, to obtain quantitative data, all steps in the sampling process need validation:

1. Quenching: Prevent leakage.
2. Extraction: Prevent conversions and maximize extraction.
3. Measurement: Minimize matrix effects and avoid bias by compounds other than the respective analyte.

1.1 Application Examples of ID-MS

ID-MS can basically be applied to all MS platforms and organisms. Following solutions have been worked out to (1) measure metabolites in newly implemented pathways and (2) measure metabolite levels of organisms in complex media.

Especially for the analysis of newly implemented pathways a "standard" ^{13}C extract might not contain the metabolites of interest. In this case, the ^{13}C extract has to be enriched with the respective metabolites, which is most easily achieved by using the production mutant also for the ^{13}C extract generation. In some cases, also the protocol for ^{13}C cultivation needs to be adapted to increase the content of relevant metabolites. For example in case of penicillin pathway analysis, ^{13}C-labeled PAA needs to be added to the medium [18].

In case of organisms that require complex media, the production of ^{13}C-labeled cell extract is very costly (if not impossible) as all carbon-containing components need to be added as labeled species. Bennett et al. [19] produced ^{13}C-labeled extract of mammalian fibroblasts with a medium containing fully labeled glucose and glutamine. The authors observed that "a nontrivial amount of certain metabolites persist in the fully unlabeled form." Based on a ratio analysis of "unlabeled contamination" the calculations can take the impurities into account [19].

A ^{13}C reference from another organism can be an interesting alternative solution for most metabolites. Especially, there is a commercially available ^{13}C extract from algal cells [20]. The authors state that the commercially available U-^{13}C-labeled algal cells are a convenient source for the generation of isotope-labeled internal standards.

The recent developments and the wider application of ID-MS protocols in diverse microbial organisms, but also mammalian cells, show the success of the approach to provide quantitative metabolome measurements.

2 Materials

2.1 Rapid Sampling and Metabolite Extraction

- Methanol (analytical grade), Ethanol (analytical grade), both from Sigma-Aldrich, Buchs, Switzerland.
- Polypropylene (PP) and polystyrene (PS) test tubes (14 mL, 17 mm diameter), 2 mL Eppendorf vials, 2 mL screw-cap vials.

2.2 LC-MS/MS-Based Analysis for Glycolytic Intermediates

2.2.1 Equipment for LC-MS/MS Analysis

- HPLC: Alliance HT pump 2795 system (Waters, USA).
- MS: Quattro LC™ triple quadrupole (Waters, UK).
- 515 HPLC pump (Waters, USA).
- 2-position-Ten port stainless steel LabPro Selector valve (Rheodyne, USA).
- Two way LabPro Solvent select (Rheodyne, USA).
- ASRS® 300 4 mm self-regenerating suppressor (Dionex, USA).

2.2.2 Materials for LC-MS/MS Analysis

- HPLC propylene screw top 300 μL vials with cap and bonded pre-slit PTFE/silicon septa (Waters, USA).
- Sodium hydroxide (J.T. Baker, The Netherlands).
- Methanol, HPLC grade (Sigma Aldrich, Germany).
- LC column for glycolytic metabolites: IonPac AS11 (Dionex, Sunnyvale, USA) 4 mm inner diameter column to 250 mm in length.
- Guard column: A XTerra® MS C18 column 3 mm inner diameter and 20 mm length (Waters, Ireland).

2.3 LC-MS/MS-Based Analysis for Nucleotides

2.3.1 Equipment for LC-MS/MS Analysis

- HPLC: Alliance HT pump 2795 (Waters, USA).
- MS: Quattro Ultima Pt triple quadrupole mass spectrometers (Waters, UK).
- 515 HPLC pump (Waters, USA).
- Ten port Selector valve (Waters, USA).

2.3.2 Materials for LC-MS/MS Analysis

- LC column for analysis of nucleotides: XBridgeTM C18 column packed with 3.5 μm particles, 2.1 mm inner diameter, 150 mm in length (Waters, Ireland).

- Guard column: XBridge MS C18, packed with 3.5 μm particles, 2.1 mm inner diameter, and 10 mm length (Waters, Ireland).
- Dibutylamine acetate concentrate, DBAA (Sigma-Aldrich, Japan).
- Acetonitrile, HPLC grade (Sigma Aldrich, Germany).
- HPLC propylene screw top 300 μL vials with cap and bonded pre-slit PTFE/silicon septa (Waters, USA).

2.4 GC-MS-Based Analysis for Glycolytic, Pentose Phosphate Pathway, and TCA Intermediates

2.4.1 Equipment for GC-MS Analysis

- GC-EI-MS (GC: 7890A, MS: 5975C Agilent Technologies, Santa Clara, CA, USA) with PTV injection system (Cooled Injection System CIS4, Gerstel GmbH, Mülheim an der Ruhr, Germany).
- Optional: Backflush system.

2.4.2 Materials for GC Method

- Straight glass liner with glass wool (Gerstel GmbH, Mülheim an der Ruhr, Germany).
- Zebron ZB-50 column (30 m × 250 μm, 0.25 μm film thickness; Phenomenex, Torrance, CA, USA).
- 10 μL syringe with Teflon tipped plunger (Agilent Technologies, Santa Clara, CA, USA).
- O-Methoxyamine-hydrochloride (MOX, purum, Sigma-Aldrich, Buchs, Switzerland).
- Trimethylchlorosilane (TMCS) (Thermo Scientific, Bellafonte, PA, USA).
- *N*-Methyl-*N*-trimethylsilyltrifuoroacteamide (MSTFA) in 1 mL ampules (Thermo Scientific, Bellafonte, PA, USA).
- Pyridine (HPLC grade 99.9 %, Sigma-Aldrich, Buchs, Switzerland).
- Isooctane (puriss) (Merck, Darmstadt, Germany).
- Ethylacetate (puriss) (Sigma-Aldrich, Buchs, Switzerland).
- Glass vials, silanized (1.5 mL, Grace, Deerfield, IL, USA) and inlets (0.1 mL micro-insert, Grace, Deerfield, IL, USA).
- Screw caps with natural rubber/Teflon septum (Grace, Deerfield, IL, USA).

3 Methods

3.1 Rapid Sampling

1. Prepare:
   $n\times$ "sampling tubes" (PP) with 5 mL 100 % v/v MeOH. Number and weigh them, and store at −40 °C.
   $2\times n$ "washing tubes" (PS) with 5 mL 100 % v/v MeOH for the washing step; store at −40 °C.

$n\times$"ethanol tubes" (PS) with 5 mL 75 % v/v EtOH (68 % m/m) for the extraction step; store at 4 °C.

2. Set centrifuge to −20 °C; the rotor for test tubes is cooled down to −40 °C. Set cryostat at −40 °C, and thaw ^{13}C-labeled extract. Place "sampling tubes" and "washing tubes" in the cryostat. Heat water bath to 95 °C.
3. Sample 1.2 mL broth into each "sampling tube."
4. Weigh each tube (wipe off ethylene glycol with a dry tissue).
5. Centrifuge suspension at 2,000 × *g* for 5 min at −20 °C.
6. Decant supernatant and resuspend pellet in 5 mL from the "washing tubes."
7. Centrifuge again and decant supernatant. Repeat **steps 5** and **6**.
8. Add ^{13}C extract to each pellet (typically 120 μL).

3.2 Metabolite Extraction

1. Place "ethanol tubes" in the water bath (4 min, 75 °C).
2. Transfer the boiling ethanol into the "sampling tube," vortex (10 s), place the tube in the water bath (95 °C) as quickly as possible, and boil for 3 min.
3. Place "sampling tube" back in the cryostat
4. Place the tubes in the RapidVap. Close the tubes with perforated caps.
5. Evaporate to dryness.
6. Resuspend the dried sediment in 600 μL MilliQ water.
7. Vortex thoroughly and transfer to 1.5 mL Eppendorf tubes.
8. Centrifuge at 15,000 × *g* for 5 min at 1 °C.
9. Transfer the supernatants to 0.2 μm Durapore PVDF centrifuge filters.
10. Centrifuge at 15,000 × *g* for 5 min at 1 °C.
11. Transfer the filtrate into screw-cap sample vials.
12. Store at −80 °C until analysis.

3.3 Preparation of Standards

For the calibration lines, standards of each metabolite should be injected over a wide concentration range (typically 0.1–100 μM). Most convenient is to prepare stock solutions, from these generate a standard mix solution, and distribute the standard mix over a series of aliquots (store at −80 °C). For the analysis take one aliquot and prepare a dilution series. From our experience, a close to logarithmic scheme works best (*see* Table 1 for a pipetting scheme). The concentration in the mix solution should be adjusted such that the expected measurements fall into the medium range of the calibration line.

Make sure that the ^{13}C-labeled cell extract that is used is the same as the one added to the analyzed samples (including the same amount of freeze/thaw cycles, similar standing time, etc.).

Table 1
Recommended dilution pipetting scheme: the standard mix is prepared in a close to logarithmic concentration series

Std	Standard mix (μL)	MilliQ (μL)
S1	1	499
S2	2	498
S3	5	495
S4	10	490
S5	20	480
S6	40	460
S7	60	440
S8	100	400
S9	200	300
S10	500	0

1. Prepare a dilution series from a standard stock mix solution (*see* Table 1 for a scheme).
2. Transfer a fixed amount to an analysis vial ($V_{std} = 100$ μL).
3. Add a fixed amount of ^{13}C extract ($V_{13C,std} = 20\ \mu L$).

3.4 LC-MS/MS Glycolytic Intermediates

Several HPLC methods are available for glycolytic intermediates, e.g., [21, 22]. Here one method is discussed exemplarily. Liquid chromatographic separation of glycolytic metabolites is based on anion-exchange chromatography adapted from the method described by van Dam et al. [23].

A guard column is placed prior to analytical column to trap hydrophobic contaminants. The analytical column is maintained at 25 °C while the auto sampler temperature is set to 4 °C. The sample injection volume is set to 10 μL.

1. A sample volume of min 40 μL is required (vial with insert).
2. Prepare four solvents: A: 5 % methanol; B: 0.0006 M NaOH; C: 0.06 M NaOH; and D: 0.3 M NaOH (*see* **Note 1**).
3. Set the flow rate of the LC to 1 mL/min with A: 95 % and B: 5 %.
4. Set the flow rate of the HPLC pump for the methanol inflow (80 % v/v) to the ion suppressor to 0.7 mL/min.
5. Set the pump for solvent selector to 0.9 mL/min. Solvent selector switches between solvents of 80 % v/v methanol and 5 % v/v methanol.
6. Switch the ion suppressor on (auto suppression external water mode, the ion suppressor removes the sodium ions from the

effluent of the analytical column). The final flow to the MS is reduced to ~100 μL/min (T union flow splitter).

7. The chromatography is started with solvent selector in position A (the guard column is switched in front of the analytical column).
8. Inject 10 μL of sample.
9. A NaOH gradient is generated using four solvent mixtures: The gradient is programmed as follows:
   - *t*= 0–1.5 min; A: 95 %, B: 5 %, C and D: 0 %.
   - *t*= 1.51 min; A: 95 %, C: 5 %, and B and D: 0 %. Switch guard column switches to B position (solvent selector sets a flow of 80 % methanol through the guard column, the guard column decoupled from the analytical column).
   - *t*= 1.51–17 min; increase the solvent C to 25 % using an exponential gradient (curve 7, concave close to linear). The exponential gradient improves the separation of the isomers of G6P, which are F6P, glucose-1-phosphate, fructose-1-phosphate, mannose-1-phosphate, and mannose-6-phosphate.
   - *t*= 17–22 min; increase solvent C linearly to 100 %.
   - *t*= 22–24 min; C: 100 %.
   - *t*= 24–24.2 min; increase solvent D linearly to 100 %.
   - *t*= 24.2–27 min; D: 100 %.
   - *t*= 27–34 min; A: 95 % and B: 5 %.
   - Switch guard column switch to position A. The system is now ready for a new injection (*see* **Note 2**).

Mass spectrometer: The MS system is equipped with an ESI source. The capillary voltage is 2,700 V in negative ionization mode. Nitrogen is used as nebulizer gas at 88 L/h and as desolvation gas at 555 L/h. The source block temperature is 90 °C and desolvation temperature is 250 °C (*see* **Note 3**).

MRM transitions: Reactions are optimized for each metabolite using standards, which are directly infused into the MS. A list of reactions for unlabeled metabolites can be found in [23]. The corresponding ^{13}C-labeled isotopologues can be determined knowing the number of carbon atoms in precursor and fragment ions [12, 24].

3.5 LC-MS/MS with Ion Pair Mobile Phase: Nucleotides

The ion pair method for separation of nucleotides is adopted from method described by Seifar et al. [25]. In this method the analytical column is maintained at 25 °C and the auto sampler temperature is set to 4 °C. The sample injection volume is set to 5 μL. A guard column is placed prior to the analytical column to protect analytical column from particles and contaminants.

1. Transfer 50 μL of sample to an HPLC vial (Waters, Teflon).
2. Add 1 μL of 50 mM ion-pair reagent DBAA to the sample vial (*see* **Note 4**).
3. Prepare two solvents: A: 5 % acetonitrile and 5 mM DBAA and B: 84 % acetonitrile and 5 mM DBAA. Set the flow rate of the LC to 1 mL/min with A: 90 % and B: 10 %.
4. Set the HPLC pump for 50 % v/v acetonitrile to a flow of 0.1 mL/min (located between the analytical column and the electrospray of the MS).
5. Using a T union flow splitter the final flow to the MS is reduced to 0.1 mL/min.
6. The chromatography is started with selector valve in position 2 (effluent from analytical column diverted to the waste).
7. Inject 5 μL of sample into the LC column.
8. The following gradient program is used:
   - t = 0–2 min; A: 90 % and B: 10 %. At t = 1 min, the selector valve is set to position 1 (the effluent from the analytical column mixes with the 0.1 mL/min flow) (*see* **step 2**).
   - t = 2–12 min; solvent B increases linearly to 60 %.
   - t = 12–15 min; B: 60 %.
   - t = 15–15.5 min; linear gradient to A: 90 % and B: 10 %. Switch selector valve to position 2 (LC effluent to waste).
   - t = 15.5–25 min; hold A: 90 % and B: 10 %.

Mass spectrometer: ESI capillary voltage is 2,700 V in negative ionization mode. Nitrogen is used as nebulizer gas at 50 L/h and as desolvation gas at 660 L/h. Argon gas is used as collision gas. The collision cell pressure is 2.83×10^{-3} mbar. The source block temperature is 120 °C and desolvation gas temperature is 300 °C.

MRM transitions: Reactions, collision energy, and cone voltage are optimized for each metabolite using standards that are directly infused. A list of reactions for nucleotides is available in [25].

3.6 GC-MS for Glycolytic, Pentose-Phosphate, and TCA Intermediates

For GC-MS analysis chemical derivatization is required. The following steps are advised:

1. Lyophilization: Transfer 100 μL of sample into a glass sample vial, close with a screw cap with Teflon septum (punch holes with a syringe needle), and freeze-dry overnight. Make sure that the samples do not start melting, by, e.g., using a cold (−80 °C) block.
2. Dissolve O-methoxyamine hydrochloride (MOX, stored in desiccator) at 20 g/L in (anhydrous) pyridine; shake until MOX is completely dissolved (prepare fresh).

3. Add 50 μL of 20 g/L MOX solution and incubate for 50 min at 70 °C.
4. Open a fresh MSTFA ampule and add 50 μL of TMCS (*see* **Note 8**).
5. Add 80 μL of MSTFA-TMCS and incubate for 50 min at 70 °C.
6. Transfer 100 μL of the derivatized sample in a glass vials and centrifuge for 1 min at 10,000 ×*g*.
7. Transfer 70 μL to a GC vial with inlet and close (tight) with a screw cap.

Injection and PTV inlet program: Set initial temperature to 70 °C, first ramp at 720 °C/min to 220 °C, and hold for 5 min. Second ramp, 720 °C/min to 300 °C, and hold for 5 min (*see* **Note 5**). Inject 1 μL in split-less mode (for metabolites present in high concentration in split mode, e.g., Tre split 1/40). Depending on the sample load, the column needs to be replaced or shortened reqularly (*see* **Note 6**). Wash syringe 3× with isooctane, and again 5× with ethyl acetate and 2× with isooctane after injection (*see* **Note** 7).

Oven program: Helium flow at 1 mL/min. Initial temperature, 70 °C, and hold for 1 min. Increase temperature with a ramp of 10 °C/min until 320 °C; no hold. At the end of a GC run the column is backflushed with 5 column volumes at 320 °C.

MS settings: Transfer line 250 °C, source temperature 230 °C, and quadrupole temperature 150 °C. The measured fragments can be found in [26] for pentose-phosphate pathway intermediates and in [24] for glycolytic intermediates.

3.7 Calculation of Intracellular Metabolite Levels

The calculation of the intracellular metabolite levels is independent of the applied analysis method (GC- or LC-MS). Essential is to use the original, aqueous sample amounts. The following symbols will be used in the calculations:

$$
\begin{aligned}
&V_{\text{sample}} \text{ broth sample } (\text{mL}) \\
&c_X \text{ biomass concentration in sample } (\text{g}_{\text{DW}} / \text{L}) \\
&V_{13\text{C,sample}} \text{ volume of } {}^{13}\text{C extract added to the cell pellet } (\mu\text{L}) \\
&V_{13\text{C,std}} \text{ volume of } {}^{13}\text{C extract added to the standard mix } (\mu\text{L}) \\
&c_{\text{sample}} \text{ concentration of } {}^{12}\text{C compound in the extract sample } (\mu\text{M}) \\
&c_{\text{std}} \text{ concentration of } {}^{12}\text{C compound } (\mu\text{M}) \\
&V_{\text{std}} \text{ volume of } {}^{12}\text{C standard } (\mu\text{L}) \\
&c'_{\text{std}} \text{ concentration of } {}^{12}\text{C compound } \left(\text{based on } V = V_{\text{std}} + V_{{}^{13}\text{C,std}}\right) (\mu\text{M}) \\
&A_{\text{std}}^{12\text{C}} \text{ area of } {}^{12}\text{C compound } (-) \\
&A_{\text{std}}^{13\text{C}} \text{ area of } {}^{13}\text{C compound } (-)
\end{aligned}
\tag{1}
$$

1. Integrate the peak areas and plot the ratio A^{12C} / A^{13C} against the concentrations of ^{12}C compound in the calibration standards c_{std}'.
2. Calculate the slope (a) and y intercept (b) by linear (weighted) regression:

$$\begin{pmatrix} a \\ b \end{pmatrix} = \left(\mathbf{X}^T\mathbf{W}\mathbf{X}\right)^{-1}\mathbf{X}^T\mathbf{W}\mathbf{y}, \quad \mathbf{X} = \begin{pmatrix} c_{std,1}^{12C} & 1 \\ c_{std,2}^{12C} & 1 \\ \vdots & \vdots \\ c_{std,n}^{12C} & 1 \end{pmatrix}, \quad \mathbf{y} = \begin{pmatrix} \frac{A_{std,1}^{12C}}{A_{std,1}^{13C}} \\ \frac{A_{std,2}^{12C}}{A_{std,2}^{13C}} \\ \vdots \\ \frac{A_{std,n}^{12C}}{A_{std,n}^{13C}} \end{pmatrix} \tag{2}$$

For the weighting matrix different assumptions on the error model can be used: (absolute) error independent of the measurement signal (equal weighting, **W**, becomes an identity matrix), linear increase in variance, weight $W_{i,i} = 1/y_i$, or linear increase in standard deviation, $W_{i,i} = 1/y_i^2$. From our experience, the latter one leads to the most reliable results.
3. Calculate the compound concentration in the sample by

$$c_{sample} = \left(b + a\left(A_{sample}^{12C} / A_{sample}^{13C}\right)\right)\frac{V_{12Cstd} + V_{13Cstd}}{V_{12Cstd}} \tag{3}$$

4. Calculate the intracellular metabolite levels as µmol/g_{DW}, thereby taking into account the broth sampling amounts and dilution due to ^{13}C addition to the standards:

$$c_{sample/X} = \frac{c_{sample}}{m_{sample}\, c_X} \tag{4}$$

3.8 Linearity in Concentration and Mass Isotopomer Ratio

When using ID-MS-based quantification, the linear range of the mass isotopomer ratio is crucial. Only within the linear dynamic range peak area ratios will correctly represent concentrations. Quadrupole MS systems usually provide a broad signal range of about 4–5 orders of magnitude. This is required, also because the intracellular metabolites have large concentration differences [3]. To test the linearity, the natural labeling content can be exploited. Especially for derivatized molecules, the natural isotope distribution leads to mass isotopomer (isotopologue) distributions that can be measured. By injection of standards and samples at different concentrations (4–5 orders of magnitude), the linearity of mass

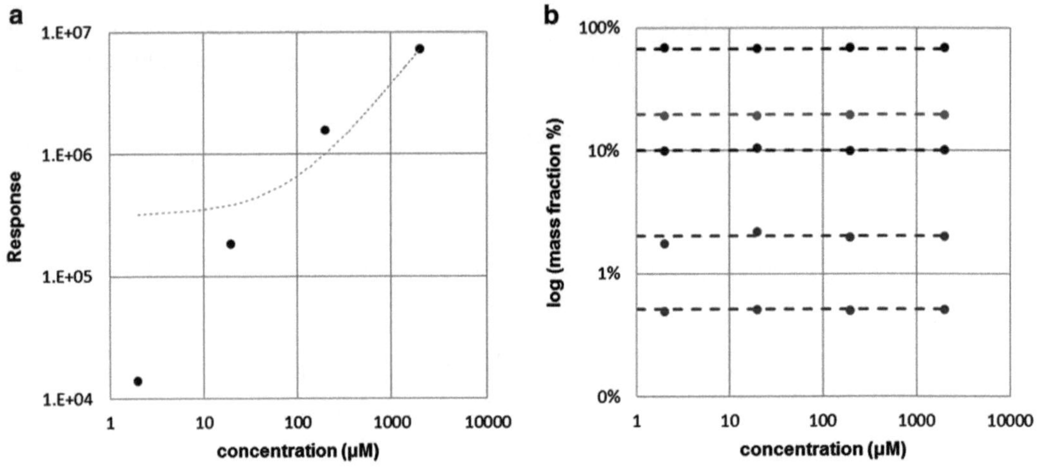

Fig. 3 (**a**) Response (area, au) over four orders of magnitude for a standard of malate. The *dashed line* indicates a linear regression (note that the scale is double logarithmic). (**b**) Measured mass isotopomer fractions over four orders of magnitude (*points*) in comparison to the expected (calculated) distribution of malate. From *top* to *bottom* the fractions of m + 0, m + 1, m + 2, m + 3, and m + 4

isotopomers can be determined. As an example, the standard of malate does not show linearity in the concentration-area response, but the isotopomer ratio is constant over four orders of magnitude (Fig. 3). This observation indicates that already for defined standards a high variability in derivatization efficiency, injection, and/or ionization is present.

With a high linear range, the amount of ^{13}C extract can be reduced—usually, a ratio of 1:5 is applied (^{13}C cell extract: sample volume). Depending on the metabolite concentrations in the sample, the peak ratio will differ significantly. In case the ratio is outside the range of 1:100 and 100:1 the amount of ^{13}C extract should be adjusted. Especially for, e.g., extracellular samples the ratio can be adapted to cover, e.g., very low trehalose or high glucose concentrations.

3.9 Preparation of ^{13}C Cell Extract

For the ^{13}C cell extract preparation the content in ^{12}C has to be minimized [2, 12]. Therefore, the cultivation is performed as fed-batch, starting with a low cell density (small inoculum) until a high cell density, fully grown on ^{13}C carbon source is reached. To balance the amount of intracellular metabolites, cells are harvested at different time points. Additionally, for the last harvest a glucose pulse is applied to increase the amount of glycolytic intermediates in the extract. The different cell extracts are pooled, measured, and diluted and aliquots of 1.2 mL are stored at –80 °C.

4 Notes

1. To secure precise gradients, a total of four solvent mixtures are used with different NaOH concentrations. Additionally, the very high final NaOH concentration improves the lifetime of the column (good washing properties). For the preparation of the solvents use fresh MilliQ water and make sure that a minimum of CO_2 dissolves (esp. prevent bubbles, turbulences, etc. during preparation). As the solvent mixtures are always in exchange with the environment, carefully monitor for the appearance of additional peaks—the time to prepare new solvent mixtures.
2. Because of the complex setup of valves, be careful when for example stopping a sample run. Make sure that when you restart the system, the valves are properly set. Otherwise the ion suppressor or even the MS system could become polluted with high salt concentrations.
3. In case of decreased signals, it is advised to first clean the cone and if necessary the ESI source. The column can be regenerated using 0.3 M NaOH (solvent D), e.g., increase the time of the last interval in case of memory effects. In this case, use an extra valve to divert the eluent to waste before entering the ion suppressor (NaOH concentration is too high for the suppressor).
4. In our experience, the addition of ion-pairing reagent to the sample improves the signal intensity as less adducts are observed during ionization. Regular cleaning of the cone and ESI sources improves the reproducibility. In case of peak broadening, the column can be cleaned with ammonium phosphate. In case of observing retention time shift, the length of the last gradient (column conditioning, A: 90 %, B: 10 %) can be increased to cover ten times column volume (to about 34 min).
5. The measurement quality is improved by the use of a PTV injection system especially for less stable compounds. Using a backflush system, the measurement time can be shortened and the lifetime of the column is extended.
6. For reliable operation, cut 10 cm of the column after about 100 samples and at the same time change the liner (best with glass wool).
7. To minimize blockage of the syringe plunger, use gastight syringes (prevents evaporation of MSTFA behind the flushed zone of the syringe). Break and blockage of the needle can be minimized by fresh caps if a sample is injected several times and the use of long-life septum or a septumless head (SLH) at the injection system.
8. For a reproducible derivatization, make sure to use fresh MSTFA (max 1 week old) and minimize the water content of

the hygroscopic chemicals (MOX, TMCS: store in desiccator). The solvent tends to react with the rubber of the sample septum—make sure to use septa with Teflon protection. Note that after injection, the Teflon layer is destroyed and the solvent starts to react with the rubber. In case the sample should be injected multiple times, we advise to renew the cap after, e.g., two injections. Cooling of the sample reduces the evaporation and reaction with the rubber material.

References

1. de Koning W, van Dam K (1992) A method for the determination of changes of glycolytic metabolites in yeast on a subsecond time scale using extraction at neutral pH. Anal Biochem 204:118–123
2. Wu L, Mashego MR, van Dam JC, Proell AM, Vinke JL, Ras C, van Winden WA, van Gulik WM, Heijnen JJ (2005) Quantitative analysis of the microbial metabolome by isotope dilution mass spectrometry using uniformly ^{13}C-labeled cell extracts as internal standards. Anal Biochem 336:164–171
3. Canelas AB, Ras C, ten Pierick A, van Dam JC, Heijnen JJ, van Gulik WM (2008) Leakage-free rapid quenching technique for yeast metabolomics. Metabolomics 4:226–239
4. Canelas AB, ten Pierick A, Ras C, Seifar RM, van Dam JC, van Gulik WM, Heijnen JJ (2009) Quantitative evaluation of intracellular metabolite extraction techniques for yeast metabolomics. Anal Chem 81:7379–7389
5. Bolten CJ, Wittmann C (2008) Appropriate sampling for intracellular amino acid analysis in five phylogenetically different yeasts. Biotechnol Lett 30:1993–2000
6. de Jonge LP, Douma RD, Heijnen JJ, van Gulik WM (2012) Optimization of cold methanol quenching for quantitative metabolomics of Penicillium chrysogenum. Metabolomics 8:727–735
7. Gale EF (1947) Assimilation of amino acids by bacteria. I. Passage of certain amino acids across the cell wall and their concentration in the internal environment of Streptococcus faecalis. J Gen Microbiol 1:53–76
8. Bolten CJ, Kiefer P, Letisse F, Portais J-C, Wittmann C (2007) Sampling for metabolome analysis of microorganisms. Anal Chem 79: 3843–3849
9. Maharjan RP, Ferenci T (2003) Global metabolite analysis: the influence of extraction methodology on metabolome profiles of Escherichia coli. Anal Biochem 313:145–154
10. Rabinowitz JD, Kimball E (2007) Acidic acetonitrile for cellular metabolome extraction from Escherichia coli. Anal Chem 79:6167–6173
11. Hajjaj H, Blanc PJ, Goma G, Francois J (1998) Sampling techniques and comparative extraction procedures for quantitative determination of intra- and extracellular metabolites in filamentous fungi. FEMS Microbiol Lett 164:195–200
12. Mashego MR, Wu L, Van Dam JC, Ras C, Vinke JL, Van Winden WA, Van Gulik WM, Heijnen JJ (2004) MIRACLE: mass isotopomer ratio analysis of U-13C-labeled extracts. a new method for accurate quantification of changes in concentrations of intracellular metabolites. Biotechnol Bioeng 85:620–628
13. Strassburg K, Walther D, Takahashi H, Kanaya S, Kopka J (2010) Dynamic transcriptional and metabolic responses in yeast adapting to temperature stress. OMICS 14:249–259
14. Nasution U, van Gulik WM, Kleijn RJ, van Winden WA, Proell A, Heijnen JJ (2006) Measurement of intracellular metabolites of primary metabolism and adenine nucleotides in chemostat cultivated *Penicillium chrysogenum*. Biotechnol Bioeng 94:159–166
15. Bennett BD, Kimball EH, Gao M, Osterhout R, Van Dien SJ, Rabinowitz JD (2009) Absolute metabolite concentrations and implied enzyme active site occupancy in Escherichia coli. Nat Chem Biol 5:593–599
16. Taymaz-Nikerel H, de Mey M, Ras C, ten Pierick A, Seifar RM, van Dam JC, Heijnen JJ, van Gulik WM (2009) Development and application of a differential method for reliable metabolome analysis in Escherichia coli. Anal Biochem 386:9–19
17. Lafaye A, Labarre J, Tabet J-C, Ezan E, Junot C (2005) Liquid chromatography-mass spectrometry and 15N metabolic labeling for quantitative metabolic profiling. Anal Chem 77:2026–2033
18. Seifar RM, Zhao Z, van Dam J, van Winden W, van Gulik W, Heijnen JJ (2008) Quantitative

analysis of metabolites in complex biological samples using ion-pair reversed-phase liquid chromatography-isotope dilution tandem mass spectrometry. J Chromatogr A 1187:103–110

19. Bennett BD, Yuan J, Kimball EH, Rabinowitz JD (2008) Absolute quantitation of intracellular metabolite concentrations by an isotope ratio-based approach. Nat Protoc 3: 1299–1311
20. Vielhauer O, Zakhartsev M, Horn T, Takors R, Reuss M (2011) Simplified absolute metabolite quantification by gas chromatography-isotope dilution mass spectrometry on the basis of commercially available source material. J Chromatogr B Analyt Technol Biomed Life Sci 879(32):3859–3870
21. Coulier L, Bas R, Jespersen S, Verheij E, Van Der Werf MJ, Hankemeier T (2006) Simultaneous quantitative analysis of metabolites using ion-pair liquid chromatography-electrospray ionization mass spectrometry. Anal Chem 78:6573–6582
22. Oldiges M, Lutz S, Pflug S, Schroer K, Stein N, Wiendahl C (2007) Metabolomics: current state and evolving methodologies and tools. Appl Microbiol Biotechnol 76:495–511
23. van Dam JC, Eman MR, Frank J, Lange HC, van Dedem GWK, Heijnen JJ (2002) Analysis of glycolytic intermediates in Saccharomyces cerevisiae using anion exchange chromatography and electrospray ionization with tandem mass spectrometric detection. Anal Chim Acta 460:209–218
24. van Dam JC, Ras C, ten Pierick A (2011) Analysis of glycolytic intermediates with ion chromatography-and gas chromatography-mass spectrometry. In: Metz TO (ed) Metabolic profiling. Human Press, New York, USA
25. Seifar RM, Ras C, van Dam JC, van Gulik WM, Heijnen JJ, van Winden WA (2009) Simultaneous quantification of free nucleotides in complex biological samples using ion pair reversed phase liquid chromatography isotope dilution tandem mass spectrometry. Anal Biochem 388:213–219
26. Cipollina C, ten Pierick A, Canelas AB, Seifar RM, van Maris AJ, van Dam JC, Heijnen JJ (2009) A comprehensive method for the quantification of the non-oxidative pentose phosphate pathway intermediates in Saccharomyces cerevisiae by GC-IDMS. J Chromatogr B 877: 3231–3236

Chapter 7

Determining the Biomass Composition of a Sponge Holobiont for Flux Analysis

Jabin Watson, Bernard Degnan, Sandie Degnan, and Jens O. Krömer

Abstract

The first step on the path of flux analysis of a new organism with little available literature is the determination of the biomass composition. Once the content of the macromolecular components (protein, RNA, DNA, carbohydrates, lipids) and their composition is known, this composition can be converted into a biomass equation. The biomass equation is an important part of metabolic flux analysis. This equation provides the information about the precursor and energy needs for growth. In many experiments the determination of the growth rate is the simplest flux to be determined, yet this rate determines the net fluxes of a whole range of anabolic pathways in the system and often is used as the objective function in FBA analysis. The challenge for the scientist is to create a biomass equation that represents the organisms of choice under the conditions studied. This chapter outlines basic protocols that can be applied to the quantification of the macromolecular components, using the marine demosponge *Amphimedon queenslandica* as a case study. As is true for all other sponges and indeed marine animals, *A. queenslandica* is a holobiont, comprising an animal host plus symbiotic and other associated microbial cells. We show how this complexity can be overcome by developing a fast, yet robust, method for biomass quantification of sponges using the displacement volume. The analytical protocols we describe herein are widely applicable not only to other organisms sampled from complex environments but also to cell cultures. The second part of the chapter highlights the procedures needed to convert a macromolecular composition into a biomass equation.

Key words Biomass composition analysis, RNA, DNA, Lipids, Carbohydrate, Protein, Sponge skeleton, Displacement, Dry weight, Biomass equation

1 Introduction

Genome-scale metabolic reconstructions are powerful tools for understanding metabolic processes, and have become the standard for studying metabolic networks at a system level [1]. To convert a reconstruction into a functioning model for metabolic flux analysis, the biochemical composition of the organism must be determined [2]. Quantifying the macromolecules of an organism allows the formulation of a biomass equation, which in turn enables the testing of a metabolic model by predicting growth rates under specific conditions. These predictions can be then tested in vivo.

Jens O. Krömer et al. (eds.), *Metabolic Flux Analysis: Methods and Protocols*, Methods in Molecular Biology, vol. 1191, DOI 10.1007/978-1-4939-1170-7_7, © Springer Science+Business Media New York 2014

The major macromolecules in a cell are DNA, RNA, proteins, carbohydrates, and lipids. By quantifying the macromolecule composition, the precursor molecules that are required for biomass synthesis can be identified. This also allows calculation of the energy requirements for cell maintenance and growth. Although flux analysis has been extensively used on single-species cultures [3–5], its application to more complex systems including two and more species is just beginning [6]. This chapter provides basic protocols that can be used to determine the macrocomposition of biomass in a complex biological system, exemplified here by a sponge holobiont. The protocols that we describe are applicable to any such characterization, from organisms in complex open environmental systems, to cell cultures.

The organism we chose to analyze as our case study was the marine sponge *Amphimedon queenslandica* (Niphatidae, Haplosclerida, Demospongiae, Porifera), a sessile invertebrate that encrusts complex substrata in tropical coral reef lagoons. At the time of writing, this is the only sponge to have a fully sequenced, assembled, and annotated genome [7]. As is the case for nearly all sponges, the *A. queenslandica* body plan consists of an irregular canal system through which water is pumped by groups of ciliated cells (*see* Fig. 1). The form of an individual sponge is shaped by various environmental factors, including prevailing currents and availability and shape of the substrate [8, 9]. At the cellular level, the sponge synthesizes a skeleton of silica spicules and a flexible extracellular matrix (ECM) that includes a diversity of proteins and carbohydrates and a sponge-specific collagen, spongin [10]. The cellular biomass of the *A. queenslandica* holobiont primarily includes both eukaryotic sponge cells and prokaryotic bacterial cells that appear to maintain a symbiotic relationship. Both cell types are easily dislodged from the ECM with minimal physical force.

Long-term cell cultures for sponges have not been established (but *see* refs. 11 and 12 as examples of sponge cell culture methods). This is also the case for other marine invertebrates. Thus, the biomass equation must be generated from, and the subsequent flux analysis be performed on, whole or partial samples of living sponge. Living material that has been directly collected from the field will be inherently variable with respect to cell and ECM composition and density. In organisms that host microbial symbionts, this variability includes the ratio of sponge to non-sponge cells. To address this natural variation, analyses need to be performed on multiple samples from multiple individuals. In this case study, we set a minimum sample size of 20 pieces of sponge per assay; 5 subsamples were collected from 4 different sponges for each of the biochemical assays. To undertake flux analyses on living *A. queenslandica*, we need to be able to relate measurements made on dried samples (i.e., dry weight values) to the live sponge. Given the complexity of the internal aquiferous canal system in sponges, we chose to

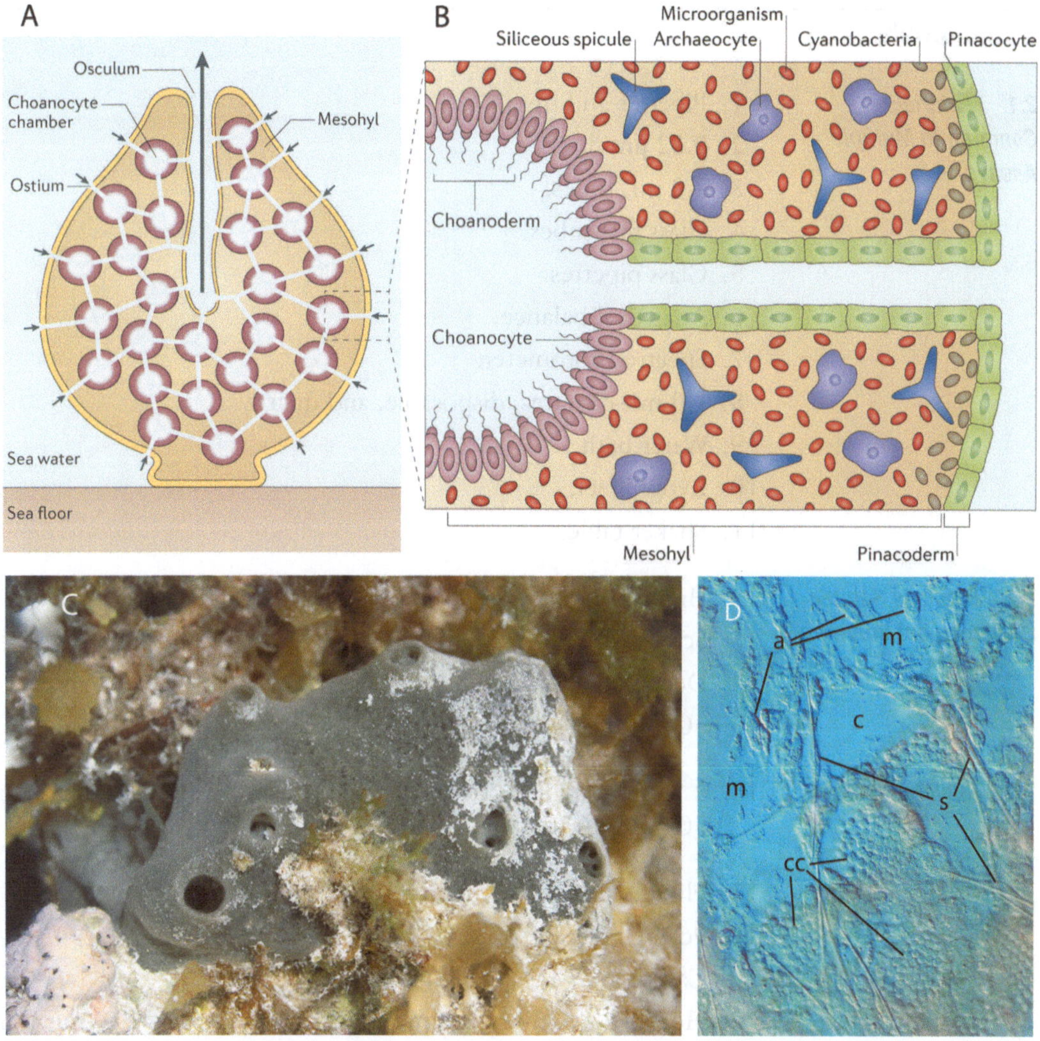

Fig. 1 Body plan and cellular arrangement of *Amphimedon queenslandica*. (**a**) Generalized sponge body plan showing directional water flow mediated by chambers of flagellated choanocyte cells. Connectivity between the inhalant ostia and exhalent oscula is irregular. (**b**) Cellular arrangement showing interacting microbial and sponge cell populations in the mesohyl. Spicules form part of the supporting skeleton. (**c**) *Amphimedon queenslandica* has an irregular body plan and incorporates foreign material as it grows. (**d**) This light microscopy image of a live *A. queenslandica* juvenile demonstrates the irregular biomass arrangement at the cellular level. The densely packed cells at the bottom of the image are dominated by choanocyte chambers (cc) and contrast to the low cell density of the extracellular matrix (m). Archeocytes (a) and sclerocytes (s) are distributed throughout the sponge. Panels **a** and **b** reproduced from [23], with permission from Nature Publishing Group

relate dry weight to displacement volume. This would allow us to measure the starting biomass with minimal disturbance to the metabolic state. Further, to ensure that values obtained from small subsamples are applicable to larger fragments or whole sponges, we determined whether the relationship between dry weight and displacement volume is linear.

2 Materials

2.1 Materials Common to Multiple Assays

1. Pipettes and tips.
2. 0.22 μm filters.
3. 2.0 mL tubes.
4. Glass test tubes.
5. Glass pipettes.
6. Analytical balance.
7. Spectrophotometer.
8. 1.0 mL cuvettes, disposable, and quartz.
9. Water bath.
10. Lyophilizer.
11. Shaker table.

2.2 Biomass Sample Displacement Volume

12. *Amphimedon queenslandica*.
13. Scalpel blades.
14. Deep petri dishes.
15. 50 mL measuring cylinder.

2.3 Skeleton Weight

16. 2 % SDS (sodium dodecyl sulfate): Dissolve 1.0 g of SDS in 50 mL of water.

2.4 DNA

17. Fluorescent plate reader.
18. 96-Well plates.
19. Calf thymus DNA (Sigma D1501).
20. Hoechst 33258 concentrated stock solution, 10 mg/mL (Invitrogen, Life Technologies catalog number H3569).
21. 1 M Tris–HCl (hydroxymethyl aminomethane) (pH 8.0): The molecular weight of Tris–HCl is 121.4 g/mol. Make 500 mL of 1 M Tris–HCl (pH 8.0) by adding 60.57 g to 400 mL of MilliQ water. Adjust the pH to 8.0 with HCl (hydrochloric acid). Make to 500 mL with MilliQ water.
22. 0.5 M EDTA (Ethylenediaminetetraacetic): The molecular weight of EDTA is 372.2 g/mol. Make 100 mL of 0.5 M EDTA by adding 18.61 g of EDTA to 90 mL of MilliQ water. Adjust the pH to 8 using NaOH (sodium hydroxide) pellets (*see* **Note 1**). Make to 100 mL with MilliQ water.
23. TE (pH 8.0): To make 100.0 mL, combine 1.0 mL of 1 M Tris–HCl (pH 8.0) and 0.2 mL of 0.5 M EDTA and make up to 100.0 mL with MilliQ water.
24. Fluorometry buffer: 2 M NaCl (sodium chloride) and 50 mM Na_3PO_4 (sodium phosphate). The molecular weight of NaCl is

58.44. To make 500 mL of fluorometry buffer, add 58.44 g to 450 mL of MilliQ water and stir to dissolve. The molecular weight of Na_3PO_4 is 163.94 g/mol. Add 4.099 g and stir to dissolve. Make to 500 mL with MilliQ water. Adjust the pH to 7.4.

25. Rehydration buffer: Dissolve 0.5 g of lysozyme in 50 mL TE (pH 8.0) (*see* **Note 2**).

2.5 RNA

26. $HClO_4$ (perchloric acid) solutions made from 70 % stock (*see* **Note 3**).
27. 3.0 M $HClO_4$: To make 25.0 mL, slowly add 6.468 mL of stock $HClO_4$ to 6.5 mL of MilliQ water. Adjust the final volume to 25 mL with MilliQ water.
28. 0.5 M $HClO_4$: Slowly add 8.625 mL of stock $HClO_4$ to 50 mL of MilliQ water. Adjust the final volume to 200 mL with MilliQ water.
29. 0.3 M KOH (potassium hydroxide): Add 1.683 g of KOH to 100 mL of water and mix until dissolved.

2.6 Carbohydrate

30. Concentrated sulfuric acid.
31. Phenol (80 % w/v).
32. Iced water.
33. Water bath set to 25 °C.
34. Analytical grade glucose.
35. Glucose standard stock solution: Add 0.01 g of glucose to a 15 mL tube and record the weight of glucose using an analytical balance. Add 10 mL of MilliQ water, recording the weight of water added. Glucose has a molecular weight of 180.16 g/mol. Calculate the concentration of glucose in the stock solution.

2.7 Protein

36. 1 M NaOH (sodium hydroxide). The molecular weight of NaOH is 39.996 g/mol. To make 500 mL add 19.9985 g to 450 mL of MilliQ water. Mix to dissolve the pellets. Make the total volume to 500 mL with MilliQ water.
37. Bovine serum albumin (BSA) 2 mg/mL (Thermo Scientific product number 23209).
38. Coomassie Plus™ (Bradford) Assay Kit (Thermo Scientific product number 23236).
39. Water bath set to 95 °C.
40. Iced water.

2.8 Lipid

41. 15 mL glass tubes with lids.
42. Hexane.
43. Isopropanol.

44. Make a 3:2 hexane:isopropanol solution. Each sample will require 5 mL.
45. Make a 7:2 hexane:isopropanol solution. Each sample will require 5 mL.
46. 0.47 M $NaSO_4$ (sodium sulfate): To make 100 mL of 0.47 M $NaSO_4$, add 6.676 g to 100 mL of MilliQ water.
47. Nitrogen gas supply, with manifold and flow control valves.

3 Methods

3.1 Displacement Volume

Displacement volume was used to normalize live sponge biomass to dry weight composition. This measurement can be taken on live samples before flux analysis.

1. Cut sponge into similar-sized pieces using a scalpel blade (*see* **Note 4**).
2. Filter seawater through a 0.22 μm filter
3. Place 5 mL of filtered seawater into the measuring cylinder, ensuring that the meniscus is in line with the 5 mL mark.
4. Place the piece of sponge into the measuring cylinder (*see* **Note 5**).
5. Place a petri dish on the analytical balance and zero it.
6. With a pipette, transfer the displaced volume to the petri dish on the analytical balance, by carefully returning the meniscus of the measuring cylinder to the 5.0 mL line.
7. Record the displacement weight.
8. Convert to a volume, correcting for the specific density of the seawater.

3.2 Dry Weight

Dry weights were obtained by freeze drying (lyophilization). The application of a vacuum to frozen samples allows the water molecules to sublimate from the solid to the gas phase. This is a gentle process that leaves molecules in their conformational states. Samples that are to be analyzed for RNA or lipid content should be lyophilized in a 15 mL glass tube.

1. Sponge pieces were cut and the displacement volume was recorded.
2. Label and weigh a tube and lid for each sample (*see* **Note 6**).
3. Transfer each sponge piece into a tube and weigh.
4. Lyophilize each sample overnight.
5. Measure and record the weight of the lyophilized sample and tube.

6. Store the lyophilized samples in a −80 °C freezer.
7. To calculate the dry weight, subtract the weight of the empty tube and lid from the weight of the tube, lid, and dry sample.
8. Correct the dry weight for salt content. First calculate the amount of water that was evaporated by subtracting the weight of the tube, lid, and dry sponge piece from the weight of the tube, lid, and wet sponge piece. Multiply the amount of water evaporated by the specific gravity of the seawater to estimate the weight of salt left behind (*see* **Note 7**). Subtract the weight of the salt from the weight of the dry sponge piece.
9. Normalize the salt-corrected dry weight to the displacement of the starting biomass.

3.3 Dry Weight and Displacement Volume Relationship

To analyze the relationship between dry weight and displacement volume, 50 different-sized pieces of sponge were collected. The displacement volume ranged from 0.0865 to 3.404 mL. The above methods were used to analyze ten pieces from five different sponges. The linearity of the relationship was plotted and resulted in an r^2 value of 0.91 (*see* Fig. 2).

3.4 Skeleton Weight

We defined the sponge skeleton as the silica spicules and the ECM combined. This skeleton plays a structural role but is here assumed not to be turned over metabolically. Instead it can be treated as a metabolic sink during the modelling. Treating the sponge pieces with a detergent to lyse the cellular component allowed us to weight the remaining skeleton and calculate the amount of biomass that was present.

1. Record the displacement and dry weight of each sample using the method described above.
2. Remove the freeze-dried samples from the −80 °C freezer and allow to return to room temperature (*see* **Note 8**).
3. Weigh and record the weight of each tube with the sample in it (*see* **Note 9**).
4. Add 3.0 mL of 2 % SDS to each tube.
5. Place on a shaker table for 2 h.
6. Thoroughly wash each piece and tube with RO water to remove all SDS and cellular debris.
7. Examine each piece with a microscope to ensure that only the skeleton remains.
8. Return each piece to its tube and lyophilize overnight.
9. Weigh each piece of lyophilized skeleton.

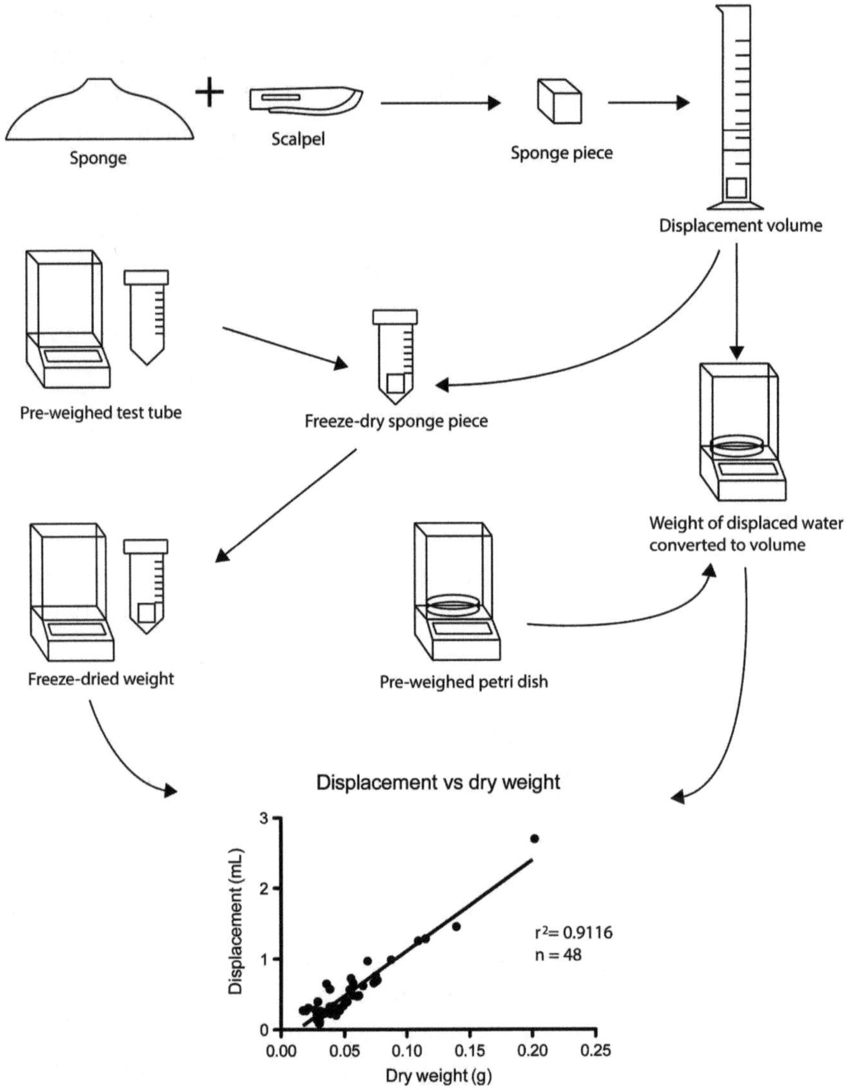

Fig. 2 This graphical representation outlines the workflow used to generate the relationship between the displacement volume and the dry weight of pieces of sponge

3.5 DNA

To quantify the DNA content, we used the fluorescent dye Hoechst 33258. Hoechst 33258 is a bis-benzimidazole compound that binds non-intercalatively to the minor groove of double-stranded DNA, with high specificity to the adenine and thymine regions [13]. It has peak excitation at 350 nm and emission at 461 nm [14]. Samples of unknown concentration are assayed against a standard curve that is generated using DNA of known concentration (*see* **Notes 10** and **11**). High-molecular-weight DNA is required for making the standard curve because Hoechst 33258 does not bind efficiently to small DNA fragments [13]. Hoechst 33258 can be used with whole-cell lysates as it does not bind to protein, or significantly to RNA [13].

1. Reconstitute standard stock DNA by adding 100 μg of calf thymus DNA to 1.0 mL of TE buffer (pH 8.0). Leave overnight at 2–8 °C to completely dissolve the DNA.
2. Measure the concentration of the standard stock DNA using a spectrophotometer at 260 nm. Multiply the reading by 50. This gives a concentration in μg/mL.
3. Make the standards with a series of 1:1 dilutions of the standard DNA and TE (pH 8.0). Perform five serial dilutions, recording the weight at each step for accurate calculation of the DNA concentration. This will give six known concentrations for the standard curve, in addition to a blank.
4. Remove the freeze-dried samples from the –80 °C freezer and allow to return to room temperature (*see* **Note 8**).
5. Weigh and record the weight of each tube with the sample in it (*see* **Note 9**).
6. The freeze-dried samples need to be rehydrated for analysis. Add 3.0 mL of rehydration buffer and record the weight.
7. Incubate at 37 °C for 1 h.
8. Make up a Hoechst working solution. Add 50 μl of Hoechst 33258 (10 mg/mL) to 50 mL of the fluorometry buffer. Keep solution covered with aluminum foil.
9. Prepare the 96-well plate. For each standard and sample, prepare three wells. Transfer 100 μL of the Hoechst working solution into each allocated well. Include a single well for the blank. Weight the plate after each addition and record the exact weight of working solution added.
10. Mix each sample and standard before adding 10 μL to an allocated well. Weigh and record the weight of the plate after each addition.
11. Mix the plate and place in the plate reader.
12. Measure at 350 nm excitation and 461 nm emission.
13. Subtract the blank value from all measurements.
14. Calculate the exact concentrations of the standards using the recorded weights at each step.
15. Make the concentration standard curve.
16. Calculate the concentrations of each replicate before calculating the average concentration of the sample (*see* **Note 12**).
17. Calculate the concentration of the initial rehydration reaction using the recorded weights (*see* **Note 11**).
18. Calculate the amount of DNA using the concentration and rehydration reaction volume.
19. Normalize the amount of DNA to the displacement volume of the initial biomass.

3.6 RNA

In this assay, an alkali is used to hydrolyze the sample before the cellular components are separated by their solubility in an acidic solution. The UV absorbance is then used to quantify the amount of RNA in solution. The addition of potassium hydroxide hydrolyzes the RNA into monomers. These RNA monomers are acid soluble. The hydrolysis of carbohydrates does not produce compounds that affect the absorbance at 260 nm, while the hydrolysis of protein does produce compounds that interfere with the absorbance reading. This production of interfering compounds is restricted by the use of low-molarity alkali and a reaction time of 1 h [15]. The main proportion of protein is then precipitated when the reaction is made acidic. This acidification also precipitates the DNA fraction that remained stable under the initial alkaline conditions. The precipitated protein and DNA are then removed by centrifugation and the amount of RNA quantified by measuring absorbance at 260 nm.

1. Remove the freeze-dried samples from the −80 °C freezer and allow to return to room temperature (*see* **Note 8**). These are the samples that were prepared and lyophilized in glass tubes (*see* **Note 6**).
2. Weigh and record the weight of each tube with the sample in it (*see* **Note 9**).
3. Add 3 mL 0.3 M KOH to each tube and incubate for 60 min in a water bath at 37 °C to hydrolyze each sample. Mix the solution every 15 min.
4. Cool the solution to room temperature.
5. Add 1 mL of 3 M $HClO_4$ and mix before recording the weight.
6. Centrifuge at 8,000 × *g* for 10 min at 4 °C.
7. Transfer the supernatant to a labeled glass tube that can hold at least 15 mL.
8. Resuspend the precipitate in 4 mL cold 0.5 M $HClO_4$. Record the weight added.
9. Centrifuge at 8,000 × *g* for 10 min at 4 °C.
10. Remove the supernatant and add it to the test tube that contains the supernatant from **step** 7.
11. Repeat **steps 8–10**, combining the $HClO_4$ supernatants.
12. Add 0.5 M $HClO_4$ to the tube containing the combined supernatants, to a total volume of 15 mL.
13. Centrifuge the tube at 8,000 × *g* for 10 min at 4 °C.
14. Transfer to a quartz cuvette.
15. Measure the absorbance at 260 nm.
16. Multiply the absorbance by 40 to quantify the amount of RNA in μg/mL.
17. Calculate the total volume of the cumulative amounts added, taking into account the specific gravity of each solution.

18. Calculate the total amount of RNA using the concentration and total reaction volume.
19. Normalize the amount of RNA in each sample to the displacement volume of the initial biomass.

3.7 Carbohydrate

Quantification of the total carbohydrate content is done using the phenol method first described by DuBois et al. (1956), which has since had several modifications to improve accuracy [16–18]. The addition of concentrated sulfuric acid dehydrates the carbohydrates into their furfural derivatives [17]. Once the reaction has been cooled, phenol is added that condenses with the furfurals to produce a yellow color [16]. Absorbance is measured at 488 nm and carbohydrate concentration is calculated using a standard curve made with glucose.

1. Make a standard series by diluting the glucose standard stock solution with MilliQ water. Add 0, 1, 5, 10, 25, 50, 75, 100, 125, and 150 μL of the glucose standard stock solution to labeled test tubes. Make the total volume of each standard up to 1.0 mL with MilliQ water. Weigh and record the weights of both glucose stock solution and water as they are added. Calculate the concentration of each standard.
2. Remove the freeze-dried samples from the −80 °C freezer and allow to return to room temperature (*see* **Note 8**).
3. Weigh and record the weight of each tube with the sample in it (*see* **Note 9**).
4. Add 1.0 mL of MilliQ water and record the weight (*see* **Note 13**).
5. Add 5.0 mL of concentrated H_2SO_4 to each of the samples and standard tubes, and invert several times to mix. Record the weight of H_2SO_4 added.
6. Let the reaction stand for 10 min.
7. Rapidly cool the tubes by placing in iced water for 1 min.
8. Add 50 μL of phenol to each of the samples and standard tubes, and mix.
9. Record the weight of phenol added.
10. Incubate the tubes for 30 min in a water bath set to 25 °C.
11. Measure the absorbance at 488 nm (*see* **Notes 14** and **15**). Take three replicate readings for each standard and sample.
12. Average the three readings for the replicate standard solutions.
13. Calculate the absorbance versus concentration standard curve.
14. Average the three replicate readings per sample.
15. Compare the average absorbance of each sample to the standard curve to calculate the concentration of carbohydrates in each sample.

16. Calculate the total volume of the cumulative amounts added, taking into account the specific gravity of each solution.
17. Calculate the total amount of carbohydrates using the concentration and total reaction volume.
18. Normalize the amount of carbohydrates of each sample to the displacement volume of the initial biomass.

3.8 Protein

The Bradford assay is used to quantify the protein content. Sodium hydroxide is used to digest the cellular biomass, as the high concentrations of detergents normally required for cell lysis are not compatible with the Bradford assay [19]. The Bradford assay exploits the change from red to blue of the dye, Coomassie Brilliant Blue 250, upon binding to protein. The reaction is complete in 2 min, and is sensitive [19].

1. Make a set of standard solutions by serially diluting the 2 mg/mL BSA with MilliQ water. Weigh and record the volumes added. Calculate the concentration of each standard.
2. Remove the freeze-dried samples from the –80 °C freezer and allow to return to room temperature (*see* **Note 8**).
3. Weigh and record the weight of each tube with the sample in it (*see* **Note 9**).
4. Add 3.0 mL of 1 M NaOH to each standard and sample tube and record the weight.
5. Incubate in a water bath for 5 min at 95 °C.
6. Cool the reaction in iced water.
7. Vortex the tube and centrifuge for 5 min at 2,000 ×*g* to pellet any particulates.
8. For each standard and sample, label three tubes.
9. Invert the Coomassie working reagent bottle several times before adding 1.5 mL to each tube. Record the weight added.
10. Add 50 μL of BSA standard or sample supernatant to the Coomassie working reagent and mix. Weigh and record the volume added (*see* **Note 10**). A dilution step may be performed here by substituting 1 M NaOH for a fraction of the 50 μL.
11. Incubate for 10 min (*see* **Note 16**).
12. Measure the absorbance at 595 nm, using MilliQ water as a blank.
13. Subtract the absorbance value of the blank from the absorbance of the standards and samples.
14. Calculate the absorbance versus concentration standard curve.
15. Compare the absorbance readings to the standard curve to calculate the concentration of protein in each replicate.

16. For each replicate, calculate the total volume of the cumulative amounts added, taking into account the specific gravity of each solution.
17. Calculate the total amount of protein in each replicate using the concentration and total reaction volume.
18. Calculate the amount of protein in each sample by averaging the three replicates.
19. Normalize the amount of protein to the amount of starting biomass used in the assay.

3.9 Lipids

Quantification of lipids is done gravimetrically. The lipids are extracted with a 3:2 hexane:isopropanol solution overnight [20]. Cellular components other than lipids are removed by the addition of an aqueous phase. The phases are then separated by centrifugation and the upper hexane layer is transferred. The aqueous layer is washed with 7:2 hexane:isopropanol to recover the total lipid fraction [21]. The hexane is then evaporated under nitrogen and the remaining lipids quantified by weighing.

1. Remove the freeze-dried samples from the −80 °C freezer and allow to return to room temperature (*see* **Note 8**). These are the samples that were prepared and lyophilized in glass tubes (*see* **Note 6**).
2. Weigh and record the weight of each tube with the sample in it (*see* **Note 9**).
3. Add 1.0 mL of MilliQ water.
4. Add 5.0 mL of 3:2 hexane:isopropanol.
5. Place the tubes on a slow rotation wheel and extract overnight.
6. Add 5.0 mL of 0.47 M $NaSO_4$ and centrifuge at 4,000 × *g* for 10 min.
7. For each sample, label and weigh a glass tube with lid included.
8. Pipette the top layer of hexane into the glass tube.
9. Add 5 mL of the 7:2 hexane:isopropanol solution to the $NaSO_4$ solution in the initial tube and mix by inverting for 5 min.
10. Centrifuge at 4,000 × *g* for 10 min.
11. Pipette off the hexane supernatant and add to the hexane supernatant in the glass tube from **step 9**.
12. Evaporate the hexane under a slow flow of nitrogen.
13. Immediately cap the tube while still under nitrogen and weigh it. The increase in weight of the tube is the weight of the lipids present in the sample.
14. Normalize the amount of lipid in the sample to the displacement volume of the starting biomass.

4 Converting the Composition into a Biomass Equation

Now that we have presented essential protocols to measure the macromolecular composition of the biomass, this information needs to be converted into a biomass equation, or a set of molar drain fluxes that allow the growth rate to be used to estimate the molar flux through a pathway. This is a challenging task that requires a couple of organism-specific assumptions and it is beyond the scope of this chapter to address all of them needed for *A. queenslandica*, but we highlight the procedure using DNA as an example. Generally, to complete the whole biomass equation will require the average amino acid composition of the proteins, the sugar composition of the carbohydrate, the fatty acid composition of the lipids, and the nucleotide composition of DNA and RNA. Furthermore some assumptions about the localization of the components must be taken (e.g., distribution of carbohydrates in storage compounds versus cell walls). Nevertheless it is useful to run through the procedure with one component (DNA) as an example as it should highlight the most important steps, especially the conversion of weights and percentages into molar coefficients:

1. Obtain the GC content of the DNA (*see* **Note 17**) in *A. queenslandica*; it is 31.14 %.
2. Translate the GC content into 15.57 % guanine, 15.57 % cytosine, 34.43 % adenine, and 34.43 % thymine.
3. Calculate molecular weights of the monophosphate-deoxyribonucleotides dxMP (*see* **Note 18**):

   $M_{dAMP} = 313$ g/mol; $M_{dTMP} = 304$ g/mol; $M_{dGMP} = 329$ g/mol; $M_{dCMP} = 289$ g/mol.
4. Calculate the average molecular weight of one DNA nucleotide: $(0.3443 \times 313) + (0.3443 \times 304) + (0.1557 \times 329) + (0.1557 \times 289) = 308.65$ g/mol.
5. Convert the measured DNA content from g_{DNA}/g_{CDW} into mol_{dxMP}/g_{CDW}. In our sponge example the measured DNA content (Table 1) was determined to be 0.031 $mg_{DNA}/ml_{displacement}$. We need to apply a further conversion (*see* **Note 19**): 1.0 mL displacement equates to 0.091 g dry sponge. 64 % of that is skeleton (assumed DNA free), while 36 % is non-skeleton tissue or cell dry weight (CDW). Therefore $0.36 \times 0.091 = 0.03276$ $g_{CDW}/ml_{displacement}$. This brings the DNA content to 0.946 mg_{DNA}/g_{CDW} and 3.065 $\mu mol_{dxMP}/g_{CDW}$, respectively.
6. Using the ratios derived from the GC content from above we can calculate the individual nucleotide yield coefficients for *DNA synthesis* to be 0.477 $\mu mol_{dGTP}/g_{CDW}$; 0.477 $\mu mol_{dCTP}/g_{CDW}$; 1.055 $\mu mol_{dATP}/g_{CDW}$; and 1.055 $\mu mol_{dTTP}/g_{CDW}$ (*see* **Note 20**).

Table 1
Macromolecular composition of *Amphimedon queenslandica*

Component	mg/mL displacement	Standard deviation	Relative error	*n*
Dry weight	92.49	10.22	11.05	42
Skeleton	62.68	10.93	17.44	21
Lipid	10.90	1.79	16.42	19
Protein	8.28	1.14	13.76	20
Carbohydrate	1.66	0.43	25.96	20
RNA	0.198	0.037	18.78	20
DNA	0.031	0.006	19.81	20

7. If a comprehensive genome-scale model is present, one can stop here and use the yield coefficients provided above. If the model does not contain the nucleotide biosynthetic pathways, these yields can be further fragmented into the drains of central metabolites as outlined in the next step.
8. dATP: Ribose-5P + 7 ATP + 2 GLN + CO_2 + GTP + NADPH + 2 ASP + 2 Formyl-THF + GLY → dATP + 6 ADP + AMP + 2 GLU + $NADP^+$ + 2 FUM + GDP + 2 THF + PPi + 4 Pi (*see* **Note 21**).
9. dGTP: Ribose-5P + 8 ATP + 3 GLN + CO2 + NADPH + ASP + 2 Formyl-THF + GLY + NAD^+ → dGTP + 6 ADP + 3 GLU + 2 AMP + NADP + FUM + 2 THF + NADH + 2 PPi + 3 Pi.
10. dCTP: HCO_3^- + 2 GLN + 6 ATP + ASP + UQ + Ribose-5P + NADPH → dCTP + 2 GLU + 5 ADP + AMP + UQH_2 + CO_2 + 3 Pi + PPi + $NADP^+$.
11. dTTP: HCO_3^- + 2 GLN + 8 ATP + ASP + UQ + Ribose-5P + NADPH + 5,10 methylene-THF → dTTP + 2 GLU + 7 ADP + AMP + UQH_2 + CO_2 + 3 Pi + PPi + $NADP^+$ + DHF + NH_3.
12. Multiply the stoichiometric coefficients in the above equations with the individual dxTP yield coefficient outlined above and you can calculate the total drain of precursors that DNA synthesis brings about (in μmol/g_{CDW}). In our example this results in the following:

    3.06 Ribose-5P + 22.5 ATP + 6.605 GLN + 1.1 GTP + 3.06 NADPH + 4.12 ASP + 3.06 Formyl-THF + 1.53 GLY + 1.05 5,10-methylene-THF + 1.5 UQ + 1.53 HCO_3^- + 0.5 NAD^+ → DNA + 19 ADP + 3.5 AMP + 6.6 GLU + 3.06 $NADP^+$ + 3 FUM + 1.05 GDP + 3.06 THF + 3.54 PPi + 10.25 Pi + 1.05 NH3 + 1.05 DHF.

13. In analogy to **step 7**, one could stop on this level. If a model, however, is simpler than this equation, the equation can be further condensed: For instance, using the following reactions:

    GDP + ATP → GTP + ADP; GLU + NH3 + ATP → GLN + ADP; AMP + ATP → 2 ADP the above term becomes:

    3.06 Ribose-5P + 33.7 ATP + 5.55 NH3 + 3.06 NADPH + 4.12 ASP + 3.06 Formyl-THF + 1.53 GLY + 1.05 5,10-methylene-THF + 1.5 UQ + 1.53 HCO_3^- + 0.5 NAD^+ → DNA + 33.7 ADP + 3.06 $NADP^+$ + 3 FUM + 1.05 GDP + 3.06 THF + 3.54 PPi + 10.25 Pi + 1.05 DHF.

14. Multiplication of the equation with the growth rate gives fluxes in (in μmol/(g_{CDW} × h)).

This section hopefully highlights essential steps in assembling the appropriate equations with yield coefficients and how to get from mass measurements to molar yield coefficients. It also highlights that this is a complex task that needs to be done for each biomass component separately using the approach described above. The level on which the fragmentation of the reactions stops will always depend on the model. As such, the more comprehensive the model, the less work needs to be put into the condensation of reactions into common precursors. The drawback: More complicated models may only be used for certain approaches; for instance a genome-scale model is well suited for flux balance analysis, while ^{13}C fluxomics with a full genome-scale isotopomer model is currently difficult to analyze. Therefore the user will have to condense reactions manually as described above to a point that is suitable for the flux analysis that is being conducted.

5 Notes

1. EDTA will not dissolve until the pH reaches 8. Stirring with a magnetic stirrer and applying gentle heat (50 °C) will increase the rate of dissolution.
2. The freezing and hydration process was enough to lyse the sponge cells but not the bacterial cells. We added lysozyme to the TE used to rehydrate the sample to enzymatically lyse the bacterial fraction. Note that lysozyme does not lyse Archaea because of different cell wall composition to bacteria. *A. queenslandica* has a low abundance of Archaea.
3. Perchloric acid is highly reactive with metals and organic matter. Only use glass pipettes and tubes.
4. Change scalpel blade very frequently. The skeleton of *A. queenslandica* is made from silica. This causes the scalpel to rapidly lose its edge. Avoid placing pressure on the blade to

prevent the loss of cellular material that is easily dislodged. Be careful to avoid foreign material such as macroalgae or inorganic debris, which is often incorporated into the sponge biomass as it grows.

5. Excess water must be removed when transferring the sponge piece from the deep petri dish to the measuring cylinder. Briefly touch the bottom of the sponge piece to the edge of the petri dish. Water should be left in the aquiferous system to prevent trapped air affecting the displacement value.
6. Use glass tube for RNA and lipid samples. Hexane will react with plastic tubes.
7. The evaporation of seawater leaves salt that significantly contributes to the dry weight. We assumed that the intracellular salt content was the same as seawater for our calculations.
8. To avoid condensation, do not open the tubes until the sample is at room temperature.
9. Remove any water that has condensed on the outside of the tube while it was warming to room temperature.
10. Each standard curve has a linear range within which it is accurate. Sample concentrations that fall outside this range must be diluted as appropriate to ensure an accurate analysis. Use samples that are uniform in size and assay test samples prior to analyzing the sample set. Published data for a closely related organism may serve as a guide for expected concentrations.
11. Small differences in the amounts of solutions used during an assay can produce cumulative error that is significant. To ensure accuracy, all weights of solutions added to an assay are determined gravimetrically on an analytical balance. The exact volume added can then be calculated using the specific gravity of the solution.
12. Do not average the readings and then calculate concentration. The replicate wells will have different concentrations due to differences in the amount of stock solution and diluent. Rather average the calculated concentrations of the replicates.
13. Water is important for the generation of heat essential for the reaction to be completed.
14. A dilution may be required, depending on the organism being analyzed. The reaction with the sponge tissue resulted in an almost black solution. We made a dilution of 1 in 50 with concentrated sulfuric acid. Record weights when performing a dilution step for accurate concentration calculations.
15. Dependent on the reaction, filtration of the acid solution may be required to remove particulate material that may interfere with spectrophotometer readings.

16. It is important to incubate all reactions for the same amount of time, as the absorbance continues to change.
17. The GC content for the genome should be available from the genome sequence. If not sequenced, use the GC content of a closely related species.
18. dxMP stands for dAMP, dTMP, dGMP, or dCMP. Remember that the dxMPs in DNA are polymerized, which happens through the loss of PPi from dxTPs. Using the molecular weight of dxTPs we need to subtract the molecular weight of PPi ($P_2O_7H_4$; $M = 178$ g/mol) to obtain the molecular weight of one dxMP in the chain. For example $M_{dATP} = 491.18$ g/mol (fully protonated); incorporated in the chain $M_{dAMP} = 313.18$ g/mol ($C_{10}H_{12}N_5O_5P$).
19. For single-cell organisms it should usually be possible to obtain $mg_{macrocomponent}$ per g_{CDW} straight from the described assays.
20. For the yield coefficients the triphosphates of the nucleotides are given, since they need to be synthesized for DNA polymerization and will be the drain points in the network.
21. The stoichiometric coefficients are derived from Chapter 3.6 in [22].

References

1. Palsson B (2009) Metabolic systems biology. FEBS Lett 583(24):3900–3904
2. Feist AM et al (2008) Reconstruction of biochemical networks in microorganisms. Nat Rev Microbiol 7:129–143
3. Holms H (1996) Flux analysis and control of the central metabolic pathways in Escherichia coli. FEMS Microbiol Rev 19:85–116
4. Blank LM et al (2005) Large-scale ^{13}C-flux analysis reveals mechanistic principles of metabolic network robustness to null mutations in yeast. Genome Biol 6(6):R49
5. Quek LE et al (2010) Metabolic flux analysis in mammalian cell culture. Metab Eng 12(2):161–171
6. Zomorrodi AR, Maranas CD (2012) OptCom: a multi-level optimization framework for the metabolic modeling and analysis of microbial communities. PLoS Comput Biol 8(2):e1002363. doi:10.1371/journal.pcbi.1002363
7. Srivastava M et al (2010) The *Amphimedon queenslandica* genome and the evolution of animal complexity. Nature 466(7307):720–726
8. Hooper JN, Van Soest R (2006) A new species of Amphimedon (Porifera, Demospongiae, Haplosclerida, Niphatidae) from the Capricorn-Bunker Group of Islands, Great Barrier Reef, Australia: target species for the "sponge genome project". Zootaxa 1314:31–39
9. Palumbi SR (1986) How body plans limit acclimation: responses of a demosponge to wave force. Ecology 67(1):208–214
10. Schutze J et al (2001) Molecular evolution of the metazoan extracellular matrix: cloning and expression of structural proteins from the demosponges Suberites domuncula and Geodia cydonium. J Mol Evol 53(4–5):402–415
11. Müller WEG et al (2004) Sustainable production of bioactive compounds by sponges—cell culture and gene cluster approach: a review. Mar Biotechnol 6:105–117
12. Caralt S et al (2007) Cell culture from sponges: pluripotency and immortality. Trends Biotechnol 25(10):467–471
13. Green MR, Sambrook J (2012) Molecular cloning. Cold Spring Harbor Laboratory Press, New York
14. Cesarone CF et al (1979) Improved microfluorometric DNA determination in biological material using 33258 Hoechst. Anal Biochem 100:188–197
15. Fleck A, Munro HN (1962) The precision of ultraviolet absorption measurements in the Schmidt-Thannhauser procedure for nucleic acid estimation. Biochim Biophys Acta 55:571–583
16. DuBois M et al (1956) Colorimetric method for determination of sugars and related substances. Anal Chem 28(3):350–356

17. Rao PP, Pattabiraman TNT (1989) Reevaluation of the phenol-sulfuric acid reaction for the estimation of hexoses and pentoses. Anal Biochem 181(1):18–22
18. Taylor K (1995) A modification of the phenol/sulfuric acid assay for total carbohydrates giving more comparable absorbances. Appl Biochem Biotechnol 53(3):207–214
19. Bradford MM (1976) A rapid and sensitive method for the quantitation of microgram quantities of protein utilizing the principle of protein-dye binding. Anal Biochem 72: 248–254
20. Carnicer M et al (2009) Macromolecular and elemental composition analysis and extracellular metabolite balances of *Pichia pastoris* growing at different oxygen levels. Microb Cell Factories 8(1):65
21. Hara A, Radin NS (1978) Lipid extraction of tissues with a low-toxicity solvent. Anal Biochem 90(1):420–426
22. Michal G, Schomburg D (2012) Biochemical pathways, 2nd edn. Wiley, Hoboken, NJ
23. Hentschel U et al (2012) Genomic insights into the marine sponge microbiome. Nat Rev Microbiol 10:641–654

Chapter 8

Successful Downsizing for High-Throughput ^{13}C-MFA Applications

Birgitta E. Ebert and Lars M. Blank

Abstract

^{13}C label-based metabolic flux analysis is a powerful technique for the determination of intracellular reaction rates and is used in such different research fields as quantitative physiology, metabolic engineering, and systems biology. Metabolic fluxes can be determined at high quality using (pseudo)-steady-state cultures and advanced mathematical models for data interpretation. Here, we describe a protocol for parallel metabolic flux analysis that consists of downsized microbial (yeast) cultivation, miniaturized sample preparation, and semiautomated analytics and data evaluation. With this protocol dozens of metabolic flux analyses can be carried out in 1 week, thereby enabling for example the analysis of genetic and environmental perturbations on the operation of metabolic networks.

Key words ^{13}C metabolic flux analysis, Quantitative physiology, Parallel cultivation, Microtiter plates, METAFoR, Bacteria, Yeast, Metabolic network, Metabolic engineering

1 Introduction

^{13}C metabolic flux analysis (MFA) allows the determination of intracellular reaction rates in complex metabolic networks. This quantitative method is now employed in all living cells, from bacteria and simple eukaryotes like yeast to plants and mammalian cells. The experimental limitations such as sample size, required isotopic steady state, and data interpretation restricted to experts are pushed by a combination of analytical, experimental, and software developments that drive the general applicability of this technique. The use of gas chromatography-coupled mass spectroscopy allowed a significant reduction in sample size, when compared to NMR methods [1]. The flux determination at non-steady state is under development [2], promising shorter and less expensive experiments and rich information content. The availability of software packages that facilitate data interpretation [3–5] enabled the widespread use and recently, a software tool automating the MFA

Jens O. Krömer et al. (eds.), *Metabolic Flux Analysis: Methods and Protocols*, Methods in Molecular Biology, vol. 1191, DOI 10.1007/978-1-4939-1170-7_8, © Springer Science+Business Media New York 2014

calculations has been developed that enables high-throughput intracellular flux calculations [6].

Here, we present the downsizing and parallelization of ^{13}C metabolic flux analyses for submerged microbes using the example of baker's yeast. Depending on the growth requirements, the protocol can be easily adapted to other yeast and unicellular fungi and bacteria. The main focus of the experimenter has to be on the quantitative reproducibility of the growth experiments. We here describe in detail for yeast what it takes to carry out the consistent cultivation, and then explain the workload-reduced sample preparation, high-quality labelling pattern determination and data evaluation. Data evaluation is described for the use of the software Flux-P [6]. Flux-P is an automated version of the tool FiatFlux [4], which is based on the metabolic flux ratio analysis approach [7]. Flux-P uses the Bio-jETI (Java Electronic Tool Integration) technology, a framework for service integration and workflow development in the bioinformatics domain [8], to make the single analysis steps, carried out manually in FiatFlux, available as a collection of platform-independent remote services. The Java Application Building Center (jABC) framework [9] provides the graphical user interface for Bio-jETI and supports the orchestration of user-specific MFA workflows for Flux-P.

The description is a hands-on protocol of the analysis, breaking down the workflow into simple single tasks in order to ease the application by non-experts and to allow first flux distribution calculations. While this works out for general flux analyses, in some cases a fine-tuned optimization is necessary for which the user has to be aware of the details of the (mathematical) operations. The description of these details is outside the scope of this chapter. The interested user can find this information in [4] and [10]. The hands-on protocol might promote studies with an increased number of experiments and might lower the entry barrier for newcomers to this exciting technique.

2 Materials

2.1 Reagents

1. ^{13}C-labelled glucose: [U-^{13}C] glucose (e.g., Sigma-Aldrich), [1-^{13}C] glucose (e.g., Omicron Biochemicals, South Bend, IN (>99 %)) (*see* **Note 1**).
2. (Naturally labelled) glucose.
3. HCl: 6 M aqueous solution.
4. Acetonitrile: High purity.
5. *N*-(*tert*-butyldimethylsilyl)-*N*-methyl-trifluoroacetamide (e.g., Macherey-Nagel GmbH & Co. KG, Germany).

6. Verduyn minimal medium: 5 g/L $(NH_4)_2SO_4$, 3 g/L KH_2PO_4, 0.5 g/L $MgSO_4 \cdot 7H_2O$, 4.5 mg/L $ZnSO_4 \cdot 7H_2O$, 0.3 mg/L $CoCl_2 \cdot 6H_2O$, 1.0 mg/L $MnCl_2 \cdot 4H_2O$, 0.3 mg/L $CuSO_4 \cdot 5H_2O$, 4.5 mg/L $CaCl_2 \cdot 2H_2O$, 3.0 mg/L $FeSO_4 \cdot 7H_2O$, 0.4 mg/L $NaMoO_4 \cdot 2H_2O$, 1.0 mg/L H_3BO_3, 0.1 mg/L KI, 15 mg/L EDTA, 0.05 mg/L biotin, 1.0 mg/L calcium pantothenate, 1.0 mg/L nicotinic acid, 25 mg/L inositol, 1.0 mg/L pyridoxine, 0.2 mg/L *p*-aminobenzoic acid, and 1.0 mg/L thiamine [11].
7. Buffered Verduyn minimal medium: Same composition as given above but buffered with 100 mM KH-phthalate ($pH \leq 5.9$) or 100 mM morpholinepropanesulfonic acid (MOPS) ($pH > 5.9$).
8. KOH: 1 M aqueous solution.
9. HCl: 1 M aqueous solution.

2.2 Analytical Devices, Equipment, and Software

1. Square deep-well plates with conical bottom (System Duetz, Enzyscreen, The Netherlands); the plates shall be acid treated to reduce leaching of plasticizers and equipped with lids that guarantee well position-independent aeration (e.g., sandwich covers, System Duetz, Enzyscreen, The Netherlands; *see* **Note 2**).
2. Incubation shaker with orbital shaking; oxygen transfer and reduced gradients because of mixing are guaranteed by using an orbital shaker with 50 mm amplitude at 300 rpm (*see* ref. 12 and **Note 3**).
3. Drying oven for biomass hydrolysis.
4. Heating block or Thermoshaker.
5. Gas chromatography-mass spectrometry (GC-MS) instrument for the quantification of ^{13}C incorporation into proteinogenic amino acids, e.g., GC 3800 combined with an MS/MS 1200 (*see* **Note 4**) unit equipped with a FactorFour VF-5 ms column (5 % phenyl, 95 % dimethyl-polysiloxane; all units: Bruker, Bremen, Germany).
6. Enzymatic kits for glucose, acetate, ethanol, and glycerol quantification, alternatively a high-pressure liquid chromatography (HPLC) instrument suited for the analysis of these organic compounds.
7. Flux-P, available for academic, noncommercial use [6] for users having a FiatFlux license (available free of charge for academic use, *see* http://www.imsb.ethz.ch/researchgroup/nzamboni/research/Software). Flux-P consists of a server running the underlying analysis software and the client-side workflows. On the server side, the software requires a Unix-based operating system (Linux, Unix, Solaris, Mac OS X), a recent Java Runtime Environment (JRE), MATLAB R2011a

(The Mathworks, Inc., Natick, MA, USA) or later (including the MATLAB Optimization and NetCDF Toolboxes), and the Flux-P jETI server. On the client side, the software requires a recent JRE (*see* **Note 5**), the jABC and Bio-jETI release, version 3.8.1 or later (available from Programming Systems, TU Dortmund, Germany. Bio-jETI. Available online: http://ls5-www.cs.tu-dortmund.de/projects/biojeti/download.php). The Flux-P workflows are platform independent and can be run on, e.g., Windows 7, Ubuntu Linux, or Mac OS X.

8. OMIX [13] (http://www.13cflux.net/) for the visualization of metabolic fluxes (optional).

3 Methods

3.1 Experiment Design

The experimental setup has to be designed to be suited for the scientific questions to be answered. Especially, the careful determination of the isotopic tracer that allows the best resolution of the metabolic network under investigation is of importance (*see* **Notes 1, 6–8**). A detailed description of the design of isotopic tracer experiments can be found in Chapter 6 of part I of this book.

3.2 Growth Medium Preparation

The Verduyn medium can be partially prepared in advance. Especially, concentrated multicomponent trace element (100-fold) and the vitamin solutions (1,000-fold) can be prepared in a 1 L volume and stored in aliquots at −20 °C. The medium preparation needs then only three salts, the buffer, 10 mL of the 100-fold trace element solution, and 1 mL of the 1,000-fold vitamin solution. The entire mixture is then sterile filtered and stored at 4 °C. As rather little medium is required (maximal 120 mL for a 96-deep-well plate), we add the sterile filtered labelled substrate freshly. Ideally, the amount of water is chosen to make up 1 L when glucose is added to the medium.

3.3 Precultivation

Prepare a preculture in minimal medium and cultivate at growth conditions identical to the planned labelling experiment (*see* **Notes 9** and **10**).

3.4 Parallel Cultivation

Parallel cultivation is realized using 96-deep-well plates. We assume pseudo-steady-state conditions during the exponential growth phase that allows steady isotopic labelling of the biomass [7, 14] (*see* **Note 6**).

1. Prepare minimal medium as described in Subheading 3.2.
2. The medium can be buffered with 100 mM KH-phthalate to reduce pH changes (*see* **Note 11**). Use KH-phthalate buffers for pH values in the range of 2.2–5.9. Above 5.9 MOPS can

be used as buffer reagent. Adjust the pH with 1 M HCl or 1 M KOH (*see* **Note 12**).

3. The carbon source glucose can be added at concentrations between 0.5 and 20 g/L. Glucose and if required antibiotics can be filter-sterilized before addition to the growth medium.
4. ^{13}C labelling of proteinogenic amino acids is achieved either by growth on a glucose mixture of 80 % (w/w) unlabelled and 20 % (w/w) uniformly labelled [U-^{13}C] glucose or 100 % [1-^{13}C] glucose.
5. For batch growth experiments in deep-well plates 1.2 mL of growth medium is used (*see* **Notes 13–17**).
6. Harvest cells from overnight cultures by centrifugation and wash the cell pellet using carbon source-free minimal medium to remove residual unlabelled carbon sources. Inoculate the 96-deep-well plate to a maximum OD_{600} of 0.03 (*see* **Note 7**).
7. Fix the 96-well plate in the incubation system; incubate at 30 °C, 300 rpm, and an amplitude of 50 mm (*see* **Note 3**).
8. Cell growth is monitored by following optical density changes at a wavelength of 600 nm (OD_{600}).

3.5 Cell Harvest

1. Cells are harvested during the exponential growth phase to guarantee pseudo-steady-state conditions (approximate OD_{600} of 1 corresponding to 0.5 g_{CDW}/L (*see* **Note 18**)). Transfer culture from well to sample tube and centrifuge for 1 min in a tabletop centrifuge at maximum speed (*see* **Note 19**). Remove residual medium by washing the pellet with water (*see* **Note 20**).

3.6 Determination of Extracellular Substrates and Metabolites

1. Determine substrates and products (e.g., glucose, acetate, ethanol, and glycerol) in the supernatant by HPLC or commercial enzymatic kits.
2. For use of enzymatic kits, the usual 3 mL cuvette protocols have to be adapted to 200 μL final volume protocols to allow the use of regular clear (polystyrol) microtiter plates. In this downscaled format, the use of enzymatic kits takes full advantage of the microtiter plates, as it allows semiautomation of concentration determination. Inexpensive non-sterile plates can be used for this analytical step, (*see* **Note 21**).
3. The use of a multichannel pipette, ideally automated, increases reproducibility and lowers the workload, especially if the growth medium has to be diluted, to match the linear range of the enzyme assay(s). Dilutions are again made in non-sterile microtiter plates (*see* **Note 22**).
4. Organic acids, mainly intermediates of the TCA cycle like succinate and malate, but also pyruvate and others, are produced by yeast as minor by-products. Also for some of these organic acids

enzyme assays exist. If required, organic acids can be determined more comprehensively via HPLC. Depending on the vial, ideally with conical micro inserts, sample volumes below 50 μL can be used, with injection volumes of 10 μL (or less).

5. The substrate uptake and product secretion rates required for the flux analysis are determined during the exponential growth phase. The differences between the substrate and product concentrations at the beginning of the culture and the time of harvest are determined and the validity of an exponential growth model is assumed (*see* **Note 23**).

3.7 Sample Preparation for GC-MS Analysis

1. For the analysis of the labelling patterns of proteinogenic amino acids, as proxies for the labelling patterns of central carbon metabolism intermediates, the harvested cells are used for sample preparation.
2. Biomass is transferred into 1.5 mL Eppendorf cups for full biomass hydrolysis (*see* **Note 24**). Aim to add about 0.3 mg dry weight (*see* **Note 25**).
3. The biomass is hydrolyzed by adding 6 M HCl at a volume of 150 μL followed by an incubation of 16 h at 105 °C (*see* **Note 26**).
4. The Eppendorf cups are kept closed by a custom-made Eppendorf cup rack that is equipped with a screwable plate that enforces pressure on the lids of the Eppendorf cups, thereby guaranteeing that no liquid is lost (*see* **Note 27**).
5. The hydrolysate is dried in a heating block at 85 °C for 6 h (*see* **Note 28**).
6. The dried hydrolysate is dissolved using 15 μL of acetonitrile (*see* **Note 29**). For silylation add *N*-(*tert*-butyldimethylsilyl)-*N*-methyl-trifluoroacetamide (*see* **Note 30**) to an equal volume of 15 μL. Derivatization shall be carried out in a heating block set at 85 °C for 1 h with little or no shaking. Particles are removed by centrifugation and the resulting supernatant is transferred to a GC vial (*see* **Note 31**).

3.8 GC-MS Analysis

1. For GC-MS analysis a device with an accuracy of <1 mol % has to be used. This accuracy is generally matched by quadrupole MS, while time-of-flight (TOF) detectors have usually higher errors. Use a system that allows for separation of 16 detectable amino acids (*see* **Note 32**). Apply a constant helium flow rate of 1 mL/min (99.999 % or higher quality).
2. The split ratio can be chosen according to the sample concentration. We often used a split ratio of 1:20 and an injection volume of 1 μL at an injector temperature of 250 °C. The temperature of the GC oven is kept constant for 2 min at 150 °C and afterwards is increased to 250 °C with a gradient

of 3 °C/min. The temperatures of the transfer line and the source are set to 280 °C and 250 °C, respectively.

3. Perform ionization by electron impact (EI) ionization at 70 eV.
4. For increased sensitivity, define a selected ion monitoring (SIM) frame for every amino acid [15].
5. Check for detector overload by analyzing the GC-MS raw data (*see* **Note 33**).
6. Save the GC-MS data in .cdf (common data format) format for evaluation with Flux-P.

3.9 Data Evaluation and Interpretation

For the complete Flux-P analysis the following is required: (1) GC-MS data of the hydrolyzed biomass in .cdf format (*see* **Note 34**); (2) extracellular rates, that is, growth rate, glucose uptake rate, and optional (by)-product formation rates; and (3) the scan ranges of the analyzed amino acids for the automated identification of these analytes. In the following the workflow for the automated calculation of net flux distributions of multiple datasets and their visualization in OMIX, an editor for drawing and analyzing metabolic reaction networks, is described.

1. Prepare one comma-separated (.csv) file with the required information of all experiments as listed in Table 1. The first row of this file has to contain the column headers; additional lines contain the data of the single experiments (*see* **Notes 35–39**).
2. Create a text file specifying the scan ranges of the measured amino acids. List the amino acid in three number code followed by the start and end scan number divided by space, e.g., ALA 23 48.
3. Create a folder containing the created .csv, .txt and the .cdf files, to be analyzed.
4. Start the jABC modelling framework.
5. Choose the workflow model for batch processing of ^{13}C-labelling datasets from the project browser on the upper left-hand side (Fig. 1). In the demo project provided with Flux-P, this is Example 2 (*see* **Note 40**). The model is activated by double mouse click after which the workflow appears on the canvas. The icons on the canvas represent different functions or services, the so-called service-independent building blocks (SIBs), which are connected with labelled branches according to the flow of control.
6. Activate the SIB "netFlux" by single mouse click. In the inspector pane particular properties of this SIB appear. The netFlux SIB entry "ext Fluxes" has to be adapted to allow proper assignment of the extracellular fluxes specified in the .csv file. If for example only growth rate and glucose uptake have been measured, the syntax of the string has to be

Table 1
Specification of entries of the data file for ^{13}C data batch processing

Header[a]	Entry	Remark
cdffile	Name of netCDF file	Enter only the name without .cdf ending. Only word characters, i.e., [a-zA-Z_0-9]+), are allowed
Description	Short description of experiment	Optional[b]
Strain	Name of the investigated strain	Optional[b]
Model	Specify model number for use of the metabolic flux ratio analysis; if not specified otherwise the corresponding metabolic network model is used to calculate the net fluxes (*see* **Notes 35** and **38**)	1 *Bacillus* 2 *Escherichia* 3 *S. cerevisiae* 4 *Zymomonas* 5 *Pseudomonas* 6 *Rhodobacter* 7 *Corynebacterium* 8 *Thiobacillus*
fraction_1_13C	Molar fraction of 1-^{13}C glucose (*see* **Note 36**)	Value has to be in the interval [0;1]
purity_1_13C	Purity of 1-^{13}C glucose	Default value is 0.99
fraction_U_13C	Molar fraction of U-^{13}C glucose (*see* **Note 36**)	Value has to be in the interval [0;1]
purity_of U_13C	Purity of U-^{13}C glucose	Default value is 0.99
fraction_of_unlabeled_biomass	Fraction of unlabelled biomass introduced with the inoculum (*see* **Note 37**)	
GCMS_error	Measurement error of the GC-MS device	
growth_rate[c]	Specific growth rate [1/h]	
growth_rate_error[c]	Standard deviation of the growth rate [1/h]	
glucose_uptake[c]	Glucose uptake rate [mmol/g_{CDW}/h]	
glucose_uptake_error[c]	Standard deviation of the glucose uptake rate [mmol/g_{CDW}/h]	
cofactor_balance	Select, which cofactor mass balance is excluded from the analysis (*see* **Note 39**)	0 disclose ATP balance 1 disclose NADH balance 2 disclose ATP and NADH balance

[a]Only word characters, i.e., [a-zA-Z_0-9]+), are allowed; do not use space
[b]Type " " to submit no information
[c]Column headers of extracellular rates can be named freely. Here only growth rate and glucose rate are listed; additional rates are specified accordingly

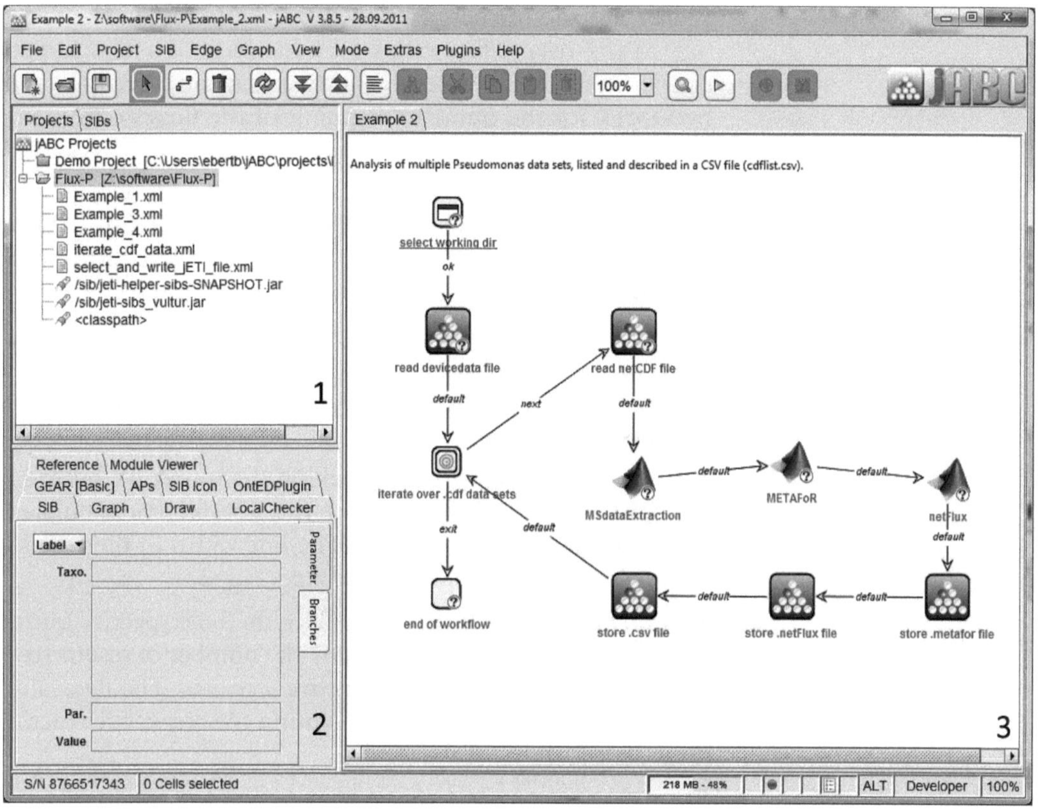

Fig. 1 The jABC user interface with its main components: project and SIB browser (1), inspector pane (2), and graph canvas (3). The displayed model shows a workflow for automated batch processing of flux analysis datasets

```
mu
${cdfdata.growth_rate}
${cdfdata.growth_rate_error} GLC
${cdfdata.glucose_uptake_rate}
${cdfdata.glucose_uptake_rate_error},
```

with "mu" denoting the growth rate. For substrate uptake and product formation the identifier of the respective metabolic network is used; here, GLC denotes glucose.

7. Execute the model by pressing the button "Run" on the toolbar.
8. In the opening window specify the working directory containing the GC-MS raw data (.cdf) files, the text file, and the .csv file containing the scan ranges of the analytes and the experimental data, respectively.
9. For every processed dataset Flux-P saves three files into the working directory. The files are saved under the same name as the .cdf file extended with the suffix .ratio, .netto, and .csv, respectively. These files contain the metabolic flux ratios

(.ratio file) and the net reaction rates (.netto file) in close analogy to the results files created by FiatFlux [4]. The .csv file contains the net flux rates in a format that can directly be read by OMIX for the optional mapping of the fluxes onto metabolic network maps.

10. Open the OMIX file of the *S. cerevisiae* metabolic network map, provided with Flux-P, which includes the OVL (Omix Visualization Language) script for the flux distribution visualization.
11. Start the OVL script by clicking on the view tool (the "eye" icon) on the "Edit" toolbar. Two buttons appear in the OVL trigger toolbar, with which visualization of single or multiple datasets are possible.
12. *Visualization of single result*: Click on the respective button in the OVL trigger toolbar. Define the .csv file containing the results in the opening window. The line width of the reaction arrows is adjusted to the specific flux (Fig. 2a).
13. *Visualization of multiple results*: Click on the respective button in the OVL trigger toolbar. Specify the number of results to be visualized and provide the respective .csv files. The flux values are assigned to the arrows representing the respective reaction (Fig. 2b).

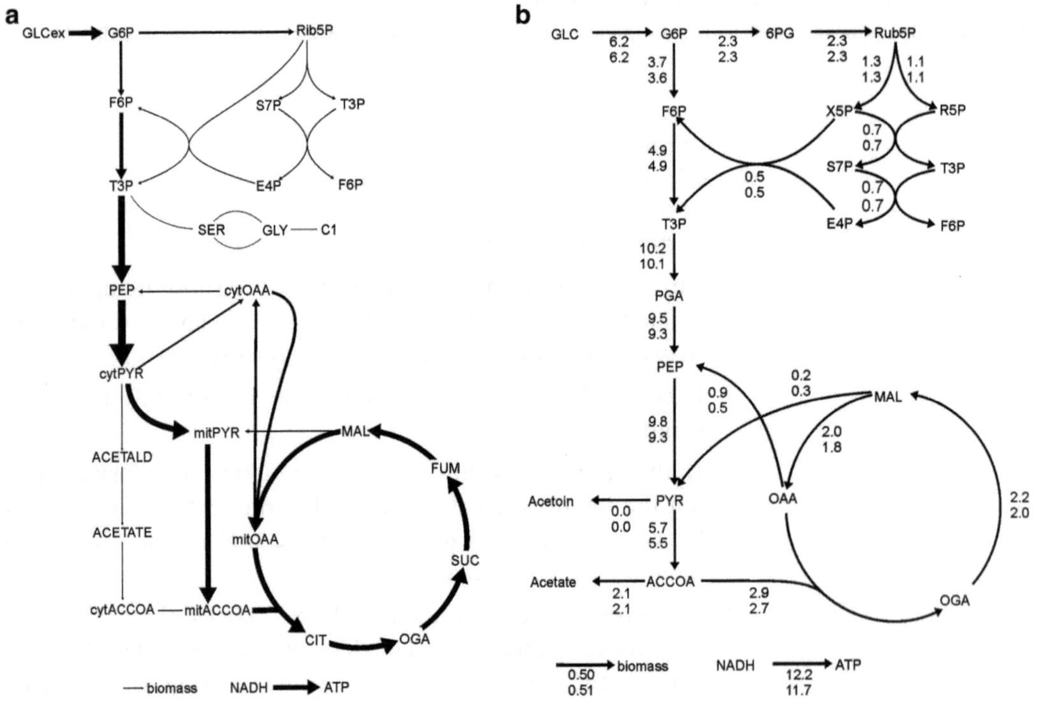

Fig. 2 Flux distribution visualizations with OMIX [13] of single (**a**) and multiple (**b**) datasets

4 Notes

1. The important parameters of stable isotope-labelled chemicals are the purity and the price. While it is advisable to choose high purity, one has to be aware that a purity of 99 % corresponds to a purity of $0.99^6 \sim 94.1$ % of [U-^{13}C] glucose per molecule, as in all six carbon atoms 1 % is not labelled. Regarding price, it is advisable to call several suppliers as the market price seems to vary. The price per gram scales with the amount ordered and drops significantly at larger quantities.
2. Enzyscreen sandwich covers have 4–5 layers made of a steel lid, two filter units, a stainless steel foil with pinholes that controls the headspace refreshment rate and evaporation, and a silicone layer with 96 holes that hermetically seals the mini-reactors. This system efficiently prevents (cross)-contamination and permits very good exchange rates of headspace air while limiting evaporation.
3. Many bench-top optical shakers have an amplitude of 25 mm. We therefore used a stackable shaker, which is available from all major distributors.
4. A single-quadrupole MS is sufficient for the amino acid analysis.
5. The JRE on the server and on the client side do not have to be of the same version.
6. ^{13}C-based metabolic flux analysis based on the analysis of proteinogenic amino acids requires steady growth conditions for several generations to ensure that the biomass approaches an isotopically steady state. This can be achieved either in batch cultivations allowing exponential growth for at least five generations or in continuously growing cultures, for which isotopically steady state is assumed after 3–5 volume exchanges. The cultivation mode of choice depends on the scientific question to answer, i.e., if it requires maximal (= batch growth) or (carbon) limited growth as given in chemostats.
7. Initial OD and glucose concentration have to be chosen so that in the mid-exponential growth phase the amount of unlabelled biomass is less than 3 %. For example, if the initial biomass concentration is 0.03 g_{CDW}/L, sample at a cell concentration of 1 g_{CDW}/L. For a biomass yield of $Y_{x/s} = 0.3$ g/$g_{Glucose}$, add two- to threefold $(1 - 0.03\ g_{CDW}/L)/0.3\ g/g_{Glucose} \approx 6.5 - 10\ g_{glucose}/L$.
8. The calculability of intracellular fluxes depends on the ^{13}C tracer used and has to be chosen to suit the specific needs of the respective experiment. For growth on glucose two main labelling strategies exist, suited to resolve intracellular fluxes of

different parts of the central carbon metabolism network: a mixture of naturally and fully labelled glucose and 100 % [1-^{13}C] glucose. The best resolution of central carbon metabolism fluxes is achieved by combining both ^{13}C carbon tracers or combining experiments performed with either of the two labelling strategies.

9. Cultivation conditions deviating from the main culture or even the use of complex medium for the preculture shall be omitted to avoid a long lag phase of the main culture.
10. The dilution of the ^{13}C signal by unlabelled biomass introduced with the inoculum can be reduced by using ^{13}C-labelled glucose for the preculture, as well.
11. In our yeast growth experiments with initial glucose concentrations of 5 g/L and a starting pH of 5.0 the pH changes were less than 0.05 [16].
12. Do not use NaOH for pH adjustment, as baker's yeast is sodium sensitive.
13. With this maximal volume, further analytics are possible. Lower volumes can be used for shorter experiments or end-point assays.
14. Using nonoptical deep-well plates requires sampling, dilution, and measurements in regular well plates. Depending on the optical density as much as 100 μL and as little as 10 μL are withdrawn from the experiment using a multichannel pipette.
15. Alternatively, optical plates can be used that allow online monitoring of biomass concentration and therefore growth. While many well plate readers are now equipped with a motor and mechanics that allow the required continuous shaking, the limitation of optical density determination via spectroscopy due to a small linear region cannot be overcome.
16. Alternative, dedicated equipment with shaker and reader function is available. We used intensively the Bioscreen (Oy Growth Curves Ab Ltd, Helsinki, Finland), which is affordable, sensitive, and robust. Limitations are well-to-well differences because of aeration via the side openings of the lid, and the resulting short experiment time without significant evaporation. An alternative to optical density determination is light scattering measurement [17], which is installed in the automated cultivation device Biolector [18] from m2p-labs GmbH (Baesweiler, Germany). In our hands, this device allows reproducible growth experiments as it overcomes some of the limitations encountered during cultivation in microtiter plates, including intense aeration, well-to-well differences, (partially) evaporation, and biomass concentration determination.
17. Sedimentation of cells disturbs experiments. To avoid sedimentation alternative microtiter plates with either flat bottom

or flower shape (m2p-labs GmbH) can be used. Alternatively, a 1.5 mm glass bead can be added to each well that adds additional mixing and thereby circumvents sedimentation.

18. The correlation between OD and cell dry weight (CDW) depends on the spectrophotometer used and therefore has to be established for every instrument.
19. If the growth rate in the parallel experiments varies little, the entire microtiter plate can be harvested with a centrifuge equipped with a rotor for microtiter plates.
20. Although turnover times of proteinogenic amino acids are relatively high, centrifuge immediately and at 4 °C. The pellet can be stored at −20 °C or further processed at once.
21. Some very cheap plates are not suited for optical analysis, as material inhomogeneities can disturb the signal.
22. Ethanol readily evaporates and therefore requires especially careful pipetting. Minimize evaporation by taking care that ethanol samples are pipetted in the fluid and not freely through the air.
23. Long lag phases (i.e., time after inoculum without growth) have to be investigated separately, as the exponential growth assumption does not hold. Often however, hardly any substrate uptake or product accumulation is observed, allowing using the described protocol from the time on when the cells start to grow.
24. Depending on the amount of available biomass, the format for full biomass hydrolysis is chosen. From miniaturized cultivations (100 μL), biomass is transferred directly into GC vials (conical microvials), while biomass from deep-well cultures (1–1.2 mL) are transferred into 1.5 mL Eppendorf cups. Note that we observed more rapid resuspension of the dried biomass hydrolysates in glass vials compared to plastic cups.
25. Significant lower biomass amounts might cause low GC-MS signals and therefore low quality of isotope determination. High biomass amounts require extra care in sample drying as a thick crust can form, which is not readily dried, and further sample preparation, as flocks and particles can exist. These particles have to be removed via centrifugation as they can block the syringe and pollute the GC inlet and the column.
26. The hydrolysis time can be varied to as short as a few hours (4 h works, but also shorter times might be possible). The time of hydrolysis becomes an important parameter if one also wants to determine amino acid concentrations. We observed for example that the amount of alanine, valine, and leucine increased with time (up to 16 h, decreasing again after 24 h), while for example the amount of detectable methionine decreases, with increased time of hydrolysis (4-h hydrolysis showed the highest signal).

27. Likewise, tight closure of the sample tubes can be achieved by clip on clamps.
28. The drying process can be accelerated through application of a constant nitrogen gas stream; avoid prolonged drying periods as these can complicate later resuspension. The heating block should be placed under a hood because of HCl evaporation. The dried sample can be stored at room temperature for several months.
29. In the literature tetrahydrofuran (THF) and dimethylformamide (DMF) and other solvents are reported. While in our hands THF does not readily dissolve the hydrolysate, both THF and DMF might have an increased health risk when compared to acetonitrile. In any case, a hood has to be used for sample preparation.
30. The silane is readily destroyed in the presence of water. Therefore, we minimize the exposure of the silane to air (i.e., water) by quickly transferring the reagent to the vials. Ideally the silane is aliquoted and only few times exposed to air. A septum and syringe can also be used to reduce air exposure. We further use GC vials dried by placing them into a heating block (85 °C) for several minutes.
31. The derivatized amino acids can be stored for days at room temperature if the sample is water free. When storage of the sample is required, after injection into the GC-MS, the GC vial has to be recapped to avoid sample evaporation via the hole in the septum. Storage at 4 °C is not recommended, as we have seen precipitations.
32. While glutamine and asparagine are deaminated to glutamate and aspartate, respectively, cysteine and tryptophan are degraded during acid hydrolysis.
33. Detector overload will always affect the most abundant ion the most and will lead to a skewed distribution. To prevent detector overload check the manufactory instructions of your GC-MS device. Data quality checks can also be performed with appropriate software, e.g., iMS2FLUX [19].
34. Because the current version of Flux-P does not allow checking the data quality of the GC-MS data, these checks have to be done externally, e.g., via FiatFlux or iMS2Flux [19].
35. Flux-P provides (1) equations for the calculation of metabolic flux ratios custom for the species *Escherichia*, *Bacillus*, *Zymomonas*, *Pseudomonas*, *Rhodobacter*, *Corynebacterium*, *Thiobacillus*, and yeast and central carbon metabolism models for the aforementioned microbial species.
36. Currently Flux-P is restricted to the analysis of experiments of the mixtures of these two ^{13}C-labelled glucose isotopomers.

37. This information is needed to correct the labelling data for natural abundance of heavy isotopes and ^{13}C dilution by the unlabelled biomass introduced with the inoculum. As these calculations are based on the usage of TBDMS as derivatization agent, Flux-P analysis is currently restricted to accordingly derivatized amino acids.
38. While the addition of new metabolic flux ratios is intricate, custom models can easily be generated. Model modifications might for example be necessary to account for inactive/active pathways (e.g., the glyoxylate shunt, which is often assumed to be inactive during growth on glucose), or additional heterogeneous enzymatic reactions. *See* the Flux-P manual for instructions [6].
39. As the ATP requirements for cell maintenance are generally unknown, most often the mass balance of ATP is not used for the flux calculation.
40. Flux-P is provided with four demo workflows, which enable the calculation of (1) metabolic flux ratios and net fluxes of single (2) and multiple datasets, (3) evaluation of two complementary datasets, e.g., with data of growth experiments with differently labelled glucose mixtures, and (4) analysis of datasets using customized metabolic network models.

Acknowledgement

This work was supported by the Cluster of Excellence "Tailor-Made Fuels from Biomass," which is funded by the Excellence Initiative of the German federal and state governments to promote science and research at German universities.

References

1. Fischer E, Sauer U (2003) Metabolic flux profiling of *Escherichia coli* mutants in central carbon metabolism using GC-MS. Eur J Biochem 270:880–891
2. Nöh K, Gronke K, Luo B, Takors R, Oldiges M, Wiechert W (2007) Metabolic flux analysis at ultra short time scale: isotopically non-stationary ^{13}C labeling experiments. J Biotechnol 129:249–267
3. Quek LE, Wittmann C, Nielsen LK, Krömer JO (2009) OpenFLUX: efficient modelling software for ^{13}C-based metabolic flux analysis. Microb Cell Fact 8:25
4. Zamboni N, Fischer E, Sauer U (2005) FiatFlux—a software for metabolic flux analysis from ^{13}C-glucose experiments. BMC Bioinformatics 6:209
5. Wiechert W, Mollney M, Petersen S, de Graaf AA (2001) A universal framework for ^{13}C metabolic flux analysis. Metab Eng 3:265–283
6. Ebert BE, Lamprecht A-L, Steffen B, Blank LM (2012) Flux-P: automating metabolic flux analysis. Metabolites 2:872–890
7. Sauer U, Lasko DR, Fiaux J, Hochuli M, Glaser R, Szyperski T, Wuthrich K, Bailey JE (1999) Metabolic flux ratio analysis of genetic and environmental modulations of *Escherichia coli* central carbon metabolism. J Bacteriol 181:6679–6688
8. Lamprecht AL, Margaria T, Steffen B (2008) Seven variations of an alignment workflow—an illustration of agile process design and management in Bio-jETI. Bioinformatics Res Appl 4983:445–456

9. Steffen B, Margaria T, Nagel R, Jörges S, Kubczak C (2007) Model-driven development with the jABC. In: Bin E, Ziv A, Ur S (eds) Hardware and Software, Verification and Testing. Springer Berlin, Heidelberg, pp 92–108
10. Nanchen A, Fuhrer T, Sauer U (2007) Determination of metabolic flux ratios from ^{13}C-experiments and gas chromatography-mass spectrometry data: protocol and principles. Methods Mol Biol 358:177–197
11. Verduyn C, Postma E, Scheffers WA, Vandijken JP (1992) Effect of benzoic acid on metabolic fluxes in yeasts—a continuous culture study on the regulation of respiration and alcoholic fermentation. Yeast 8:501–517
12. Duetz WA, Ruedi L, Hermann R, O'Connor K, Büchs J, Witholt B (2000) Methods for intense aeration, growth, storage, and replication of bacterial strains in microtiter plates. Appl Environ Microbiol 66:2641–2646
13. Droste P, Miebach S, Niedenfuhr S, Wiechert W, Nöh K (2011) Visualizing multi-omics data in metabolic networks with the software Omix—a case study. Biosystems 105:154–161
14. Fischer E, Zamboni N, Sauer U (2004) High-throughput metabolic flux analysis based on gas chromatography-mass spectrometry derived ^{13}C constraints. Anal Biochem 325: 308–316
15. Wittmann C (2007) Fluxome analysis using GC-MS. Microb Cell Fact 6:6
16. Blank LM, Kuepfer L, Sauer U (2005) Large-scale ^{13}C-flux analysis reveals mechanistic principles of metabolic network robustness to null mutations in yeast. Genome Biol 6:R49
17. Samorski M, Muller-Newen G, Büchs J (2005) Quasi-continuous combined scattered light and fluorescence measurements: a novel measurement technique for shaken microtiter plates. Biotechnol Bioeng 92:61–68
18. Kensy F, Zang E, Faulhammer C, Tan RK, Büchs J (2009) Validation of a high-throughput fermentation system based on online monitoring of biomass and fluorescence in continuously shaken microtiter plates. Microb Cell Fact 8:31
19. Poskar CH, Huege J, Krach C, Franke M, Shachar-Hill Y, Junker BH (2012) iMS-2Flux—a high-throughput processing tool for stable isotope labeled mass spectrometric data used for metabolic flux analysis. BMC Bioinformatics 13:295

Chapter 9

Labelling Analysis for ^{13}C MFA Using NMR Spectroscopy

Paula Jouhten and Hannu Maaheimo

Abstract

NMR spectroscopy is an efficient method for analyzing ^{13}C labelling of cellular metabolites. The strength of it is especially the ability to provide direct quantitative positional information on the ^{13}C labelling status of carbon atoms in metabolites. NMR spectroscopic methods allow also for detection of contiguously ^{13}C-labelled fragments in the carbon backbones of the metabolites. Furthermore, the recent developments of NMR spectroscopy hardware have substantially improved the sensitivity of the methods. In this chapter we describe a method for analyzing the ^{13}C labelling of the biomass amino acids for metabolic flux analysis, sample preparation for NMR spectroscopy, acquiring and processing the NMR spectra, and extracting the ^{13}C labelling information from the NMR data. Different NMR methods are applied depending on the ^{13}C labelling strategy chosen. These strategies include uniform ^{13}C labelling, positional ^{13}C labelling, or a combination of both. Not only the preparation of sample for analysis of ^{13}C labelling in proteinogenic amino acids in biomass is described, but also the necessary modifications to the method when analysis of ^{13}C labelling in free metabolic intermediates is of interest. Finally the strategies for using the different NMR-detected ^{13}C labelling data in ^{13}C MFA are discussed.

Key words NMR, ^{13}C, MFA, Metabolic flux, HSQC, [^{13}C, ^{1}H] COSY, ZQF-TOCSY

1 Introduction

NMR spectroscopy is an efficient method for obtaining quantitative data on the positional fractional ^{13}C enrichment of metabolic intermediates or end products. In addition, observation of the ^{13}C-^{13}C scalar couplings in the NMR spectra provides a means for monitoring cleavage and formation of the covalent bonds of the carbon backbone of the substrate molecule by alternative pathways.

For ^{13}C MFA, two alternative ^{13}C labelling strategies, or sometimes a combination of them, can be used [1]. Uniform ^{13}C labelling is achieved by feeding the cells with a mixture of uniformly ^{13}C-labelled (10–20 %) and non-labelled (with natural ^{13}C abundance of about 1.1 %) carbon source compound. In this approach, different carbon-carbon bond cleavage and reformation reactions of alternative pathways are manifested in the ^{13}C-^{13}C scalar coupling fine structure of the ^{1}H-^{13}C HSQC signals [2] and can be quantified.

Jens O. Krömer et al. (eds.), *Metabolic Flux Analysis: Methods and Protocols*, Methods in Molecular Biology, vol. 1191, DOI 10.1007/978-1-4939-1170-7_9, © Springer Science+Business Media New York 2014

In the second ^{13}C labelling strategy, positional ^{13}C labelling, the carbon source molecules have ^{13}C label in a specific position or positions. In this type of experiment, it is the fractional ^{13}C enrichment of each (or some specific) carbon position of the metabolites, rather than the contiguous labelling of the carbon backbone of the molecule, that is to be detected. The ^{13}C labelling of a given carbon can be detected as the ^{13}C satellites of its signal in a zero-quantum filtered TOCSY (ZQF-TOCSY) spectrum [3].

In this chapter, we provide detailed instructions for ^{13}C label detection by NMR spectroscopy, including the ^{13}C labelling of bacterial or yeast cells in either continuous (chemostat) or batch cultures, sample preparation, NMR experiments, and data processing. The ^{13}C labelling is analyzed from the amino acids incorporated into the biomass proteins under conditions of metabolic steady states. This is currently the prevailing method, mainly due to its robustness and the stable nature of the analytes, but also because it provides large enough samples for even relatively insensitive methods like NMR. However, recent technological development has markedly increased the sensitivity of NMR and, on the other hand, dynamic ^{13}C labelling experiments might be of interest to the reader. Therefore, we also discuss in Subheading 4 the special requirements and sample preparation in the cases, when the labelling of soluble intermediates of the metabolic pathways is to be analyzed.

Data from steady-state ^{13}C labelling experiments discussed in this chapter can be incorporated into metabolic flux analysis in two different ways. If alternative pathways modify the carbon-carbon bonds in the metabolic intermediates in different ways, algebraic flux ratio equations that couple the relative fluxes of the pathways to the ^{13}C labelling patterns of metabolic intermediates can be derived [2]. An underdetermined metabolic flux analysis problem can be turned solvable with local flux ratios as constraints for fluxes of alternative pathways. This efficient approach was first shown for *B. subtilis* by Sauer et al. [4] and later on for yeasts [5–7]. Because of the local nature of the flux ratios, the flux distribution obtained by this approach is less prone to inaccuracies in the metabolic network model or to scarce ^{13}C data [1, 8] than the alternative approach based on global iterative fitting [9]. In global iterative fitting isotopomer mass balances and metabolite mass balances are simultaneously solved. In this nonlinear optimization problem candidate flux distributions are iteratively generated and the corresponding isotopomer distributions simulated and compared to the experimental data until the fit is satisfactory. This approach is being applied to a great variety of cell types from bacteria to mammalian cells [10–13]. A drawback of the global iterative fitting approach is that because of the nonlinearity of the computational problem it is difficult to ensure that the resulting optimal flux distribution solution is global and unique [8].

2 Materials

2.1 Cell Cultivation and ^{13}C Labelling Components

1. Fermentor equipped for continuous cultivation of cells with a possibility to change the growth medium during the cultivation. In case the labelling is carried out in a batch cultivation during the exponential growth phase, the fermentor can be replaced by a thermostated incubator shaker and culture flasks.
2. Minimal medium for cell culture: Choose a medium known to be suitable for the cells in question. The medium should not contain any other carbon sources than the compound to be used as the ^{13}C source or any amino acids (*see* **Note 1**).
3. Medium for ^{13}C labelling: Use the same medium as above, but either (1) replace 10 % (*see* **Note 2**) of the carbon source compound with its uniformly ^{13}C-labelled version, e.g., [U-^{13}C]Glc (uniform ^{13}C labelling), or (2) replace a fraction or all of the carbon source compound with its positionally ^{13}C-labelled version (depending on the scientific question). It is a good idea to record a ^{1}H NMR spectrum of the medium in order to determine the exact percentage of the labelled compound, because the amount of crystal water in sugar varies and makes the exact preparation of a mixture with a given ratio difficult.

2.2 Sample Preparation Components

1. 6 M HCl for hydrolysis of the biomass, 7 ml per sample.
2. Cold (−40 °C) methanol for quenching the metabolism (only needed in case labelling of the intermediary metabolites is to be analyzed).
3. 75 % (v/v) boiling absolute ethanol containing 0.25 M Hepes, pH 7.5 for extracting the metabolites (only needed in case labelling of the intermediary metabolites is to be analyzed).
4. Rotary vacuum evaporator and vacuum-grade 100 ml round-bottom flasks (one per sample) with ground glass joints fitting the evaporator. If necessary, an adapter with ground glass joints can be used. Use of a bump trap between the flask and the condenser is highly recommended.
5. 10 ml screw-cap Pyrex® glass tubes (or other heat- and HCl-resistant, sealable tubes).
6. 15 ml plastic tubes.
7. Deuterium oxide (D_2O) (99.9 % atom % D) with approx. 0.05 wt.% 3-(trimethylsilyl)-propionic-2,2,3,3-d4 acid, sodium salt (TSP) (*see* **Note 3**).
8. 0.2 μm syringe-driven filters, 10 ml syringes.
9. Freeze drier system.

2.3 NMR Spectroscopy Components

1. High-field NMR spectrometer (400 MHz or above), preferably equipped with a 5 mm inverse detection probe head.
2. 5 mm NMR tubes (one per sample): Use high-quality tubes from a known vendor, e.g., Wilmad or Norell.
3. Software for processing the NMR spectra and for integration of the ^{13}C multiplet fine structure. Although the NMR spectrometer software can be used, for more accurate results software with more advanced deconvolution/fitting routines (e.g., PERCH NMR Software, *see* **Note 4**) is recommended.

3 Methods

3.1 Cell Cultivation and ^{13}C Labelling

In this method the cells are ^{13}C labelled in a continuous cultivation in a fermentor and the ^{13}C labelling is detected from the amino acids of the intracellular proteins. This kind of labelling is carried out under conditions of metabolic steady state, which has to be carefully monitored during the whole labelling period. *See* Subheading 4 for labelling in batch cultivation or for ^{13}C labelling analysis from intermediary metabolites. The same procedure can be applied regardless whether the uniform ^{13}C labelling or the positional ^{13}C labelling strategy is used.

1. Start a continuous chemostat cultivation of the cells on a minimal medium in a fermentor in a standard manner. Monitor the CO_2 production rate and preferably also other parameters like redox potential of the culture broth in order to detect when the system has reached the metabolic steady state. Once a stable steady state has been reached, move to **step 2** (*see* **Notes 5–7**).
2. Switch the fermentor feed to the ^{13}C label-containing medium (either with 10 % of uniformly labelled carbon source compound or a fraction of positionally ^{13}C-labelled carbon source, depending on the labelling strategy chosen). Write down the exact time of starting the labelling period (*see* **Note 8**). Continue the labelling for 1.5 residence times (*see* **Notes 8–11**) with constant monitoring of the metabolic state. Note that if the CO_2 production is monitored by MS, the ^{13}C labelling will have an effect on the MS signal, as typically only $^{12}CO_2$ is detected.
3. Harvest the cells by centrifuging them down and removing the supernatant or by filtering. Write down the exact time of harvesting (=end of the ^{13}C labelling period). The biomass can be freeze-dried for storage or transportation. If the sample preparation is to be continued immediately, freeze-drying is not necessary (*see* **Notes 12** and **13**).

3.2 Sample Preparation for NMR Spectroscopy

1. Suspend the biomass (freeze-dried or wet) into 7 ml of 6 M HCl and move the suspension to a screw-capped glass tube. Carefully close the cap and incubate at 110 °C for 20 h (*see* **Notes 14–16**).

2. Remove the sample from the oven and let it cool down in a refrigerator or on ice.
3. Carefully open the cap of the tube and move the dark-colored hydrolysate into a round-bottom flask. Keep the possible other samples in cold (*see* **Note 17**). Remember that you are handling a strong acid solution; work in a fume hood and wear necessary personal protection. Carefully observe the round-bottom flask for any flaws, such as star cracks, before using it. Attach the flask to the bump trap of the rotary evaporator and evaporate to dryness. Observe the sample while the vacuum is building up; at some point you will see the dissolved gases to boil out. After that you can warm the flask by a 40–50 °C water bath to facilitate the evaporation. When the sample is almost dry (the thick syrup is difficult to dry completely and it is not necessary) rinse the hydrolysis tube with 3 ml of water and add the water to the flask. Carefully dissolve the black syrup to the water and evaporate to dryness. Repeat the cycle of adding 3 ml of water and evaporating to dryness still twice.

   Dissolve the syrup in 2 ml of water and move the solution to a 15 ml plastic tube. Rinse the flask with 2 ml of water and move to the same tube. The solid particles probably present in the sample are not desired and can be left in the flask.
4. Repeat this procedure with the other samples you may have. Handle the samples one by one and keep the waiting samples in either ice bath or refrigerator.
5. Remove any solid particles by centrifugation and by filtering the supernatant into a 15 ml plastic tube using a syringe-driven filter and a 10 ml syringe. Rinse the syringe and the filter with 2 ml of water to the same tube. Repeat for other samples and freeze-dry the samples (*see* **Note 18**).
6. Dissolve the freeze-dried sample(s) in 500 μl of D_2O containing 0.05 wt.% of TSP (*see* **Notes 19** and **20**).
7. Move 600 μL of the sample into a 5 mm NMR tube (*see* **Note 21**). If the sample contains any solid particles, remove them by a syringe-driven filter; if there are no visible particles or if they are heavy enough to settle on the bottom of the tube, filtering is not necessary (*see* **Note 22**).

3.3 NMR Spectroscopy of Samples Labelled with a Uniformly Labelled Carbon Source

1. Acquire a standard 1D ^{1}H spectrum at 40 °C with presaturation of the water signal and store the spectrum.
2. Acquire two HSQC (heteronuclear single-quantum coherence) spectra, one for aliphatic and the other for aromatic signals (Fig. 1), using the parameters given in Table 1 (*see* **Notes 23–25**).
3. Process the two spectra using, e.g., the spectrometer software or other suitable NMR processing software. Prior to the Fourier transformation, zero-fill the data matrix in F1 (^{13}C dimension)

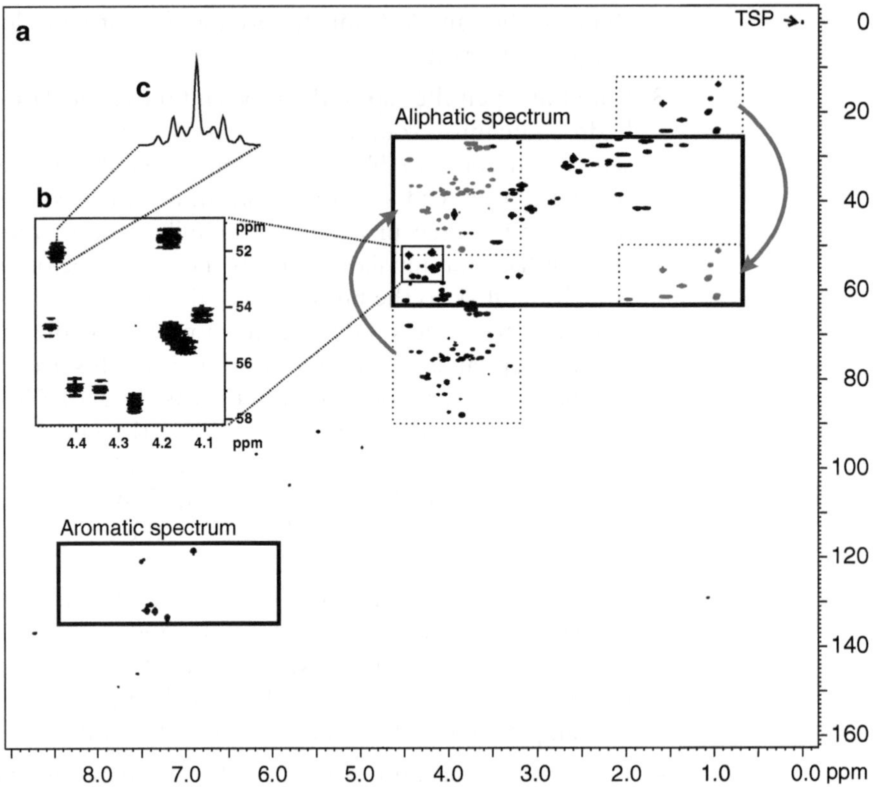

Fig. 1 NMR analysis of proteinogenic amino acids labelled with uniform ^{13}C labelling using HSQC spectra. Instead of recording one HSQC spectrum over the whole spectral area (**a**), two spectra are recorded, one for the aromatic signals and the other for the aliphatic signals. When the aliphatic spectrum is recorded using the parameters given in Table 1, the signals at either higher or lower ^{13}C frequencies are left outside the spectral window and appear as fold-back signals on the opposite (otherwise empty) side of the spectrum. This folding is illustrated by the *gray arrows* and the folded signals in their new locations are shown in *gray*. This narrowing of the spectral window in the ^{13}C dimension allows high enough point resolution, so that the ^{13}C-^{13}C scalar coupling fine structure of the cross peaks can be observed, as is illustrated in panel **b** for a selection of the Hα/Cα signals. *Inset c* shows a cross section along the ^{13}C axis of one of the signals (Asp Hα/Cα), revealing the multiplet structure (*see* also Fig. 2)

to 4096 complex data points and use a cosine window function in both dimensions.

4. Observe the fine structure of the HSQC cross peaks in the ^{13}C dimension arising from the scalar couplings to the neighboring ^{13}C-labelled carbons. For example, the fine structure in the Cα signal of Asp (Fig. 2) consists of the following components: a singlet (only Cα ^{13}C), two doublets (59.8 Hz, Cα and carboxyl ^{13}C, Cβ ^{12}C; 37.5 Hz, Cα and Cβ ^{13}C, carboxyl ^{12}C), and a quartet (all three carbons ^{13}C) (*see* **Note 26**).
5. Assign the HSQC signals based on the chemical shifts given in Table 2.

Table 1
NMR parameters for recording the aliphatic and aromatic region HSQC spectra

Parameter	Aliphatic	Aromatic
^{1}H carrier frequency	2.66 ppm	7.70 ppm
^{1}H sweep width	4.00 ppm	2.7 ppm
^{13}C carrier frequency	46.06 ppm	126 ppm
^{13}C sweep width	40.86 ppm	18.00 ppm
Number of points	1,024[a]	1,024[a]
Number of increments	1,500	800
Probe temperature	40 °C	40 °C
Relaxation delay	1 s	1 s
^{13}C decoupling during acquisition	Yes	Yes

[a]Number of points (=point resolution in ^{1}H dimension) can be even smaller, because all the information on labelling is in the indirect (^{13}C) dimension

6. Integrate the different constituent multiplets of each amino acid H/C signal. Since the multiplets are typically heavily overlapping, software capable of deconvoluting the original signals to their constituent multiplets should be used (*see* **Note 27**).
7. Estimate the error of the results. Provided that the spectral baseline is flat, the two main sources of error are the spectral noise and the inaccuracy of integration/deconvolution of the multiplet components of the cross peaks (*see* **Note 28**).

3.4 NMR Spectroscopy for Determination of Fractional ^{13}C Enrichments (Positional ^{13}C Labelling Experiments)

1. Record a 2D ZQF-TOCSY (zero-quantum filtered total correlation spectroscopy, Fig. 3, *see* **Note 29**) spectrum using a pulse sequence with a refocusing ^{13}C pulse in the middle of the increment delay (*see* **Note 30**) and DIPSI-2 mixing time of 25–80 ms (*see* **Note 31**).
2. Assign the amino acid ^{1}H signals using the ^{1}H chemical shifts from Table 2.
3. Determine the fractional ^{13}C labelling from the ratio of the ^{13}C satellites of the TOCSY signal to the total intensity of the signal (Fig. 3, *see* **Notes 32** and **33**).

3.5 Direct Local Constraints for ^{13}C-MFA from NMR Spectroscopy of Uniformly Labelled Samples

1. Calculate the relative abundances of intact carbon fragments, fragmentomers, arising from a single molecule of the carbon source compound (Fig. 2). Fragmentomers are specifically defined as molecular fragments differing only in the cleavage status of the carbon-carbon bonds. First calculate the fraction

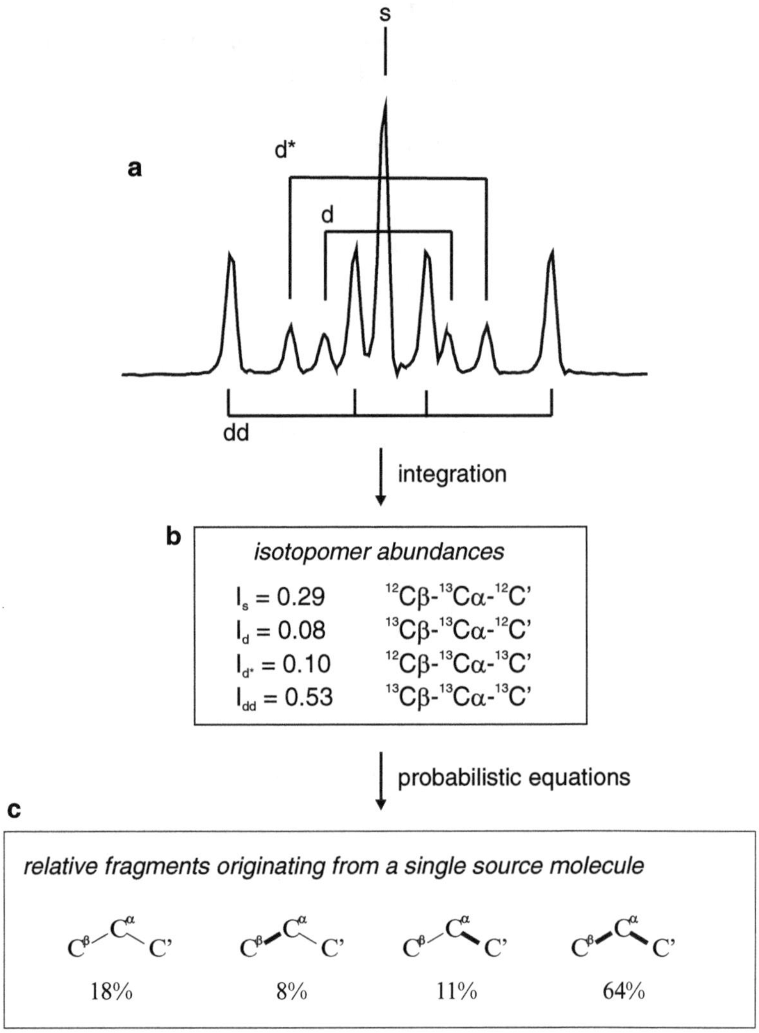

Fig. 2 Connection between the ^{13}C-^{13}C scalar coupling fine structure of the HSQC signals (**a**) and the ^{13}C labelling of the three carbon backbone fragments in question. The multiplet has four components: a singlet (s), representing the case where only Cα is ^{13}C, while the two neighboring carbons are ^{12}C; a doublet (d), where also Cβ is ^{13}C; a doublet with a larger coupling constant (d*), where Cα and the carboxyl carbon are ^{13}C; and a doublet of doublets (dd), where all the three carbons are ^{13}C. By separately integrating the four components the local isotopomer ratios can be obtained (**b**). Either these can be used as such, e.g., for the iterative fitting approaches or they can be turned to ratios of intact carbon fragments originating from the carbon source compound (indicated by *thick lines* between the carbons in panel **c**) using probabilistic equations taking into account the effect of the natural ^{13}C abundance and the statistical recombination of ^{13}C atoms

Table 2
^1H and ^{13}C chemical shifts in D_2O, pH < 1, of the key amino acid signals for ^{13}C metabolic flux analysis at 40 °C

Carbon position	δ ^1H (ppm)	δ ^{13}C (ppm)	Carbon position	δ ^1H (ppm)	δ ^{13}C (ppm)
Ala-α	4.185	51.55	Lys-β	1.989, 2.053	31.98
Ala-β	1.592	18.15 (55.36)	Lys-γ	1.560	24.30 (61.50)
Arg-β	2.006, 2.065	29.68	Lys-δ	1.756	29.00
Arg-δ	3.284	43.19	Lys-ε	3.057	41.92
Asp-α	4.446	52.03	Met-α	4.300	54.45
Asp-β	3.174	36.48	Phe-α	4.402	56.85
Glu-α	4.182	54.87	Phe-β	3.265, 3.367	38.32
Glu-β	2.229, 2.287	27.62	Pro-α	4.478	62.35
Glu-γ	2.669	32.27	Pro-β	2.202, 2.468	31.04
Gly-α	3.935	42.90	Pro-γ	2.086	26.13
His-α	4.461	54.72	Pro-δ	3.442, 3.478	49.19
His-β	3.500	27.80	Ser-α	4.263	57.45
His-$δ^2$	7.490	121.11	Ser-β	4.052, 4.121	62.09
Ile-α	4.074	60.08	Thr-α	4.067	61.16
Ile-β	2.080	38.53	Thr-β	4.440	68.09 (30.69)
Ile-$γ^1$	1.354, 1521	27.56	Thr-$γ^2$	1.381	21.70 (59.10)
Ile-$γ^2$	1.045	17.21 (54.25)	Tyr-α	4.342	56.94
Ile-δ	0.956	13.87 (51.03)	Tyr-β	3.188, 3.283	37.49
Leu-α	4.106	54.28	Tyr-$δ_x$	7.210	133.70
Leu-β	1.767, 1.868	41.61	Tyr-$ε_x$	6.900	118.85
Leu-$δ^1$	0.981	24.30 (61.70)	Val-α	4.003	61.13
Leu-$δ^2$	0.968	23.77 (61.17)	Val-$γ^1$	1.082	20.05 (57.45)
Lys-α	4.143	55.43	Val-$γ^2$	1.066	19.66 (57.06)

The chemical shifts are referenced to internal TSP (0 ppm for both nuclei). For ^{13}C chemical shifts, values in parentheses are for a spectrum with folded signals, while the values without parentheses are the real, non-folded chemical shifts

of the labelled biomass in the fermentor using first-order wash-out kinetics in case of chemostat cultures (*see* **Note 9**). In case of batch cultures all of the biomass can be assumed to be labelled if the inoculum was small (less than 1 %) compared to the amount of biomass produced by the sampling time. Then the effects of natural ^{13}C abundance (1.1 %) as well as statistical

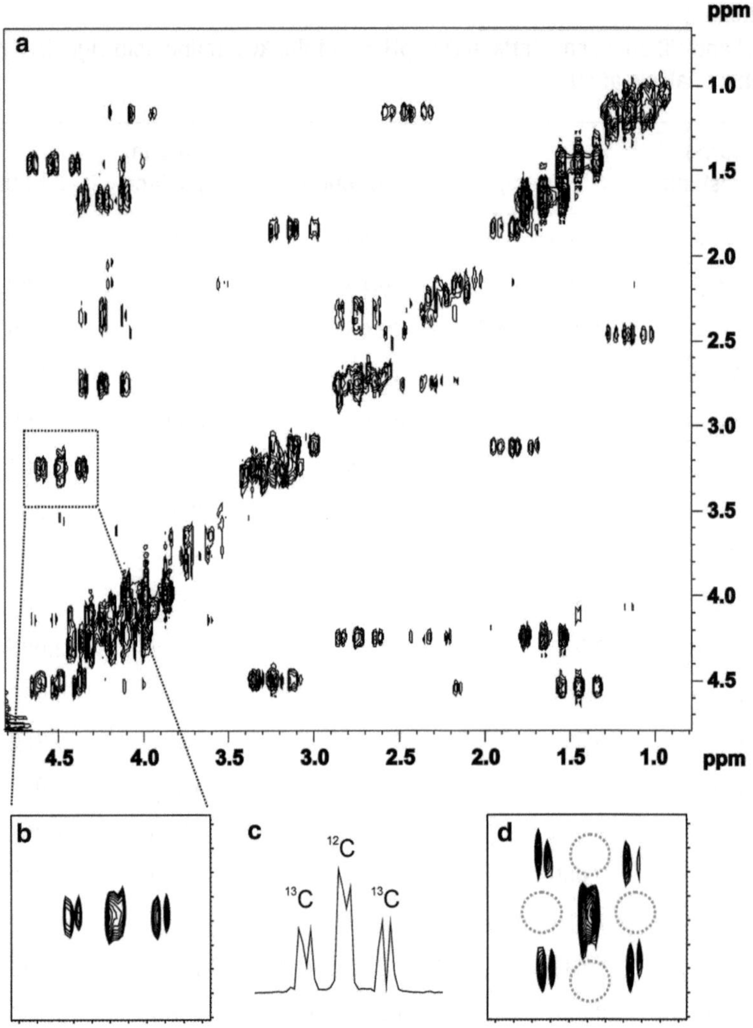

Fig. 3 NMR analysis of fractional ^{13}C enrichments of different carbons of amino acids using a ZQF-TOCSY spectrum with refocusing of ^{1}H-^{13}C scalar couplings in the indirect dimension. Panel **a** shows the aliphatic area of the spectrum and panel **b** shows an expansion of Asp Cα/Cβ cross peak, revealing the fractional ^{13}C enrichment of the Cα. The signal in the middle originates from the ^{12}C-bound protons, while the two signals on the sides (^{13}C satellites) come from the ^{13}C-bound protons. Therefore, the fractional ^{13}C enrichment of Asp Cα can be determined by taking the ratio of the integral of the ^{13}C satellites to the integral over the whole signal. The integration can be done in either 2D or by taking a cross section (panel **c**) or sum of the 1D slices. In panel **d**, the effect of not having the refocusing ^{13}C pulse is illustrated. In this case, ^{13}C satellites appear in both dimensions. While this spreads the intensity of the signal to multiple locations and makes the spectrum much more crowded, it also provides additional information on the labelling. For instance, the absence of the satellites from the locations indicated by the *dashed-line circles* reveals that in this sample if Cα is ^{12}C, also the neighboring carbons are unlabelled

recombination of the ^{13}C-labelled fragments must be taken into account. This can be done by calculating the expected ^{13}C scalar coupling fine structures for the given fragmentomers using probabilistic equations and subsequently decomposing the observed fine structure to these calculated fine structures, yielding the relative abundances of the intact carbon fragments; find the equations from reference [2] (*see* **Note 34**).

2. Construct a stoichiometric metabolic model of central carbon metabolism by listing all the reactions, the stoichiometric relations of the substrates and products, and the carbon atom transitions from substrates to products in the reactions (*see* **Note 35**). Observe that metabolic intermediates may react symmetrically, like succinate, if they possess a center of symmetry. Symmetrization should be taken into account in the carbon atom mappings of the particular reactions.
3. Propagate the fragmentomers derived from the amino acid ^{13}C labelling patterns to the precursor metabolic intermediates in the central carbon metabolism, if the amino acids are beyond the extent of the metabolic network model. Utilize the knowledge on the carbon atom transitions in the reactions of amino acid biosynthetic pathways (*see* **Note 36**).
4. Flux ratio is defined as the relative contribution of a flux to the total flux converging at a metabolic junction, i.e., at a metabolite. Derive algebraic flux ratio equations relating the relative fluxes of the alternative pathways to the mass balances of suitable fragmentomers at a metabolic junction. Derivation can be made from scratch using knowledge on how the pathways modify carbon-carbon bonds or one can pick previously derived flux ratio equations for central carbon metabolism of the model organisms on glucose, *E. coli* [2], *B. subtilis* [14], *S. cerevisiae* [6, 15], *Pichia stipitis* [16], and *Pichia pastoris* on glycerol [17], on glycerol/methanol mixtures [18], on glucose/methanol mixtures [7], and for filamentous fungus *Trichoderma reesei* on glucose and on sorbitol [19]. Suitable fragmentomers for the derivation of flux ratio equations are the ones including the carbon-carbon bond, which gets differentially modified by the alternative pathways. Often it is useful or required to derive flux ratio equations from sums of different fragmentomers to include all the information on the status of a particular carbon-carbon bond. Fragmentomer data is required for the metabolite where the fluxes converge, and for the source metabolites of the converging fluxes. Flux ratio equations can also be derived automatically from different types of ^{13}C-labelling data (*see* **Note 37**). Use the derived flux ratio equations to calculate the flux ratios from fragmentomer data (*see* **Note 38**). Use Gaussian law of error propagation to calculate experimental error estimates for the flux ratios from the error estimates of

the fragmentomers. If flr is a flux ratio calculated from the relative abundances of different fragmentomers $x_1, x_2, \ldots, x_k$ with standard deviations $s_{x1}, s_{x2}, \ldots, s_{xk}$ (Eq. 1) the standard deviation of the flux ratio is calculated from Eq. 2:

$$flr = f\left(x_1, x_2, \ldots, x_k\right) \tag{1}$$

$$s_A = \sqrt{\sum_{j=1}^{k}\left(\frac{\partial}{\partial x_j} flr\right)^2 s_{x_j}^2} \tag{2}$$

Flux ratios are as such informative of the in vivo activities of the pathways in cells. They are the result of the above-described procedures termed metabolic flux ratio (METAFoR) analysis [2, 4].

5. Use the flux ratios as additional constraints to solve the steady-state distributions of fluxes in the central carbon metabolism by means of metabolite balancing. If the flux ratio constraints render the MFA system solvable, i.e., MFA problem is no longer underdetermined, minimize the sum of square residual deviations from extracellular metabolite mass balances while fulfilling the flux ratios, stoichiometric intracellular metabolite mass balance constraints, and flux lower and upper bounds [6]. By repeatedly sampling noise with zero mean and experimentally estimated variance to the flux ratios, analyze the sensitivity of the flux solution to the noise in the flux ratio data. Alternatively the sum of square residual deviations from extracellular metabolite mass balances and experimentally determined flux ratios weighted with the experimental error estimates can be simultaneously minimized while fulfilling the stoichiometric intracellular metabolite mass balance constraints and flux lower and upper bounds [20]; *see* also **Note 39**.

If the MFA system retains unsolvable, set flux ratios as strict constraints (*see* **Note 40**), and use linear optimization to solve an optimal flux distribution. Then, sample noise to the flux ratio constraints from a normal distribution characterized by the experimental error estimates of the flux ratios. Recompute the linear optimization problem to analyze the variation in the distribution of fluxes. Since solving a linear optimization problem returns only one point solution, which is not necessarily unique, study the flux distribution solution space also by means of flux variability analysis [21].

3.6 ^{13}C-MFA by Global Iterative Fitting with Fragmentomer Data from NMR Experiments

1. Construct a metabolic network model and include carbon atom mappings into it (see instructions above) in a suitable format for a chosen software to use in ^{13}C-MFA by global iterative fitting. There are different software available, to mention i.e., 13C-FLUX [9], 13C-FLUX2 (http://www.13cflux.net),

and OpenFLUX [22]. 13C-FLUX and 13C-FLUX2 are directly NMR data compatible whereas OpenFLUX requires the user to formulate the model for NMR data outside the software (*see* **Note 41**).

4 Notes

1. It is also possible to use more than one carbon source compound simultaneously. If the uniform ^{13}C labelling strategy is used, then the uniform fractional ^{13}C labelling of each compound should be the same [18].
2. In ^{1}H-^{13}C HSQC NMR only ^{13}C carbons give a signal meaning that the higher the labelling degree, the stronger is our signal. On the other hand, the probability of finding two ^{13}C atoms covalently bound together in the case of cleaved C-C bond (=natural abundance + labelling degree) increases with increasing labelling degree and approaches the probability of finding such pairs, when the bond was not cleaved (100 %). This decreases the dynamic range of the measurement. In addition, ^{13}C-labelled compounds can be very costly making lower labelling degree desirable. In practise, labelling degree of about 10 % usually gives high enough sensitivity, but in cases where the amount of the sample is limited, one may consider using a bit higher (15–20 %) fraction of ^{13}C-labelled carbon source.
3. Preferably buy TSP containing D_2O (Aldrich), but you can also add TSP by yourself (the concentration of TSP is not very critical, but it should not give a signal larger than your amino acid signals). Once the NMR carrier frequencies have been calibrated, TSP is not necessarily needed. Note that D_2O is extremely hygroscopic, so recap the bottle immediately after taking the amount needed.
4. http://www.perchsolutions.com.
5. Because ^{13}C-labelled compounds can be very costly, it is advisable to use the smallest growth volume capable to produce the required amount (100 mg DW) of biomass.
6. The dilution rate, agitation speed, aeration, and temperature depend on the scientific question and need to be discussed with a fermentation expert.
7. The ^{13}C labelling can also be carried out in a batch cultivation, in either bioreactor or in shake flasks, during the exponential growth phase, during which the metabolism can be considered to be in a steady state. The main difference to a chemostat cultivation is that the growth rate is not limited but the organism grows on its maximal growth rate. Prior to such an experiment, it is necessary to determine the growth curve in order to

ensure that the cells are harvested during the exponential phase. In a batch cultivation the label is introduced into the medium already in the beginning of the cultivation.

8. In order to ensure that the label starts entering the fermentor as soon as possible after switching the feed, keep the tubing from the ^{13}C-labelled medium container to the fermentor as short as possible. In any case, it is necessary to determine the time it takes for the ^{13}C-labelled carbon source to reach the fermentor.

9. Residence time in continuous fermentor cultivation is the average time a cell spends in the fermentor. Residence time (R) is the inverse of the dilution rate (D):

$$R = \frac{1}{D}$$

The dilution rate of the cultivation is defined as

$$D = \frac{F_{\mathrm{out}}}{V}$$

where F_{out} is the rate of the outflow of the culture broth and V is the working volume of the chemostat.

The fraction of the biomass in the fermentor produced under labelling conditions can be estimated using first-order incorporation kinetics:

$$f = 1 - \mathrm{e}^{-Dt}$$

where f is the fraction of the labelled biomass, D is the dilution rate, and t is the labelling time.

10. Choosing the labelling time is a balance between the cost of the label and the fermentor time, and the accuracy of the results. A longer labelling period gives more accurate results as a larger fraction of the biomass gets labelled. Labelling for 1.5 residence times is a good compromise as then already over 80 % of the biomass in fermentor has been produced under labelling conditions. This has been discussed, e.g., by Gombert et al. [23] and by Solá et al. [17].

11. For a batch cultivation, start the cultivation, e.g., in shake flasks on the labelled medium and cultivate the cells for the time determined from the growth curve of a pilot experiment.

12. The cells may excrete into the cultivation broth compounds, the labelling patterns of which may also be of interest. If the compounds are not too volatile, the supernatant can be freeze-dried and analyzed by NMR. The labelling of the more volatile

compounds can be analyzed by NMR directly from the supernatant, provided their concentrations are high enough.

13. If the labelling of the metabolic intermediates is to be analyzed, it is essential to quench the metabolism (i.e., rapidly stop all metabolic reactions) upon harvesting the cells. For bacteria and yeasts this is best done by spraying the growth medium to −40 °C methanol. After this the cells can be centrifuged down in cold (−20 °C) and the supernatant removed. For other types of cells different quenching methods (e.g., using liquid nitrogen) may be required. The quenching methods have been recently reviewed [24] and detailed instructions can be found from reference [25].
14. Carefully examine the edge of the tube for flatness and the seal of the cap for any flaws. Mark the surface level of the suspension in each tub and after an hour check that the levels have not changed to make sure that the caps are not leaking.
15. With this procedure asparagine and glutamine are deaminated to aspartate and glutamate, respectively, and cysteine and tryptophane are oxidized, and their ^{13}C labelling cannot be analyzed by this method. However, in most cases they only contain labelling information already available from the other amino acids.
16. For label analysis from the intermediary metabolites, the quenched and centrifuged bacteria or yeast cells can be extracted by suspending them into 75 % (v/v) boiling absolute ethanol containing 0.25 M Hepes, pH 7.5 [26]. For other types of cells different extraction methods may be required.
17. We find that keeping the sample as cold as possible reduces the risk of bumping during the buildup of the vacuum.
18. At this stage the samples are very stable and can, e.g., be mailed to another lab or stored in cold for longer periods of time.
19. Dissolving the thick, syrup-like sample is a slow process. Either let it dissolve overnight at room temperature or, if in a hurry, place the closed tube in a boiling water bath and shake it occasionally.
20. The pH of the NMR sample should be 1 or lower. Despite the repeated cycles of dissolving in water and evaporating into dryness, there typically remains enough HCl in the sample to keep the pH down. However, it is best to measure the pH and add dilute HCl as necessary.
21. 600 μL in a 5 mm tube is 40 mm high. You can use this information to add a correct amount of D_2O in the NMR tube in case part of the sample has been lost, e.g., during filtering.
22. Already during the first days new precipitate will be formed. This is usually not a problem as the precipitate settles down on the bottom of the tube, but in order to prevent the formed

precipitate to stick to the tube walls, the tubes should be stored in a strictly upright position.

23. Exceptionally high resolution is required in the indirect (^{13}C) dimension of the spectra. Therefore it is best to run two separate spectra and narrow the spectral window of the aliphatic spectrum so that part of the signals are outside the window and appear in the spectrum as fold-back signals with an opposite sign (Fig. 1) [27].
24. Using the parameters from Table 1 and carefully setting the ^{13}C transmitter position related to the TSP signal should result in a spectrum with no splitting of peaks at the edges of the 2D spectra. However, this should be checked at the first time by acquiring a quick version of the aliphatic HSQC with a smaller number of increments. The ^{13}C scale can be calibrated relative to the TSP signal by acquiring a quick HSQC with a large enough window (or by focusing the window to the right area of the ^{13}C spectrum). Number of points can be even smaller, as the resolution in the direct dimension (^{1}H) is not critical for the ^{13}C label analysis.
25. For chemical shift calibration of the aliphatic and aromatic spectra, first calibrate the chemical shifts of the large spectrum to TSP, then observe the chemical shifts of well-separated signals in the areas of both subspectra, and use those signals as references for the narrow spectra.
26. In a ^{1}H-^{13}C HSQC spectrum signals are observed between ^{13}C atoms and their directly bound (over one covalent bond) ^{1}H atoms. The fine structure of these signals is dependent on whether or not the neighboring carbons also are ^{13}C, and therefore reflects the metabolic history of the carbon backbone of the molecule. For a carbon in the middle of the carbon chain, four different fine structures are possible. If both of the neighboring carbons are ^{12}C, then a singlet is observed. If one of the neighbors is a ^{13}C but the other one is a ^{12}C, then the scalar coupling between the two ^{13}C nuclei splits the signal to a doublet. The split of the doublet depends on the coupling constant J (given in Hz). In case the other neighboring carbon is a ^{13}C, but the previous one is not, then a doublet is again observed, but depending on the coupling constant, the spilt of this doublet may or may not be different from that in the previous case. The coupling constant between the Cα and the carboxyl carbon of amino acids is always larger than the one between Cα and Cβ. Thus for an α-carbon the two doublets can always be distinguished from each other. If all three carbons are ^{13}C, a quartet (or doublet of doublets) with the splits depending on the two coupling constants is observed. Because the fine structure of an HSQC signal reflects only the labelling

of the carbons directly bound to the observed one, the longest carbon backbone fragment for which the labelling can be observed from an HSQC signal spans three carbons. If the carbon observed is a terminal one, then only two fine structures are possible depending on whether the neighboring carbon is a ^{12}C (singlet) or a ^{13}C (doublet).

27. Spectrometer software usually have deconvolution routines, but they rarely are able to fit real multiplets but only deconvolute the original signal to singlets. Therefore we advise to use software with more advanced deconvolution/line-fitting routines, e.g., PERCH NMR Software (*see* **Note 3**). Note also that one can take a cross section of the HSQC signal along the ^{13}C axis (Figs. 1 and 2) in order to be able to do the deconvolution in one dimension (and, in some cases, to resolve otherwise overlapping signals), but a more accurate result is obtained by summing all the sections containing parts of the signal [28]. Note also that due to the isotope effects the centers of the different multiplet components are not exactly aligned.
28. The effect of the noise is different on signals or multiplet components of different amplitudes. This effect can be estimated, e.g., by defining a region of pure noise (no signals or ridges) and repeatedly adding or subtracting a sample of noise from that area from the signal in question, followed by a new integration. Most NMR software packages have this kind of routine, but it would be most convenient to carry out the noise sampling and the integration/deconvolution of the multiplet components with the same program.
29. Although a standard TOCSY sequence can be used, the spectra will contain signal distortions arising from zero-quantum transitions during the mixing period, which will compromise the accuracy of the fractional labelling determination [29], and also make the visual interpretation of the spectrum more difficult.
30. The refocusing pulse removes the ^{13}C satellites from the indirect dimension [29] (*see* Fig. 3b, d), giving several advantages: (1) only the singlet (^{12}C-bound proton) and a doublet (^{13}C-bound proton) have to be integrated and this can be carried out from a 1D slice (although summing all the slice-containing parts of the signal will give more accurate result), (2) having the ^{13}C satellites in only one dimension makes the typically already very crowded spectrum a bit less crowded, and (3) if the fractional labelling is low or the metabolite in question has a low concentration, spreading the already low intensity of the satellites to up to eight peaks instead of two may not be a good idea.

    Pulse sequences for a ZQF-TOCSY can usually be found from the standard (up-to-date) pulse sequence libraries of the

NMR spectrometer software, e.g., Bruker sequence dipsi2gpphzs, but the sequences typically do not have the refocusing ^{13}C pulse for removal of the ^{13}C satellites from the indirect dimension. A Bruker sequence with the refocusing pulse can be obtained upon request from the authors.

31. The optimal length of the mixing time (during which magnetization is spread to the whole spin system) depends on the structure of the compound in question. As a rule of thumb, a bit too long mixing time, especially when using a zero-quantum filter, should not be a problem, but it can substantially lengthen the duration of the experiment.

32. Although in most cases standard spectrometer software can be used for integration, better results, in particular in the cases of overlapping signals, can be obtained by more advanced line-fitting software, as discussed in **Note 26**.

33. This approach alone does not provide comprehensive data on the labelling of the amino acids, because the ^{13}C labelling of carbons not having directly bound protons (like carbonyl/carboxyl groups) is not detected. Data on the labelling of such carbons may be obtained if the TOCSY spectrum is recorded without the refocusing of the ^{13}C couplings in F1, *see* Fig. 3d [30], but this approach often results in very crowded spectrum. Alternatively, various NMR experiments mainly designed for protein NMR can be used as thoroughly described in reference [31]. A third option, which may be very useful in particular if the uniform and positional ^{13}C labelling are combined in the same cultivation, is to determine the fractional ^{13}C enrichment from the refocused ZQF-COSY spectrum and then scan the neighboring atoms of the ^{13}C-labelled subpopulation of carbons by HSQC [3].

34. For exact description of the probabilistic calculations *see* reference [2].

35. Curated genome-scale metabolic reconstructions provide good sources of organism-specific metabolic reactions. Carbon atom transitions can be obtained from biochemistry text books and a few databases (ARM database (http://www.metabolome.jp) [32]; KEGG ligand and rpair database [33]), or they can be directly calculated for reactions by solving the "atom mapping problem" for which Heinonen et al. have provided an efficient approach [34]. Different sources of atom mappings provide the mappings in unequal syntax. The consistency should be checked before the combination of data from different sources.

36. For common model organisms the carbon-atom mappings from amino acid to their precursor metabolic intermediates can be found in pioneering publications, e.g., *Escherichia coli*

[2], *Bacillus subtilis* [14], *Saccharomyces cerevisiae* [15], *Pichia stipitis* [16], *Pichia pastoris* [14], and *Trichoderma reesei* [19]. If the amino acid biosynthetic pathways of an organism under study are not known, path finding on genome data can be performed to resolve the carbon-atom mappings between the precursor metabolic intermediates and amino acids [35].

37. Alternatively the derivation of flux ratios from fragmentomer mass balances for metabolic junctions can be automatized for a defined metabolic network structure with carbon atom mappings implementing the method developed by Rantanen et al. [8].

38. Example of the derivation of a flux ratio from fragmentomer data:
Contribution of a pyruvate carboxylase flux to the cytosolic pool of oxaloacetate, $X_{\mathrm{Oaa_{cyt}\_from\_Pyr_{cyt}}}$, for *S. cerevisiae* during growth on glucose is derived as the pyruvate carboxylase flux divided by the sum of fluxes converging to the cytosolic pool of oxaloacetate: the pyruvate carboxylase flux and a transport flux from the mitochondrial pool of oxaloacetate. First a mass balance of x1x fragments of cytosolic oxaloacetate, $\mathrm{Oaa_{cyt}\_x1x}$, is derived, where 1 refers to an intact bond and x refers to an intact or a cleaved bond in a molecule. These fragments are chosen because TCA cycle activity produces mitochondrial oxaloacetate cleaved in this bond whereas in pyruvate the corresponding bond, according to the carbon atom mappings, is close to intact when *S. cerevisiae* grows on glucose. Since the flux from phosphoenolpyruvate to the cytosolic pool of pyruvate through pyruvate kinase and further through pyruvate carboxylase to the cytosolic pool of oxaloacetate can be assumed to be irreversible during growth on glucose, the C2-C3 fragments of phosphoenolpyruvate can be used in the mass balance equations under the particular conditions. The C2-C3 fragment of mitochondrial oxaloacetate is conserved in the TCA cycle as the C2-C3 fragment of 2-oxoglutarate and thus, it can be utilized in the mass balance equation [6], Eq. 3:

$$\begin{aligned}\mathrm{Oaa_{cyt}\_x1x} &= X_{\mathrm{Oaa_{cyt}\_from\_Pyr_{cyt}}} \times \mathrm{Pyr\_x1} + \left(1 - X_{\mathrm{Oaa_{cyt}\_from\_Pyr_{cyt}}}\right) \times \left(\mathrm{Oaa_{mit}\_x1x}\right) \\ &= X_{\mathrm{Oaa_{cyt}\_from\_Pyr_{cyt}}} \times \mathrm{Pep\_x1} + \left(1 - X_{\mathrm{Oaa_{cyt}\_from\_Pyr_{cyt}}}\right) \times \left(\mathrm{Oga\_x1xx}\right)\end{aligned} \quad (3)$$

$X_{\mathrm{Oaa_{cyt}\_from\_Pyr_{cyt}}}$ is solved from Eq. 1 and formulated in terms of amino acid fragmentomers according to the propagation of amino acid carbons to their precursors in the central carbon metabolism. Accordingly, a sum of $\{f^{(2)} + f^{(3)}\}$ fragmentomers of Phe-Cα and Tyr-Cα represents the fraction of intact C2-C3 bonds of phosphoenolpyruvate and a sum of $\{f^{(2)} + f^{(3)}\}$ Glu-Cα and Pro-Cα represents the fraction of intact C2-C3 bonds of 2-oxoglutarate. Therefore, $X_{\text{Oaa}_{\text{cyt}}\text{\_from\_Pyr}_{\text{cyt}}}$ can be solved from Eq. 4 [6]:

$$X_{\text{Oaa}_{\text{cyt}}\text{\_from\_Pyr}_{\text{cyt}}} = \left(\text{Oaa}_{\text{cyt}}\text{\_x1x} - \text{Oga\_x1xx}\right) / \left(\text{Pep\_x1} - \text{Oga\_x1xx}\right)$$
$$= \frac{\left[f^{(2)} + f^{(3)}\right]\{\text{Asp, Thr} - \text{C}\alpha, \text{Asp} - \text{C}\beta\} - \left[f^{(2)} + f^{(3)}\right]\{\text{Glu, Pro} - \text{C}\alpha\}}{\left[f^{(2)} + f^{(3)}\right]\{\text{Phe, Tyr} - \text{C}\alpha, \text{C}\beta\} - \left[f^{(2)} + f^{(3)}\right]\{\text{Glu, Pro} - \text{C}\alpha\}} \quad (4)$$

39. If five specific flux ratios are available, they can be introduced as constraints to an MFA system in FiatFlux software which supports ^{13}C-MFA for three model organisms *B. subtilis*, *E. coli*, and *S. cerevisiae* on glucose [36]. FiatFlux is in general developed for utilization of MS-data but it has been made modular. The flux ratios with their error estimates can be introduced into the second module, which solves constrained MFA problems in a user-friendly way.
40. Strict flux ratio constraints can be combined with linear optimization by defining them as linear constraints [37]. The stoichiometric matrix can be augmented with rows of linear constraints. In this way the flux ratios can be introduced also into CobraToolbox framework [38] for analysis of constraint-based models even in genome scale.
41. OpenFLUX utilizes EMU (elementary metabolite unit) framework [39]. EMU variable is generally a mass distribution vector directly compatible with MS-detected ^{13}C-labelling data. However, fragmentomer data from NMR measurements can also be simulated with the EMU framework.

Acknowledgements

This study was financially supported by the Academy of Finland through the Centre of Excellence in White Biotechnology Green Chemistry (grant 118573). The research leading to these results has received funding from the European Union Seventh Framework Programme FP7/2007-2013 under grant agreement number 222716—SMARTCELL. P.J. wants to acknowledge funding from the Academy of Finland for a postdoctoral researcher's project (grant 140380).

References

1. Zamboni N et al (2009) (13)C-based metabolic flux analysis. Nat Protoc 4:878–892
2. Szyperski T (1995) Biosynthetically directed fractional ^{13}C-labeling of proteinogenic amino acids. An efficient analytical tool to investigate intermediary metabolism. Eur J Biochem 232: 433–448
3. Massou S et al (2007) NMR-based fluxomics: quantitative 2D NMR methods for isotopomers analysis. Phytochemistry 68:16–18

4. Sauer U (1999) Metabolic flux ratio analysis of genetic and environmental modulations of *Escherichia coli* central carbon metabolism. J Bacteriol 181:6679–6688
5. Fredlund E et al (2004) Oxygen- and glucose-dependent regulation of central carbon metabolism in Pichia anomala. Appl Environ Microbiol 70:5905–5911
6. Jouhten P et al (2008) Oxygen dependence of metabolic fluxes and energy generation of *Saccharomyces cerevisiae* CEN.PK113-1A. BMC Syst Biol 2:60
7. Jordà J et al (2012) Metabolic flux profiling of recombinant protein secreting *Pichia pastoris* growing on glucose:methanol mixtures. Microb Cell Fact 11:57
8. Rantanen A et al (2008) An analytic and systematic framework for estimating metabolic flux ratios from ^{13}C tracer experiments. BMC Bioinformatics 9:266
9. Wiechert W et al (2001) A universal framework for ^{13}C metabolic flux analysis. Metab Eng 3:265–283
10. Costenoble R (2007) ^{13}C-labeled metabolic flux analysis of a fed-batch culture of elutriated *Saccharomyces cerevisiae.* FEMS Yeast Res 7: 511–526
11. Fan TW-M, Lane AN (2011) NMR-based stable isotope resolved metabolomics in systems biochemistry. J Biomol NMR 49: 267–280
12. Bartek T (2011) Comparative ^{13}C metabolic flux analysis of pyruvate dehydrogenase complex-deficient, L-valine-producing *Corynebacterium glutamicum.* Appl Environ Microbiol 77: 6644–6652
13. Niklas J (2011) Metabolite channeling and compartmentation in the human cell line AGE1.HN determined by ^{13}C labeling experiments and ^{13}C metabolic flux analysis. J Biosci Bioeng 112:616–623
13. Schneider K (2012) The ethylmalonyl-CoA pathway is used in place of the glyoxylate cycle by *Methylobacterium extorquens* AM1 during growth on acetate. J Biol Chem 287:757–766
14. Sauer U et al (1997) Metabolic fluxes in riboflavin-producing *Bacillus subtilis.* Nat Biotechnol 15:448–452
15. Maaheimo H et al (2001) Central carbon metabolism of *Saccharomyces cerevisiae* explored by biosynthetic fractional ^{13}C labeling of common amino acids. Eur J Biochem 268:2464–2479
16. Fiaux J et al (2003) Metabolic-flux profiling of the yeasts *Saccharomyces cerevisiae* and *Pichia stipitis.* Eukaryot Cell 2:170–180
17. Solà A et al (2004) Amino acid biosynthesis and metabolic flux profiling of *Pichia pastoris.* Eur J Biochem 271:2462–2470
18. Solà A et al (2007) Metabolic flux profiling of *Pichia pastoris* grown on glycerol/methanol mixtures in chemostat cultures at low and high dilution rates. Microbiology 153:281–290
19. Jouhten P et al (2009) ^{13}C-metabolic flux ratio and novel carbon path analyses confirmed that *Trichoderma reesei* uses primarily the respirative pathway also on the preferred carbon source glucose. BMC Syst Biol 3:104
20. Fischer E, Zamboni N, Sauer U (2004) High-throughput metabolic flux analysis based on gas chromatography-mass spectrometry derived ^{13}C constraints. Anal Biochem 325:308–316
21. Mahadevan R, Schilling CH (2003) The effects of alternate optimal solutions in constraint-based genome-scale metabolic models. Metab Eng 5:264–276
22. Quek LE et al (2009) OpenFLUX: efficient modelling software for ^{13}C-based metabolic flux analysis. Microb Cell Fact 8:25
23. Gombert AK et al (2001) Network identification and flux quantification in the central metabolism of *Saccharomyces cerevisiae* under different conditions of glucose repression. J Bacteriol 183:1441–1451
24. van Gulik W (2010) Fast sampling for quantitative microbial metabolomics. Curr Opin Biotechnol 21:27–34
25. van Gulik WM et al (2012) Fast sampling of the cellular metabolome. Methods Mol Biol 881:279–306
26. Gonzalez B, Francois J, Renaud M (1997) A rapid and reliable method for metabolite extraction in yeast using boiling buffered ethanol. Yeast 13:1347–1355
27. Hochuli H et al (1999) Amino acid biosynthesis in the halophilic archaeon Haloarcula hispanica. J Bacteriol 181:3226–3237
28. van Winden W et al (2001) Innovations in generation and analysis of 2D [(13)C, (1)H] COSY NMR spectra for metabolic flux analysis purposes. Metab Eng 3:322–343
29. Massou S et al (2007) Application of 2D-TOCSY NMR to the measurement of specific (13C-enrichments in complex mixtures of 13C-labeled metabolites. Metab Eng 9:252–257
30. Lane AN, Fan TW-M, Higashi RM (2008) Isotopomer-based metabolomic analysis by NMR and mass spectrometry. Methods Cell Biol 84:541–588
31. Fan TW-M, Lane AN (2011) NMR-based stable isotope resolved metabolomics in systems biochemistry. J Biomol NMR 49:267–280

32. Arita M (2003) In silico atomic tracing by substrate-product relationships in Escherichia coli intermediary metabolism. Genome Res 13:2455–2466
33. Kanehisa M et al (2006) From genomics to chemical genomics: new developments in KEGG. Nucleic Acids Res 34:D354–D357
34. Heinonen M et al (2011) Computing atom mappings for biochemical reactions without subgraph isomorphism. J Comput Biol 18:43–58
35. Pitkänen E, Jouhten P, Rousu J (2009) Inferring branching pathways in genome-scale metabolic networks. BMC Syst Biol 3:103
36. Zamboni N, Fischer E, Sauer U (2005) FiatFlux—a software for metabolic flux analysis from 13C-glucose experiments. BMC Bioinformatics 6:209
37. Choi HS et al (2007) Incorporating metabolic flux ratios into constraint-based flux analysis by using artificial metabolites and converging ratio determinants. J Biotechnol 129:696–705
38. Becker SA et al (2007) Quantitative prediction of cellular metabolism with constraint-based models: the COBRA Toolbox. Nat Protoc 2:727–738
39. Antoniewicz MR, Kelleher JK, Stephanopoulos G (2007) Elementary metabolite units (EMU): a novel framework for modeling isotopic distributions. Metab Eng 9:68–86

Chapter 10

GC-MS-Based ^{13}C Metabolic Flux Analysis

Judith Becker and Christoph Wittmann

Abstract

The in vivo analysis of metabolic fluxes has become a valuable method for the investigation of microorganisms. It turned out especially useful in industrial biotechnology for the prediction of beneficial genetic targets for rational strain optimization. Here, we describe in detail the procedure for the state-of-the-art approach of ^{13}C metabolic flux analysis comprising steady-state cultivations in a mineral salt medium with ^{13}C-labeled substrates, GC-MS measurement for labeling analysis, as well as metabolic modeling using the open-source software OpenFlux.

Key words Metabolic flux, Metabolic network, OpenFlux, Modeling, GC-MS, ^{13}C, Metabolite balancing, Tracer, In vivo pathway analysis, Central metabolism

1 Introduction

In recent years ^{13}C metabolic flux analysis has evolved as a powerful tool for in vivo determination of metabolic pathway activity. By this combined approach of metabolite balancing and the use of ^{13}C-labeled tracer substrates, the limitations and pitfalls of stoichiometric flux estimation have been overcome [1, 2]. The state-of-the-art ^{13}C metabolic flux analysis is typically based on labeling information from the amino acids of the cell protein and comprises an experimental part and a computational part [3–5]. In the experimental part, the ^{13}C labeling is introduced in the biological system by the addition of specifically labeled substrates to the cultivation broth [4]. During cell growth, the substrate is taken up and metabolized to form energy, reducing power and building blocks, e.g., amino acids for biomass formation. The distribution of the labeled carbon in the different metabolites is thereby determined by the metabolic flux distribution in the cell, as the highly specific carbon transition in each enzymatic reaction defines the fate of each carbon atom (Fig. 1a, b). By incorporation of labeled amino acids into the cell protein, the labeling pattern is stored throughout the cultivation and can be made accessible by protein hydrolysis and

Jens O. Krömer et al. (eds.), *Metabolic Flux Analysis: Methods and Protocols*, Methods in Molecular Biology, vol. 1191, DOI 10.1007/978-1-4939-1170-7_10, © Springer Science+Business Media New York 2014

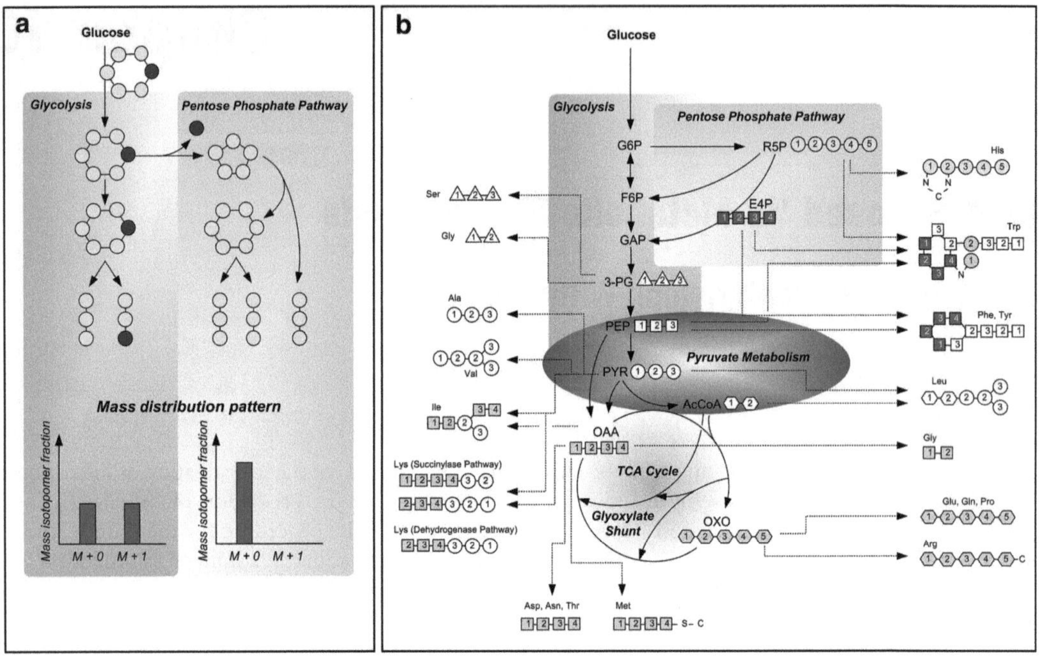

Fig. 1 Carbon transition in the central metabolism of *Corynebacterium glutamicum*. Part (**a**) illustrates the transition of the ^{13}C labeling through glycolysis and PPP, and the resulting fractional enrichment of the carbon C_3 as response to the sole use of one of these pathways, when [1-^{13}C] glucose is employed as tracer substrate. In part (**b**) the connection between proteinogenic amino acids and their respective building block(s) from the central metabolism is shown

measurement of the ^{13}C enrichment in the free amino acids [3]. Using GC-MS techniques thereby provides information on the mass isotopomer distribution of the analytes [4, 6, 7]. The labeling data are then fed into a metabolic model together with stoichiometric data from growth and product formation to compute the metabolic flux distribution. Recently, the open-source software OpenFlux has been developed which allows a straightforward model adaptation and fast calculation of the metabolic fluxes [8]. In this chapter we guide the reader through the different steps of ^{13}C-based metabolic flux analysis using the industrially relevant cell factory *Corynebacterium glutamicum*.

2 Materials

1. *Corynebacterium glutamicum* ATCC 13032.
2. Cultivation equipment (autoclave, clean bench, baffled shake flasks, shaking incubator).
3. Consumables (Eppendorf tubes, falcon tubes, Ultrafree-MC centrifugal filter devices (Amicon Bioseparations, Bedford, USA), glass vials).
4. Photometer.

5. Centrifuge for Eppendorf and falcon tubes.
6. Quadrupole gas chromatography/mass spectrometry (GC/MS) with electron impact ionization.
7. HP5-MS column (5 % phenyl-methyl-siloxane-diphenylpolysiloxane; 60 m × 0.251 mm × 0.25 μm).
8. Biochemical glucose analyzer (YSI 2700 Select, Kreienbaum, Langenfeld, Germany) or other appropriate equipment for glucose quantification (HPLC, enzyme test, etc.).
9. Incubator at 80 and 105 °C.
10. Freeze-drying equipment.
11. Personal computer including software licenses for MS Excel, Matlab 7 and Optimization Toolbox, OpenFlux, Java.
12. 99 % [1-^{13}C] glucose.
13. *N*-Methyl-*N*-*t*-butyl-dimethylsilyltrifluoroacetamide (MBDSTFA).
14. Dimethylformamide (DMF) with 0.1 % pyridine.
15. Sterile 0.9 % NaCl solution.
16. 6 M HCl.
17. 6 M NaOH.
18. Preculture agar plates: 5 g/L glucose, 5 g/L yeast extract, 10 g/L tryptone, 5 g/L NaCl, 12 g/L agar.
19. Preculture medium: 5 g/L glucose, 5 g/L yeast extract, 10 g/L tryptone, 5 g/L NaCl.
20. Solution A: 0.1 g NaCl, 5.5 mg $CaCl_2$, 20 mg $MgSO_4 \cdot H_2O$ in 55.9 mL H_2O_{dest}, sterilize by autoclaving.
21. Solution B: 0.5 g $(NH_4)_2SO_4$ in 20 mL H_2O_{dest}, sterilize by autoclaving.
22. Solution C: 1.6 g K_2HPO_4, 0.2 g KH_2PO_4 in 10 mL H_2O_{dest}, sterilize by autoclaving.
23. Solution D1: 1 g naturally labeled glucose in 10 mL H_2O_{dest}, adjust to pH 5 and sterilize by filtration (0.22 μm).
24. Solution D2: 1 g of 99 % [1-^{13}C] glucose in 10 mL H_2O_{dest}, adjust to pH 5 and sterilize by filtration (0.22 μm).
25. Solution E: 20 mg $FeSO_4 \cdot H_2O$ in 10 mL H_2O_{dest}, adjust to pH 1 and sterilize by filtration (0.22 μm).
26. Solution F: 2.5 mg biotin, 5 mg thiamin·HCl, 5 mg pantothenic acid Ca-salt in 100 mL H_2O_{dest}, sterilize by filtration (0.22 μm).
27. Solution G: 200 mg $FeCl_3 \cdot 6\ H_2O$, 200 mg $MnSO_4 \cdot H_2O$, 50 mg $ZnSO_4 \cdot H_2O$, 20 mg $CuCl_2 \cdot 2H_2O$, 20 mg $Na_2B_4O_7 \cdot 10\ H_2O$, 10 mg $(NH_4)_6Mo_7O_{24} \cdot 4\ H_2O$ in 1,000 mL H_2O_{dest}, adjust to pH 1 and sterilize by filtration (0.22 μm).
28. Solution H: 300 mg dihydroxybenzoic acid, 500 μM NaOH (6 M) in 10 mL H_2O_{dest}, sterilize by filtration (0.22 μm).

3 Methods

In this chapter, we describe the different steps required for in vivo flux determination including the experimental part with tracer cultivations, sampling and analysis of ^{13}C enrichment in proteinogenic amino acids, and the computational part with network reconstruction and flux estimation.

3.1 Tracer Cultivation with *C. glutamicum*

3.1.1 Inoculum Preparation

Inoculum preparation for the tracer cultivation includes two liquid precultures: one in complex medium and the other in the synthetic mineral salt medium that is also used for the tracer cultivation (*see* **Note 1**).

First preculture

1. Streak cells of *C. glutamicum* on agar plates and incubate for 48 h at 30 °C.
2. Pick one colony and inoculate 5 mL of complex preculture medium in a 50 mL baffled shake flask.
3. Incubate for 12 h at 30 °C on a rotary shaker (230 rpm, shaking diameter 5 cm).
4. Harvest cells by centrifugation (5 min, 8,800 × *g*, 4 °C).
5. Wash cells with 10 mL sterile 0.9 % NaCl. The washing step is required to remove complex medium residues from the preculture.
6. Resuspend cells in 2 mL sterile 0.9 % NaCl solution.
7. Determine the cell concentration of the inoculum by measuring the optical density at 660 nm (OD_{660}) in order to estimate the exact volume of inoculum that is required to start the second preculture with an OD_{660} of 0.5.

Second preculture

1. Prepare 25 mL of the synthetic mineral salt medium in 250 mL baffled shake flasks by mixing 13.9 mL solution A, 5 mL solution B, 2.5 mL solution C, 2.5 mL solution D1, 250 μL solution E, 500 μL solution F, 250 μL solution G, and 25 μL solution H.
2. Use cells from the first preculture for inoculation with an initial OD_{660} of 0.5.
3. Incubate at 30 °C on a rotary shaker (230 rpm, shaking diameter 5 cm) until cells reach an OD_{660} of ~8.
4. Harvest cells by centrifugation (5 min, 8,800 × *g*, 4 °C).
5. Wash cells with 10 mL sterile 0.9 % NaCl. The washing step is required to remove medium residues from the preculture.
6. Resuspend cells in 2 mL sterile 0.9 % NaCl solution.

7. Determine the cell concentration of the inoculum by measuring the optical density at 660 nm (OD_{660}) in order to estimate the exact volume of inoculum that is required to start the tracer cultivation with an OD_{660} of 0.05.

3.1.2 Tracer Cultivation and Cell Harvest

1. Prepare 25 mL of the synthetic mineral salt medium in 250 mL baffled shake flasks with aluminum caps by mixing 13.9 mL solution A, 5 mL solution B, 2.5 mL solution C, 2.5 mL solution D2, 250 μL solution E, 500 μL solution F, 250 μL solution G, and 25 μL solution H.
2. Inoculate the main cultivation with cells from the second preculture. Starting OD_{660}: 0.05 (*see* **Note 2**).
3. Incubate at 30 °C on a rotary shaker (230 rpm, shaking diameter 5 cm).
4. Take 1–2 mL samples within the exponential growth phase at different time points (corresponding to OD_{660} of ~5, 7, 9) (*see* **Note 3**).
5. Perform (1) determination of cell concentration by measuring the OD_{660}, (2) cell harvest by centrifugation (5 min, 8,800 × *g*, 4 °C), and (3) collection of supernatant (*see* **Note 4**).

3.2 Protein Hydrolysis

1. Carefully wash the harvested cells by complete suspension of the cell pellet in H_2O_{dest} and subsequent centrifugation (5 min, 8,800 × *g*, 4 °C).
2. Discard supernatant and repeat the washing step.
3. Resuspend cells in 1 mL H_2O and measure the OD_{660} to determine the cell concentration. Transfer an appropriate volume containing about 1 mg cell dry mass (CDM) into a centrifuge tube (*see* **Note 4**).
4. Separate cells by centrifugation (5 min, 8,800 × *g*, 4 °C) and discard supernatant.
5. Solve cells in 50 μL of 6 M HCl, carefully fix the lid of the Eppendorf tube with a clamp, and incubate for 24 h at 105 °C to hydrolyze the cell protein.
6. Wait until the hydrolyzate has cooled down to room temperature and neutralize with 6 M NaOH. This is to avoid corrosion processes caused by HCl during the freeze-drying process.
7. Remove insoluble compounds by centrifugation (1 min, 8,800 × *g*, 4 °C) using the Ultrafree-MC centrifugal filter devices.
8. Transfer the clear filtrate to a 1 mL glass vial and freeze at –70 °C in an upright position to assure that the sample will remain at the bottom of the vial.
9. Freeze-dry the frozen filtrate.

3.3 GC/MS Analysis

GC/MS analysis of the amino acids requires derivatization of the functional groups using silylation agents such as MBDSTFA (*N*-methyl-*N*-*tert*-butyldimethylsilyl-trifluoroacetamide). Ionization of MBDSTFA derivatives in electron impact ionization typically leads to [M-57], [M-85], and [M-159] fragments (Fig. 2a), whereby the [M-57] contains the complete carbon backbone of the amino acid and is hence the most valuable for metabolic flux calculation (*see* **Note 5**).

3.3.1 Derivatization

1. Add 50 μL DMF containing 0.1 % pyridine and 50 μL MBDSTFA on the freeze-dried sample.
2. Carefully cap the glass vial and incubate for 30 min at 80 °C in an oven (do not use a water bath as contamination of the sample with water might interfere with the derivatization reaction).
3. Transfer the sample into a 100 μL GC-micro insert. Take care not to transfer insoluble matter to avoid instrument damages.

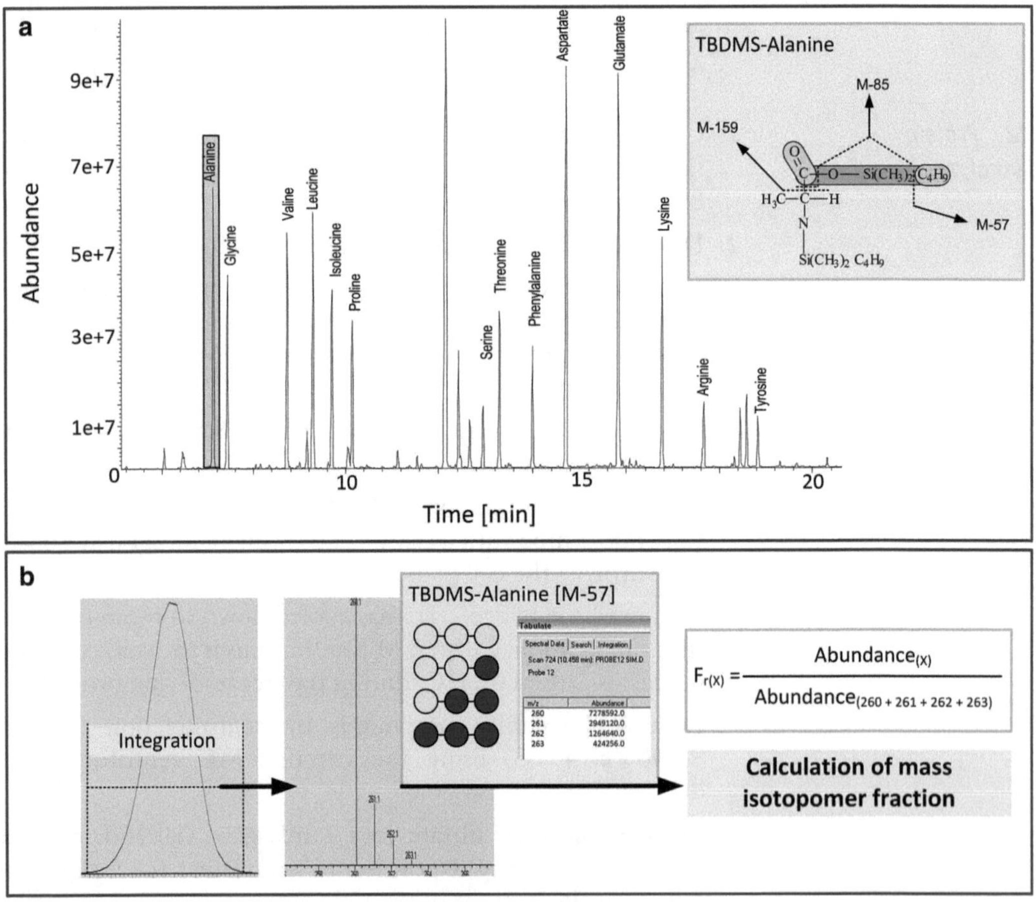

Fig. 2 GC/MS spectrum of TBDMS derivatives of proteinogenic amino acids and illustration of typical fragmentation pattern for TBDMS amino acids (**a**), and procedure for GC/MS data processing for determination of the ^{13}C labeling exemplified for the [M-57] fragment of TBDMS alanine (**b**)

3.3.2 GC/MS Measurement

The measurement procedure is based on a protocol for the determination of the ^{13}C labeling enrichment in TBDMS-derivatized amino acids in cell extracts [9]. A corresponding GC/MS spectrum is shown in Fig. 2a.

1. Instrument settings: 0.5 μL sample injection, HP5MS capillary column, electron impact ionization at 70 eV, quadrupole detector, 1 mL/min helium flow, operation temperature of 320 °C for inlet, 320 °C for interface, and 325 °C for quadrupole.
2. Temperature profile for measurement: 120 °C for 2 min, 8 °C/min up to 180 °C, 10 °C/min up to 325 °C, and hold at 325 °C for 5 min.
3. Perform measurements in scan mode (to check for sample impurities and potential signal interference), and selected ion monitoring (SIM) mode (for determination of the exact mass isotopomer distribution). For the latter, select masses for detection of the [M-57] fragment of each amino acid. For an amino acid with n carbons, $n+1$ ions should be measured referring to the non-labeled (m0), singly labeled (m1), or higher labeled molecule.
4. Data extraction from SIM mass spectra is illustrated in Fig. 2b.

3.4 Estimation of In Vivo Carbon Fluxes

3.4.1 Evaluation of Data Quality

1. Compare the labeling data of the different biomass samples that have been taken throughout the tracer cultivation at different time points (*see* **Note 3**).
   - If the labeling pattern is identical for all the samples, proceed with flux estimation.
   - If the labeling data are inconstant, repeat tracer cultivation (*see* **Note 3**).

3.4.2 Network Reconstruction and Flux Estimation

The network is based on the central metabolism of *C. glutamicum* that comprises glycolysis, pentose phosphate pathway, TCA cycle, and anaplerotic reactions. Beyond this, routes for amino acid synthesis and for biomass formation are implemented. The open-source software OpenFlux is used for metabolic modeling (*see* **Note 6** and Fig. 3).

1. Download and install the software from http://sourceforge.net/projects/openflux.
2. Write down the network stoichiometry in an Excel sheet, implement carbon transition, and define reaction type and basis as exemplified in Fig. 4 for the reaction of pyruvate dehydrogenase.
3. Implement the stoichiometric data on biomass and product formation to account for the carbon withdrawal from the central pathway intermediates (*see* **Note 4**).

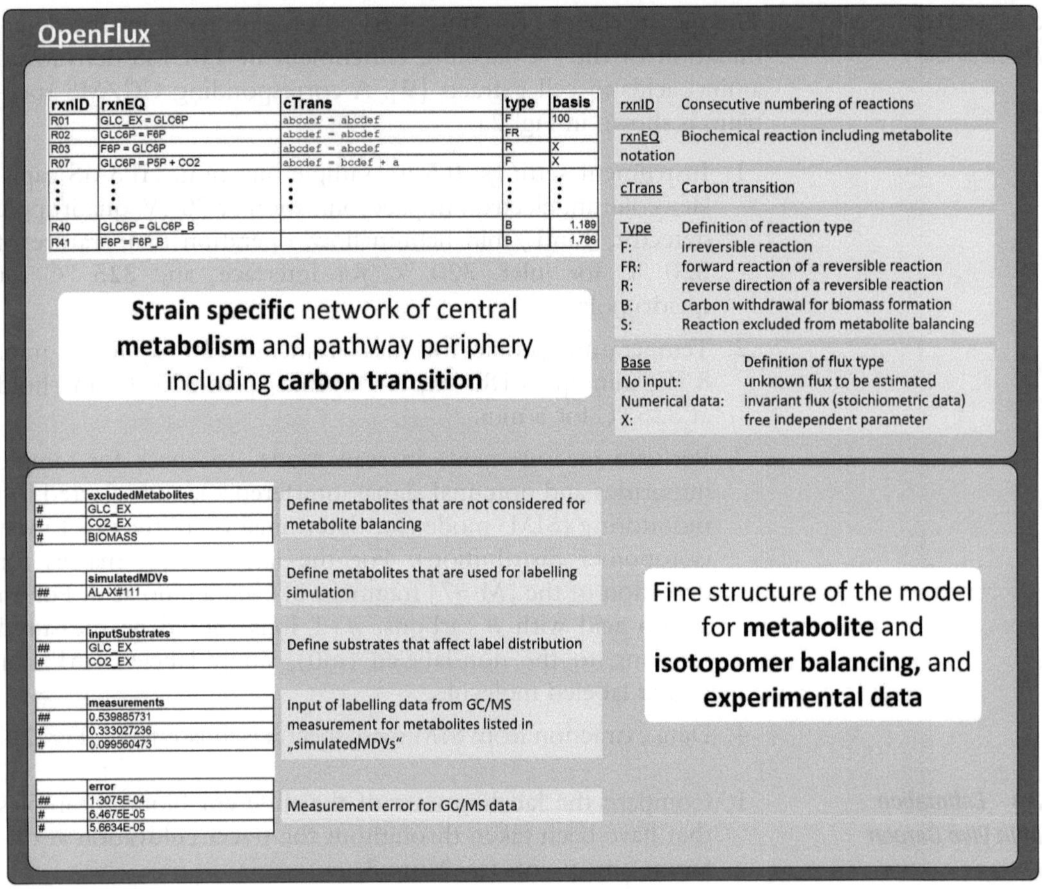

Fig. 3 Simplified scheme for model definition in OpenFlux. Programming in MS Excel is separated into two parts, whereby the *upper part* is dedicated to a user-defined network definition of the desired organism including reaction equations and carbon atom transitions. The *lower part* further narrows the structure and scope of the model by defining the metabolites that are used for metabolite and isotopomer balancing, and additionally contains the experimental data

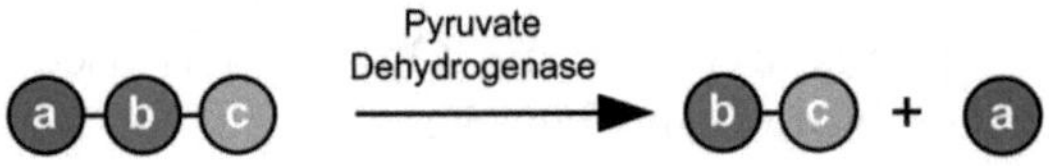

rxnEQ	cTrans	type	basis
PYR =ACCOA + CO2	abc = bc + a	F	

Fig. 4 Procedure for programming of the metabolic network exemplified for the reaction of pyruvate dehydrogenase that converts pyruvate to acetyl-CoA and CO_2. Type (F) and basis (no input) define this reaction as irreversible and unknown flux that needs to be estimated in the metabolic simulation

4. Implement labeling data from GC/MS measurement referring to the non-labeled (m0), singly labeled (m1), and doubly labeled (m2) molecule.
5. Perform network validation and run the flux estimation as given in the OpenFlux manual.

4 Notes

1. The metabolic flux studies are performed in a synthetic mineral salt medium with 99 % [1-^{13}C]-labeled glucose as sole carbon source. Using complex medium compounds would introduce other carbon sources into the system with unknown effect on the labeling pattern of the cell protein and hence interfere with the studied substrate.
2. The inoculum size should be low to minimize the effect of non-labeled biomass on the final labeling pattern of the proteinogenic amino acids. Similar to **Note 1**, the inoculum can be considered as complex compound that interferes with the tracer substrate. For the case that the inoculum will contribute to less than 1 % of the finally harvested biomass, its effect can be neglected for the flux studies.
3. For the biomass-based ^{13}C metabolic flux analysis it is crucial that the cells maintain a metabolic and isotopomer steady state throughout the whole tracer cultivation to ensure that the harvested biomass sample is representative for a single flux state [10]. Any metabolic changes throughout the cultivation will result in a changed labeling pattern of the amino acids, which will then be stored in the cell protein. If this is the case in the tracer cultivation, the ^{13}C enrichment will represent a mixture of different physiological states and will not be suitable for this flux approach. Analysis of the labeling pattern of biomass samples from different cultivation times can be used to validate the metabolic and isotopomer steady states [10].
4. The model for flux estimation includes experimental data for (1) the amino acid labeling and (2) the formation of biomass and secreted products. The latter can be obtained by determination of the biomass yield ($Y_{X/S}$) and product yields ($Y_{P/S}$). The amount of cell dry mass (CDM) can thereby be estimated from the optical density (OD_{660}) using a previously determined correlation factor for *C. glutamicum* of 0.255 g CDM per OD_{660} [11]. It should be noted that this factor depends on the photometer that is used. Glucose and eventually secreted products in the culture supernatant can be determined by a biochemical analyzer or other appropriate analytical methods (HPLC, enzyme test, etc.). From the biomass yield, the anabolic demand for the carbon precursors from the central metabolism can be calculated [12].

5. For metabolic modeling, the information on the labeling pattern of the carbon backbone of the amino acid is essential. The molecule structure of the GC/MS fragments of the TBDMS amino acids is, however, complex and contains additional constituents, e.g., Si, O, N, H, S, and C from the derivatization residue (Fig. 2a). To account for this, the raw data from GC/MS measurement have to be corrected for the natural isotopes of these elements to obtain the labeling information of the carbon skeleton of the analyte itself. Correction for isotopes can be carried out using isotope correction matrices and mathematical software [13]. The OpenFlux software package already contains an appropriate function in the model and predefined correction matrices for standard analytes [8].
6. OpenFlux is a user-friendly modeling tool to perform ^{13}C metabolic flux analysis and allows straightforward adaptation of the metabolic network structure [8]. The detailed manual thoroughly guides the user through the different programming steps, so we skipped a widespread introduction into the software. When using the software, we strongly recommend a careful study of the manual.

References

1. Becker J, Wittmann C (2012) Pathways at work—metabolic flux analysis of the industrial cell factory *Corynebacterium glutamicum*. In: Yukawa H, Masayuki I (eds) Corynebacterium glutamicum, 1st edn. Springer, Berlin
2. Wiechert W (2001) ^{13}C metabolic flux analysis. Metab Eng 3(3):195–206. doi:10.1006/mben.2001.0187
3. Kelleher JK (2001) Flux estimations using isotopic tracers: common ground for metabolic physiology and metabolic engineering. Metab Eng 3:100–110. doi:10.1006/mben.2001.0185
4. Wittmann C (2007) Fluxome analysis using GC-MS. Microb Cell Fact 6:6. doi:10.1186/1475-2859-6-6
5. Kohlstedt M, Becker J, Wittmann C (2010) Metabolic fluxes and beyond-systems biology understanding and engineering of microbial metabolism. Appl Microbiol Biotechnol 88(5):1065–1075. doi:10.1007/s00253-010-2854-2
6. Christensen B, Nielsen J (1999) Isotopomer analysis using GC-MS. Metab Eng 1(4):282–290. doi:10.1006/mben.1999.0117
7. Wittmann C (2002) Metabolic flux analysis using mass spectrometry. Adv Biochem Eng Biotechnol 74:39–64
8. Quek L-E, Wittmann C, Nielsen LK et al (2009) OpenFLUX: efficient modelling software for ^{13}C-based metabolic flux analysis. Microb Cell Fact 8:25. doi:10.1186/1475-2859-8-25
9. Wittmann C, Hans M, Heinzle E (2002) In vivo analysis of intracellular amino acid labelings by GC/MS. Anal Biochem 307(2):379–382
10. Becker J, Klopprogge C, Wittmann C (2008) Metabolic responses to pyruvate kinase deletion in lysine producing *Corynebacterium glutamicum*. Microb Cell Fact 7:8. doi:10.1186/1475-2859-7-8
11. Becker J, Klopprogge C, Schröder H et al (2009) Metabolic engineering of the tricarboxylic acid cycle for improved lysine production by *Corynebacterium glutamicum*. Appl Environ Microbiol 75(24):7866–7869. doi:10.1128/AEM.01942-09
12. Kind S, Becker J, Wittmann C (2013) Increased lysine production by flux coupling of tricarboxylic acid cycle and biosynthetic pathway—metabolic engineering of succinyl-CoA availability in *Corynebacterium glutamicum*. Metab Eng 15:184–195
13. van Winden WA, Wittmann C, Heinzle E et al (2002) Correcting mass isotopomer distributions for naturally occurring isotopes. Biotechnol Bioeng 80(4):477–479. doi:10.1002/bit.10393

Chapter 11

Dynamic Analysis of CO_2 Labeling and Cell Respiration Using Membrane-Inlet Mass Spectrometry

Tae Hoon Yang

Abstract

Here, we introduce a mass spectrometry-based analytical method and relevant technical details for dynamic cell respiration and CO_2 labeling analysis. Such measurements can be utilized as additional information and constraints for model-based ^{13}C metabolic flux analysis. Dissolved dynamics of oxygen consumption and CO_2 mass isotopomer evolution from ^{13}C-labeled tracer substrates through different cellular processes can be precisely measured on-line using a miniaturized reactor system equipped with a membrane-inlet mass spectrometer. The corresponding specific rates of physiologically relevant gases and CO_2 mass isotopomers can be quantified within a short-term range based on the liquid-phase dynamics of dissolved fermentation gases.

Key words Dissolved gas analysis, Membrane-inlet mass spectrometry, MIMS, Miniaturized reactor system, CO_2 mass isotopomer, Fermentation exhaust analysis, Metabolic flux analysis

1 Introduction

On-line fermentation gas analysis is of general interest as any biological systems consume and evolve gaseous molecules such as oxygen and carbon dioxide in aerobic fermentation processes [1]. The parameters gained from on-line gas analysis such as respiratory quotient (*RQ*), specific rates of O_2 uptake (q_{O_2}) and CO_2 evolution (q_{CO_2}) are not only important for on-line process control but also key physiological values for constraints-based metabolic flux analysis including ^{13}C-based approaches [2]. For instance, any stoichiometric models contain different intracellular fluxes that generate/consume reducing equivalents and CO_2. Thus, the parameters obtained from on-line gas analysis constrain the physiologically allowed metabolic flux ranges over redox and elemental balances. For example, the carbon balance in the below

Jens O. Krömer et al. (eds.), *Metabolic Flux Analysis: Methods and Protocols*, Methods in Molecular Biology, vol. 1191, DOI 10.1007/978-1-4939-1170-7_11,

equation contains fluxes of intracellular decarboxylation and carboxylation reactions and is constrained by the total product effluxes including CO_2 [2]:

$$\sum_{i=1}^{n} n_{S,i} \nu_{S,i} - \sum_{j=1}^{m} n_{p,j} \nu_{P,j} - \left(\sum_{k=1}^{p} \nu_{\text{decarb},i} - \sum_{l=1}^{q} \nu_{\text{carb},l} \right) = 0 \qquad (1)$$

Here, ν_S and ν_P denote the effluxes of substrates and products, respectively, typically normalized by substrate uptake q_S, and n_S and n_P the number of carbons in the corresponding compounds. The net CO_2 efflux on glucose (q_{CO_2} / q_S) equals the difference between the decarboxylation (ν_{decarb}) and the carboxylation (ν_{carb}) fluxes. A similar constrained can be made based on the degree of reduction balance [3], involving q_{O_2}. Also, the redox balance during oxidative phosphorylation can be introduced such that:

$$\frac{q_{O_2}}{q_S} - \left(\frac{1}{2} \sum_{i=1}^{n} \nu_{\text{e.prod},i} - \frac{1}{2} \sum_{i=j}^{m} \nu_{\text{e.cons},i} \right) = 0 \text{ with } RQ = \frac{q_{CO_2}}{q_{O_2}}, \qquad (2)$$

where $\nu_{\text{e.prod}}$ and $\nu_{\text{e.cons}}$ are the intracellular fluxes producing and consuming reducing equivalents such as NADH, NADPH, and quinones. For ^{13}C-based metabolic flux analysis, those constraints can be applied to the computation of feasible range of fluxes (solution space) prior to flux estimation involving ^{13}C-labeling measurements: this is typically done by solving a constrained linear or quadratic programming problem for the given stoichiometry [4].

Another useful information that can be obtained from fermentation gas analysis is CO_2 labeling. The ubiquitous presence of CO_2 in biological systems under aerobic as well as anaerobic conditions and both during growth and non-growth phases grants a wide spectrum of potential applications to itself. The theoretical concept of the so-called respirometric ^{13}C flux analysis that relies solely on ^{13}C-labeling of CO_2 evolving from multiple ^{13}C-tracers has been introduced by Yang et al. [5] and applied to study the entire central metabolic pathways in vivo [6, 7]. CO_2-labeling carries information on instantaneous metabolic phenotypes given at the time of measurement because CO_2 does not accumulate in culture broth. For instance, ^{13}C-labeling patterns of CO_2 and intracellular metabolites can be applied to dynamic profiling of non-steady-state bioprocesses such as multi-phase processes.

Typically, nuclear magnetic resonance spectroscopy or gas chromatography/mass spectrometry is employed as the analytical tools of ^{13}C-based flux analysis for isotopic labeling measurements. To the high precision analysis of exhaust or dissolved gases from a bioreactor, mass spectrometry (MS) techniques have been applied successfully [1, 8, 9]. In particular, Heinzle et al. [1] has extensively investigated both theoretical and practical aspects of mass spectrometric fermentation exhaust gas analysis, and Yang et al. [6] has developed a dynamic method to rapidly quantify cell respiration parameters and CO_2-labeling using membrane-inlet MS for dissolved gas analysis. Exhaust gas analysis can be conducted by directly introducing the fermentation outlet gas stream into MS. In comparison to this, membrane-inlet MS relies on a membrane probe which is directly immersed in the cultivation media and connected to a mass spectrometer for sample introduction. The membrane-inlet MS is widely used in various scientific areas due to its high precision, speed, and selectivity for dissolved gas analysis.

Indeed, fermentation off-gas analysis has been well studied and established as one of the standard tools for fermentation process monitoring and control. In contrast, in vivo cell respiration parameters including ^{13}C-labeling of CO_2 as a metabolic product have rarely been applied to ^{13}C flux analysis. This is often because ^{13}C experiments are frequently carried out in small-scale shake flasks without cell respiration measurements and CO_2-labeling is guessed based on material balancing during flux estimation. Hence, an analytical technique that can be applied to the acquisition of cell respiration and CO_2-labeling in vivo from small-scale ^{13}C-cultures can be desirable for more accurate ^{13}C-based flux analysis. In this regard, the current chapter aims at describing the relevant modeling approaches and experimental procedures for on-line cell respiration and CO_2 labeling analysis. Different from conventional fermentation exhaust gas analysis methods, the approach relies on a short-term analysis of dissolved gas dynamics using a miniaturized reactor system coupled with membrane-inlet MS.

To begin with, basic material balances that are applied to fermentation exhaust gas analysis need to be understood. These basic material balances are also applied more broadly to fermentation analysis.

1.1 Basic Material Balance

Let us define a well-mixed stirred tank reactor with a constant headspace and liquid-phase volume of V_G and V_L, respectively. The reactor is continuously aerated with a molar gas flow rate of G_0 for inlet stream and G_1 for outlet stream. In this case, the following balance equations can be set up for gas and liquid phase, respectively [1].

$$\left(\frac{pV_G}{RT}\right)\frac{dy_1}{dt} = G_0 y_0 - G_1 y_1 - k_L a\left(c_L^* - c_L\right)V_L \;\{\text{gas} - \text{phase}\} \quad (3)$$

$$V_L \frac{dc_L}{dt} = k_L a\left(c_L^* - c_L\right)V_L + rV_L \;\{\text{liquid} - \text{phase}\} \quad (4)$$

Here, p is the total pressure, R the gas constant, T the temperature, y the mole fraction in the gas phase c_L the liquid-phase concentration, c_L^* the dissolved gas concentration at equilibrium with the partial pressure of a certain gaseous species, k_La the mass transfer coefficient, and r the reaction rate.

For a system with volume change, e.g., a fed-batch reactor, one can apply the product rule, and the left-hand side of the each balance equation may become

$$\left(\frac{p}{RT}\right)\left(V_{\mathrm{G}}\frac{dy_1}{dt}+y_1\frac{dV_{\mathrm{G}}}{dt}\right) \text{ and } \left(V_{\mathrm{L}}\frac{dc_{\mathrm{L}}}{dt}+c_{\mathrm{L}}\frac{dV_{\mathrm{L}}}{dt}\right), \text{ respectively.}$$

However, the term dV/dt disappears in the below Eq. 7 due to inert gas balance under the liquid- and gas-phase steady-state condition.

Now, combining the above two balance Eqs. 3 and 4 and solving for the reaction rate yields the following equation:

$$rV_{\mathrm{L}} = -G_0y_0 + G_1y_1 + \left(\frac{pV_{\mathrm{G}}}{RT}\right)\frac{dy_1}{dt} + V_{\mathrm{L}}\frac{dc_{\mathrm{L}}}{dt} \quad (5)$$

For an inert gas, e.g., N_2 or Ar that is neither consumed nor produced within the system, i.e., $r=0$, the molar flow rate in outlet gas stream is given by:

$$G_1 = G_0\frac{y_{0,\mathrm{inert}}}{y_{1,\mathrm{inert}}} - \frac{1}{y_{1,\mathrm{inert}}}\left\{\left(\frac{pV_{\mathrm{G}}}{RT}\right)\frac{dy_1}{dt} - V_{\mathrm{L}}\frac{dc_{\mathrm{L}}}{dt}\right\}. \quad (6)$$

Usually, the accumulation terms of dy_1/dt and dc_{L}/dt are negligible for inert gases (liquid- and gas-phase steady-state), which holds also valid for oxygen. Thus, we get the following equation for oxygen reaction rate by combining Eqs. 5 and 6 under a steady-state condition.

$$r_{\mathrm{O}_2}V_{\mathrm{L}} = -G_0y_{0,\mathrm{O}_2} + G_0y_{1,\mathrm{O}_2}\frac{y_{0,\mathrm{inert}}}{y_{1,\mathrm{inert}}} \quad (7)$$

For CO_2 which undergoes liquid-phase reactions (*see* **Note 1**), one may need to take this into account. Using a pH-dependent conversion factor $K_{\mathrm{Conv}}^{\mathrm{pH}}$ derived from the equilibrium constants between CO_2 species (*see* **Note 2**), the liquid-phase mass balance Eq. 4 can be formulated for CO_2 such that:

$$K_{\mathrm{Conv}}^{\mathrm{pH}}\cdot V_{\mathrm{L}}\frac{d[\mathrm{CO}_2]_{aq}}{dt} = k_La_{\mathrm{CO}_2}\left([\mathrm{CO}_2]_{aq}^* - [\mathrm{CO}_2]_{aq}\right)V_{\mathrm{L}} + r_{\mathrm{CO}_2}V_{\mathrm{L}}. \quad (8)$$

The constant $K_{\mathrm{Conv}}^{\mathrm{pH}}$ also depends on the ionic strength [10], and its estimation is explained in Subheading 3.3. The liquid phase accumulation of Eq. 8 may not be negligible in Eq. 5, e.g., at a higher pH, and one may need to take this account when calculating the rate.

$$r_{CO_2} V_L = -G_0 y_{0,CO_2} + G_1 y_{1,CO_2} \frac{y_{0,\text{inert}}}{y_{1,\text{inert}}} + K_{\text{Conv}}^{\text{pH}} \times V_L \frac{d[CO_2]_{aq}}{dt} \quad (9)$$

If $[CO_2]_{aq}$ is not measured directly, one can assume a rapid equilibrium between gas and liquid phases, i.e., $[CO_2]_{aq}$ equals the equilibrium concentration $[CO_2]^*_{aq}$ which can be calculated from measured CO_2 partial pressure in the off-gas, e.g., using Henry's law or Bunsen's solubility coefficient. The time-derivative $d[CO_2]_{aq}/dt$ can be computed using finite difference or spline-based differentiation.

In Eqs. 7 and 9, r_{O_2} and r_{CO_2} equals the oxygen consumption and CO_2 evolution rate, respectively. The corresponding specific rates of q_{O_2} and q_{CO_2} are obtained by dividing the rates with cell concentration c_X, e.g., $q_{O_2} = r_{O_2} / c_X$.

As an alternative to the direct exhaust gas analysis, membrane-inlet MS can be employed, e.g., when a non-negligible time-delay is expected depending on residence time of exhaust gas in head-space region or on gas flow rate in transfer line [11]. In this case, one can measure dissolved gas concentrations to determine cell respiration rates by solely relying on Eq. 4, which however necessitates knowledge in mass transfer coefficient $k_L a$. If the liquid-phase accumulation is negligible, the rate of production or consumption equals the mass transfer rate, i.e.,

$$r = -k_L a\left(c_L^* - c_L\right) \text{ with } \frac{dc_L}{dt} \approx 0. \quad (10)$$

When dc_L/dt is not negligible, one may compute this using finite difference or spline-based differentiation, as explained above. Under this non-steady-state condition, c_L becomes time-variant and so do CO_2 isotopomer distributions. Hereto, one can implement the following short-term dynamic cell respiration and CO_2 labeling analysis using a miniaturized reactor system coupled with membrane-inlet MS.

2 Materials

To carry out dynamic measurements of cell respiration including dissolved CO_2 mass isotopomers, a system of miniaturized reactors coupled with a multichanneled membrane-inlet MS can be employed. Such a miniaturized reactor system has been introduced elsewhere [6, 7, 10]. Cells are incubated with a buffered minimal medium containing ^{13}C-labeled substrates in the reactor system, and dynamics of dissolved gaseous species are measured using

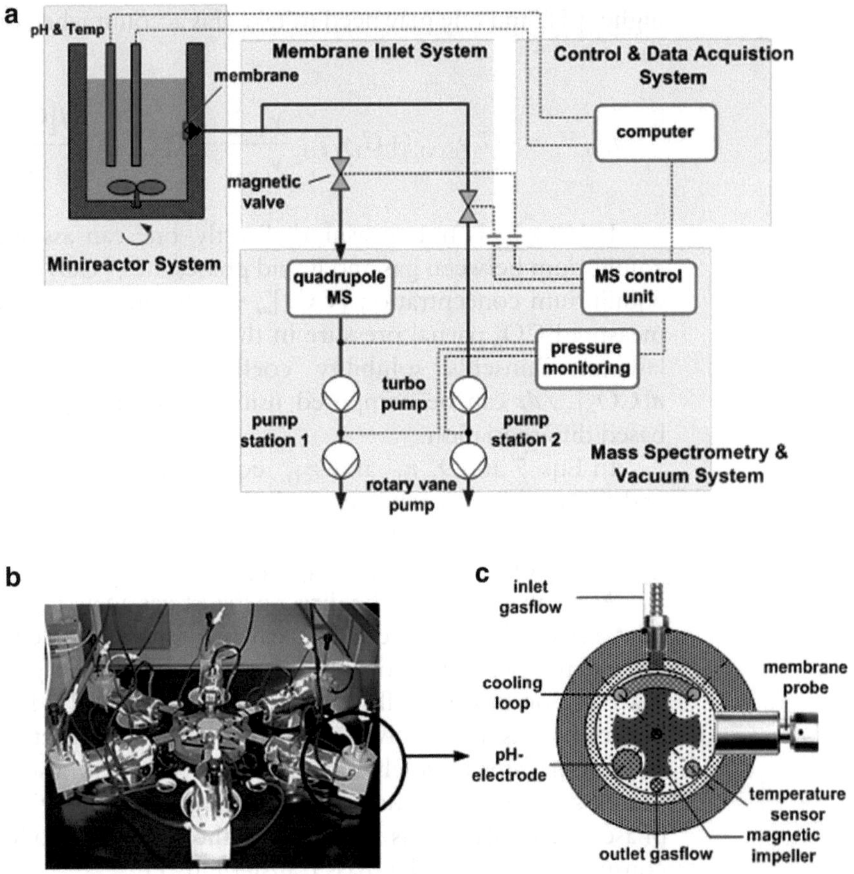

Fig. 1 Measurement device equipped with (1) miniaturized reactor system, (2) membrane inlet system, (3) mass spectrometer and vacuum system, and (4) control and data acquisition system (**a**); Miniaturized reactor system with 6 reactors coupled with multichanneled membrane-inlet system (**b**); Cross-sectional view (**b**) of a single reactor unit with equipments (**c**). Reproduced from [7] with permission from Elsevier

membrane-inlet mass spectrometry. The measurement system developed by Yang et al. [7] was specially designed for the so-called respirometric ^{13}C flux analysis, involving multiple reactors to collect CO_2 labeling information evolving from multiple ^{13}C-tracer substrates.

2.1 Measurement Device

As shown in Fig. 1a, the measurement system consists of four major units of (1) minireactor system, (2) membrane inlet system, (3) MS & vacuum system, and (4) control & data acquisition system. The details of instrumental design and mechanical set-up can be found in Yang et al. [7].

1. *Miniaturized reactor system.* Each unit of multiple miniaturized stirred tank reactors that is connected to a quadrupole MS via multichanneled membrane inlet system (Fig. 1b). Each reactor is equipped with a cooling loop, a pH-electrode,

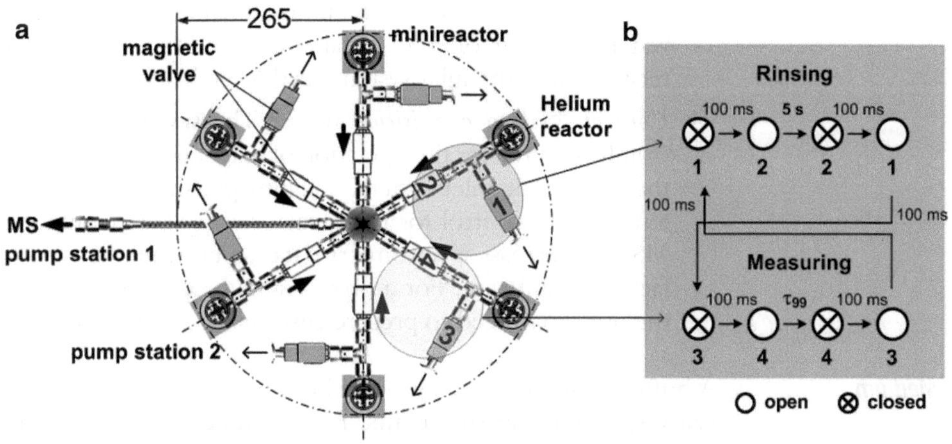

Fig. 2 Magnetic valve-controlled, multichannel membrane inlet system (**a**): The magnetic valves are electrically activated to selectively introduce dissolved gas molecules into ion source of a quadrupole MS from a reactor (*bold arrow*: into MS with pump station 1 through the *white-colored valves*) and to evacuate and hold vacuum in the inlet tube line before and after measurements (*arrow*: into pump station 2 through the *gray-colored valves*). Operating sequence of valves for rinsing and measuring processes (**b**). τ_{99} denotes the time when the MS signal reaches 99 % of its final value. Reproduced from [7] with permission from Elsevier

a temperature sensor, a magnetic impeller, and headspace and bottom aeration openings, as shown as a cross-sectional view in Fig. 1c. A recommended reactor size is about 20 mL with a working volume of 10 mL.

2. *Membrane-inlet system*. To selectively introduce dissolved gas molecules into the ion source of MS, a magnetic valve-controlled, multichanneled membrane inlet system can be applied. Yang et al. [7] introduced a 6-channel inlet system (Fig. 1b and Fig. 2a) equipped with a nonporous perfluoroalkoxyalkane membrane (25 μm thickness & 23.3 mm^2 contact area with the liquid), which has a great selective permeability for oxygen and carbon dioxide. Each channel of the membrane inlet system is equipped with vacuum fittings and two magnetic valves. *See* Subheading 3.1 for a detailed example of membrane-inlet system operation.
3. *Mass spectrometry & vacuum system*. As an adequate mass spectrometer, a quadrupole system equipped with a 90° axial secondary electron multiplier and a gas tight electron impact axial beam ion source is recommended. In addition, a vacuum system consisting of two separate turbo pump stations is desired. One pump station is need to maintain ultra high vacuum in the MS recipient (pump station 1 in Fig. 1a) and another to support a high vacuum in the inlet system as well as a fast vacuum recovery in the inlet during the magnetic valve operation. A vacuum gauge is also necessary to monitor vacuum pressure in the inlet system and protect the system from potential damages such as

membrane breakage. To minimize memory effects caused by water adhesion on the metal surface of the system, the MS recipient and the inlet system need to be heated to 120 °C.

4. *Control & data acquisition system.* The system control unit is mainly required for the operation of magnetic valves imbedded in the membrane-inlet system, vacuum pressure monitoring, and temperature control for each reactor as well as membrane-inlet MS. Online mass spectrometry data can be acquired using manufacturer's software. For a proper system operation and control, the users may need to prepare customized software tools.

2.2 Medium

A buffered minimal medium can be applied for MS calibration and cell respiration measurements. For example, the medium applied to lysine-producing *C. glutamicum* [6, 7] was composed of 8 stock solutions. 1 L medium contained: (a) 0.055 g $CaCl_2 \times 2H_2O$, 0.2 g $MgSO_4 \times 7H_2O$, 1 g NaCl in 579.5 mL deionized water; (b) 2 g K_2HPO_4, 0.25 g KH_2PO_4 in 10 mL deionized water; (c) 16 g K_2HPO_4, 2 g KH_2PO_4 in 280 mL deionized water and adjusted to pH 6.5 with HCl; (d) 5 g $(NH_4)_2SO_4$ in 100 mL deionized water; (e) 0.5 mg biotin, 1 mg thiamine·HCl in 20 mL deionized water; (f) 5 mL of 2 g/L $FeSO_4 \times 7H_2O$ (20 mg $FeSO_4 \times 7H_2O$ in 10 mL deionized water, adjusted to pH 1.0 with HCl); (g) 5 mL of 100× trace elements [12] and (h) 30 mg 3,4-dihydroxybenzoic acid in 0.5 mL deionized water. The solution (h) was prepared by adding 50 mL/L 4 M NaOH. The final pH of the example medium is 6.55 with an ionic strength 0.232 M at 30 °C. The so-called salting-out coefficient of the calibration medium calculated using the ionic composition of the medium (ionic strength) is 0.044 for O_2 and 0.031 for CO_2 [10]. Hence, the solubility of O_2 is decreased 1.044-fold, and that of CO_2 1.032-fold in comparison to pure water [10]. An extensive study regarding gas solubility in fermentation broth and salting-out effect can be found elsewhere [13].

3 Methods

For the dynamic measurements of cell respiration and evolution of CO_2 mass isotopomers from ^{13}C-tracers, cells from ^{13}C-cultures can be incubated in the miniaturized reactor system for a certain time period within which both physical and physiological steady-states are maintained. From the reactors, in vivo dynamics of dissolved gaseous species are acquired online to get rates of CO_2 evolution and O_2 uptake. To properly acquire CO_2 labeling data and cell respiration rates from the short-term dynamic experiments using a membrane-inlet MS, relevant theories as well as experimental techniques such as inlet system operation and liquid-phase material balances have to be understood in advance.

3.1 Membrane-Inlet System Operation

For a single culture, one can continuously monitor gas molecules introduced into MS through membrane-inlet from one measurement reactor cell. In case simultaneous CO_2 labeling measurements from multiple reactors are desired, this can be done best using a multi-channeled inlet system with an appropriate switching procedure that gives minimal carryover from other inlet channels. Hereto, Yang et al. [7] designed a measurement method consisting of measuring and rinsing cycles, e.g., using helium. An example configuration is shown in Fig. 2a: one of the reactors is continuously aerated with helium, and the inlet system is rinsed between each measurement cycle using the helium-aerated reactor.

Hence, the multiple reactor operation requires valve opening and closing events, which can cause vacuum pressure fluctuation. Therefore, a sufficient waiting time for vacuum recovery is necessary between each valve manipulation event. About 100 ms is reported to be sufficient when employing two turbo pumps [7], which, however, depends on the system dimension and needs to be determined experimentally. In addition to this, the time at which signals reach steady-state (τ_{99}) during a measurement event has to be determined. Yang et al. [7] found that the signal dynamics of the gaseous molecules immediately after activating an inlet towards MS exactly follow the first-order dynamics, i.e.,

$$\frac{dI_{m/z}}{dt} = \frac{1}{\tau}\left(I_{m/z} - I_{m/z}^{\text{asy}}\right). \tag{11}$$

Here, $I_{m/z}$ the ion current signal at an arbitrary m/z, $I_{m/z}^{\text{asy}}$ the corresponding steady-state signal expected for $t = \infty$ (asymptote), and τ the characteristic time constant given for the dynamics. Further, when the corresponding background signal ($I_{m/z}^{b}$) is determined, i.e., the asymptote extrapolated from the first-order dynamics during helium rinsing (decreasing signals for other gaseous species), the time constant τ_{99} at which a signal of a gaseous species ($I_{m/z}$) reaches 99 % of its maximum ($I_{m/z}^{\text{asy}}$) from its lowest signal ($I_{m/z}^{b}$) can be calculated using the following relation.

$$\tau_{99} = -\log\left(\frac{0.01 \cdot I_{m/z}^{\text{asy}}}{I_{m/z}^{\text{asy}} - I_{m/z}^{b}}\right) \times \tau \tag{12}$$

Yang et al. [7] observed that τ_{99} is not dependent on dissolved gas concentrations but rather ion-specific, e.g., 21.3 ± 0.3 s for $^{28}N_2^+$, 22.6 ± 0.9 s for $^{32}O_2^+$, 27.3 ± 0.3 s for $^{44}CO_2^+$, 27.9 ± 0.4 s for $^{45}CO_2^+$ and 28.2 ± 0.4 s for $^{46}CO_2^+$ for atmospheric air.

After determining τ_{99}, the following steps of sequential valve operation can be implemented to acquire signals from multiple reactors (Fig. 2b):

1. *Stand-by state (10 min prior to initiation)*: open all the valves connecting the measurement reactors with pump station 2 to maintain pre-vacuum in the inlet system. Keep valve 1, connecting the helium reactor with pump station 2, closed, while valve 2, connecting the helium reactor with pump station 1, is kept open.
2. *Measurement event*: Close valve 2 and open valve 1; open the valve connecting one of the measurement reactor with pump station 1 (valve 3) and close the valve connecting the reactor with pump station 2 (valve 4). Wait a few 10 s (τ_{99} in Eq. 12) for signal steady-state and collect signals for a few seconds.
3. *Helium-rinsing event*: Close valve 3 and open valve 4; open valve 2 and close valve 1. Wait for a few seconds for a sufficient inlet rinse prior to activating the next measurement event. 5 seconds of rising time may be sufficient, but this depends on dimension of inlet system and pump station efficiency.

3.2 Dynamic MS Calibration

Prior to in vivo dynamic experiments, membrane-inlet MS has to be calibrated for dissolved gas species such as CO_2 and O_2. Hereto, Yang et al. (2003) developed a rigorous method of the so-called dynamic calibration [10]. Conventional calibration for the dissolved gas analysis requires attaining steady states of 0 % and 100 % saturation, which can be very time-expensive and inaccurate. In contrast, the method of dynamic calibration has the advantage that equilibrium does not need to be attained completely for the calibration. In this dynamic method, one just relies on a short-term measurement using different calibration gases. For each dissolved gaseous species, the measured signal has a linear correlation with the dissolved gas concentration (c_L) within the dynamic range of MS such that

$$c_L = aI_{m/z} + b. \tag{13}$$

In addition, the dynamics of mass transfer between liquid and gas phase follows the asymptotically stable first-order process, i.e.,

$$\frac{dc_L}{dt} = \frac{1}{\tau}\left(c_L^* - c_L\right) \text{ or } c_L = c_L^* + \left(c_L^0 - c_L^*\right)e^{-(t-t_0)/\tau} \text{ with } c_L\left(t_0\right) = c_L^0. \tag{14}$$

Here, c_L^* denotes the dissolved gas concentration at equilibrium with the partial pressure of a certain gaseous species in a calibration mixture, $c_L{}^0$ the concentration given at time zero, and τ the characteristic time constant given for the dynamic process. Concerning the gas and liquid-phase equilibrium, refer to Schumpe et al. (1982) for an exact conversion between partial pressure and dissolved concentration using Bunsen's coefficients by considering physical conditions such as temperature and ionic strength (salting-out effect) [13].

Further, we get the corresponding first-order function given for the measured ion currents with $t_0 = 0$ by combining Eq. 14 with Eq. 13:

$$I_{m/z} = I^*_{m/z} + \left(I^0_{m/z} - I^*_{m/z}\right)e^{-t/\tau}. \tag{15}$$

By applying a nonlinear curve-fitting algorithm to measured signals $I_{m/z}$, the asymptotic signal $I_{m/z}{}^*$ corresponding to $c_L{}^*$, the time constant τ and the initial signal $I_{m/z}{}^0$ can be determined from measurements. From $I_{m/z}{}^*$ measured under different partial pressure conditions, one can get the calibration factors of a and b in Eq. 13 for the conversion of measured ion currents into concentrations. The experimental procedure of the dynamic MS calibration is as follows:

1. Fill reactors with the medium with a proper working volume (e.g., 10 mL) which is identical to the medium for in vivo measurement. A buffered, defined calibration medium is recommended, e.g., to prevent pH change and eliminate uncertainties during in vivo ^{13}C labeling measurements for CO_2.
2. Saturate the calibration medium with synthetic air containing, e.g., 1.5 % CO_2 kept at the desired cultivation temperature in the reactor.
3. Aerate reactor sequentially with (1) pure argon or helium, (2) synthetic air with 1.5 % CO_2, and (3) nitrogen with 2 % O_2, and (4) CO_2-free synthetic air from the headspace (10 VVM: gas volume flow per unit of liquid volume per minute) at a constant stirring speed, e.g., of 1,000 rpm. Each aeration cycle typically takes about 15–20 min.
4. Record time-domain dynamic changes of O_2 and CO_2 ion current signals ($I_{m/z}$) at m/z 32, 44, 45, and 46 during each cycle of aeration.
5. Conduct nonlinear curve fitting for the signals obtained at each aeration cycle using Eq. 13, and get asymptotic signals ($I_{m/z}{}^*$) measured at different partial pressure levels to get calibration factors, e.g., for O_2 and CO_2.
6. Store calibration factors and the time constant (τ) to apply those to dynamic CO_2 labeling and cell respiration measurements.

3.3 Dynamic Quantification of CO_2 Labeling & Cell Respiration

Once the membrane-inlet MS is calibrated for the dissolved gaseous ion species of interest, one can conduct cell respiration and CO_2 labeling measurements dynamically by a short-term incubation of cells with ^{13}C-tracer substrates in the miniaturized reactors. One application of those in vivo measurements is to the so-called respirometric ^{13}C metabolic flux analysis (^{13}CMFA), in which metabolic fluxes are determined from ^{13}C-labeling patterns in CO_2 evolving from ^{13}C-tracers with different labeling positions (Fig. 3).

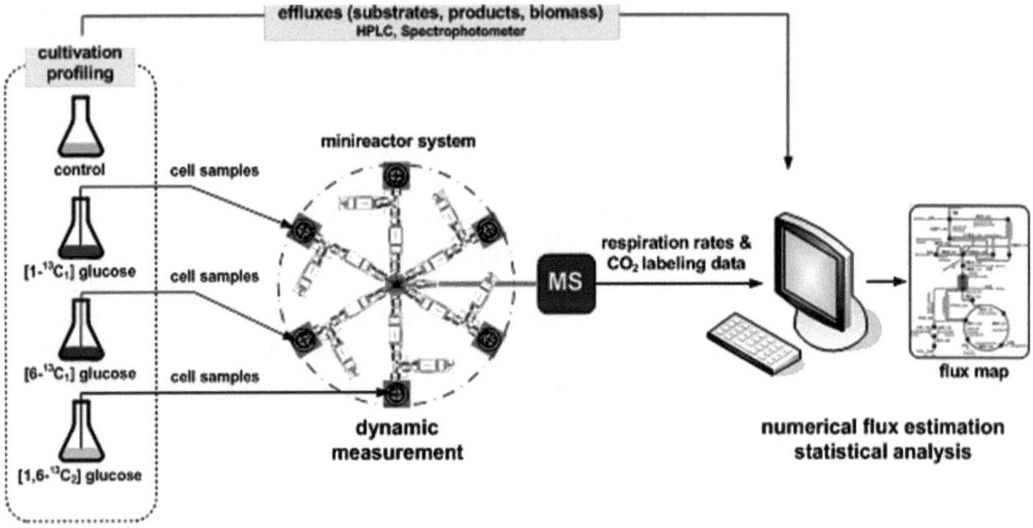

Fig. 3 Overall procedure of respirometric ^{13}C metabolic flux analysis using simultaneous CO_2 labeling measurements from multiple ^{13}C-cultures containing differently labeled ^{13}C-tracers

1. *Experimental prerequisite.* The prerequisites for the dynamic cell respiration and CO_2 labeling measurements are constant specific rates of growth, O_2 uptake, and CO_2 production (metabolic steady-state) under a constant pH and temperature condition. Thus, the design of a short-term measurement is necessary, within which both physical and biological conditions are kept invariant. Of course, the steady-state values for CO_2 mass isotopomer concentrations may not be reached within the measurement interval. These isotopic steady-state values are however, obtainable from a short-term dynamic measurement using the modeling efforts described below. To begin with, the liquid-phase mass balances for O_2 and CO_2 have to be fully understood.
2. *Liquid-phase mass balance.* For an inert gas, with respect to water, such as O_2, its accumulation in culture broth containing cells with constant rates of specific growth (μ) and oxygen uptake (q_{O_2}) can be expressed as follows in a continuously aerated reactor:

$$\frac{d[O_2]_{aq}}{dt} = k_L a_{O_2}\left([O_2]^*_{aq} - [O_2]_{aq}\right) - c_{X_0} e^{\mu(t-t_0)} \times q_{O_2} \quad (16)$$

Here, $[O_2]_{aq}^*$ is the equalibrium concentration with O_2 partial pressure in inlet gas stream, c_{X0} the initial cell concentration, and $k_L a_{O_2}$ the liquid–gas-phase mass transfer coefficient of oxygen. The above equation is valid for a metabolic steady state (balanced growth condition), e.g., cells with a constant μ in a batch culture.

In contrast to oxygen, CO_2 undergoes liquid-phase reactions as shown in Subheading 1.1 and **Note 1**, and this has to be taken into account. As also shown in Eq. 8, the material balance set up for CO_2 involves a pH-dependent conversion factor $K_{Conv}{}^{pH}$, derived from the liquid-phase reactions of CO_2 in Eq. 25. Similarly to Eq. 8, one can set up the following accumulation term for each mass CO_2 isotopomer:

$$\frac{d\,{}^{m+i}[CO_2]_{aq}}{dt} = \frac{k_L a_{CO_2}}{K_{Conv}^{pH}}\left({}^{m+i}[CO_2]_{aq}^{*} - {}^{m+i}[CO_2]_{aq}\right) + c_{X_0} e^{\mu(t-t_0)} \frac{{}^{m+i}q_{CO_2}}{K_{Conv}^{pH}}. \quad (17)$$

Here, ${}^{m+i}[CO_2]_{aq}{}^{*}$ denotes the dissolved ${}^{m+i}CO_2$ concentration at equilibrium with inlet gas stream, ${}^{m+i}q_{CO_2}$ the specific production rate of ${}^{m+i}CO_2$ and $m+i$ the mass of a CO_2 mass isotopomer with $m=44$, $i=0, 1,\ldots, n$ and $n+1$ the number of the measurable CO_2 mass isotopomers.

As shown in Eq. 25, the conversion factor $K_{Conv}{}^{pH}$ requires pH of the medium with the equilibrium constants between $[CO_2]_{aq}$, $[HCO_3^-]$, and $[CO_3^{2-}]$, which are subject to variations depending on temperature and ionic strength of the medium [10]. Instead of calculating $K_{Conv}{}^{pH}$ from the equilibrium constants, one can determine this using the time constants (τ) determined during the dynamic calibration procedure.

3. *Mass transfer coefficients.* To begin with measurements of CO_2 labeling and cell respiration, liquid–gas-phase mass transfer coefficients of CO_2 and O_2 ($k_L a_{O_2}$, $k_L a_{CO_2}$) as well as the pH-dependent conversion factor ($K_{Conv}{}^{pH}$) need to be available in advance. All these parameters can be obtained from the system calibration (Subheading 3.2) based on the fact that the ratio of mass transfer coefficients of two gas molecules is proportional to their liquid-phase diffusivity ratio, i.e.,

$$k_L a_{CO_2} = \left(\frac{D_{CO_2}}{D_{O_2}}\right)^{0.5} \times k_L a_{O_2}. \quad (18)$$

The square root of the diffusivity ratio between CO_2 and O_2 is 0.918 as calculated by Wilke–Chang's empirical correlation and 0.925 by Siddiqi–Lucas. These empirical correlations can be validated by comparing the $k_L a$ values of inert gases (e.g., N_2, He, Ar, O_2) and their diffusivity ratios. Note that the ratio finally reduces to the ratio of the liquid molar volume (υ) of a gas at its normal boiling point, i.e., $\left(D_{CO_2} / D_{O_2}\right)^{0.5}$ equals $\left(\upsilon_{O_2} / \upsilon_{CO_2}\right)^{0.274}$ by Siddiqi–Lucas' correlation, where υ_{O_2} and υ_{CO_2} is 25.6 and 34.0 cm^3/mol, respectively.

For inert gases, such as O_2, the reciprocal of the time constant (τ) in Eq. 15 directly equals $k_L a_{O_2}$. In case of CO_2, τ

equals $K_{Conv}^{pH} / K_L a_{CO_2}$. When the time constants determined for O_2 (τ_{O_2}) and CO_2 (τ_{CO_2}) are available, we get the following relationship between $K_{Conv}{}^{pH}$, $k_L a_{CO_2}$, and $k_L a_{O_2}$:

$$K_{Conv}^{pH} = \tau_{CO_2} \times k_L a_{CO_2} = \tau_{CO_2} \times \left(\frac{D_{CO_2}}{D_{O_2}} \right)^{0.5} \times k_L a_{O_2} = \frac{\tau_{CO_2}}{\tau_{O_2}} \times \left(\frac{D_{CO_2}}{D_{O_2}} \right)^{0.5}. \quad (19)$$

Hence, at a constant pH and temperature, we can get apparent values of $k_L a_{O_2}$ and $k_L a_{CO_2}$ using Eq. 18 and that of $K_{Conv}{}^{pH}$ from Eq. 19 without involving any equilibrium constants of dissolved CO_2 species that is affected by medium conditions such as ionic strength.

4. *Dynamic CO_2-labeling measurement.* During a short-term dynamic experiment, the fractional enrichment of the measurable $n+1$ CO_2 mass isotopomers at isotopic steady-state, $f_{CO_2} = \left({}^{m+0}f_{CO_2}, {}^{m+1}f_{CO_2}, \ldots, {}^{m+n}f_{CO_2} \right)^T$, may not be obtained within the time frame of the experiment. This is because the measurement time may be too short to attain stable values of each CO_2 mass isotopomer experimentally. To address this problem, the asymptotic stability of the liquid-phase material balance of CO_2 mass isotopomers can be applied. The explicit solution of differential equation (Eq. 17) with an initial condition of $[CO_2]_{aq} = [CO_2]_{aq}^0$ at t_0 is given by

$$[CO_2]_{aq} = [CO_2]_{aq}^* - \left([CO_2]_{aq}^* - [CO_2]_{aq}^0 + \frac{{}^cX_0\, {}^qCO_2\, K_{Conv}^{pH}}{k_L a_{CO_2} + K_{Conv}^{pH} \mu} \right) e^{-\frac{k_L a_{CO_2}}{K_{Conv}^{pH}}(t-t_0)} - \frac{{}^cX_0\, {}^qCO_2\, K_{Conv}^{pH}}{k_L a_{CO_2} + K_{Conv}^{pH} \mu} e^{\mu(t-t_0)}. \quad (20)$$

As t approaches infinity, the second term of Eq. 20 reaches zero and the third infinity. Hence, the dissolved concentration of an individual CO_2 mass isotopomer is unstable (time-variant, divergent) as long as $\mu \neq 0$. Let us formulate the limiting value of the time function ${}^{m+i}f_{CO_2}(t)$ in time domain such that

$$\lim_{t \to \infty} {}^{m+i}f_{CO_2}(t) = \lim_{t \to \infty} \frac{{}^{m+i}[CO_2]_{aq}}{\sum_{k=0}^{n} {}^{m+k}[CO_2]_{aq}}. \quad (21)$$

Here, ${}^{m+i}f_{CO_2}$ denotes the fractional enrichment of a measurable CO_2 mass isotopomer, and each ${}^{m+k}[CO_2]_{aq}$ and

${}^{m+i}[CO_2]_{aq}$ are time-dependent functions as Eq. 20. Now, multiplying the denominator and the nominator of Eq. 21 with $\left(\left(k_L a_{CO_2} + K_{Conv}^{pH}\mu\right) / e^{\mu(t-t_0)}\right)$, we get the limiting value of ${}^{m+i}f_{CO_2}(t)$, which equals its fractional specific production rate.

$$\lim_{t\to\infty} {}^{m+i}f_{CO_2}(t) = \frac{c_{X_0}\,{}^{m+i}q_{CO_2} K_{Conv}^{pH}}{\sum_{k=0}^{n}\left(c_{X_0}\,{}^{m+k}q_{CO_2} K_{Conv}^{pH}\right)} = \frac{{}^{m+i}q_{CO_2}}{\sum_{k=0}^{n} {}^{m+k}q_{CO_2}} \qquad \because \mu \geq 0 \qquad (22)$$

As shown, the dynamic system given for ${}^{m+i}f_{CO_2}(t)$ has an asymptotically stable point, i.e., CO_2 labeling dynamics converges to a steady-state value, which can be calculated from the specific rates of CO_2 mass isotopomer evolution of $q_{CO_2} = \left({}^{m+0}q_{CO_2}, {}^{m+1}q_{CO_2}, \ldots, {}^{m+n}q_{CO_2}\right)^T$. Hence, q_{CO_2} of each CO_2 mass isotopomer is now to be estimated by fitting in vivo concentration measurements using the model function Eq. 17 or Eq. 20.

5. *Dynamic cell respiration estimation.* Now, when online CO_2 and O_2 measurements are available, one can get cell respiration values and CO_2 labeling data, e.g., using nonlinear least-squares minimization, i.e.,

$$\min_{\Theta} \left\| c_{L,meas} - \mathbf{F}(\Theta, t) \right\|_2^2 \qquad (23)$$

Here, $c_{L,meas}$ denotes the measured dissolved gas concentration (e.g., oxygen and CO_2 mass isotopomers), $\mathbf{F}(\boldsymbol{\Theta}, t)$ the model function, $\boldsymbol{\Theta}$ the design parameters, $\Theta = \left({}^{32}q_{O_2}, k_L a_{O_2}, {}^{32}[O_2]_{aq}^0\right)^T$ for ${}^{32}O_2$ or $\Theta = \left({}^{m+i}q_{CO_2}, K_{Conv}^{pH}, {}^{m+i}[CO_2]_{aq}^0\right)^T$ for an ${}^{m+i}CO_2$ mass isotopomer. The model function equals the analytical solutions of Eqs. 16 and 17. Of course, one can implement numerical integration using an ordinary differential equation solver.

For each CO_2 mass isotopomer balance Eq. 17, $k_L a_{CO_2}$ can be expressed using $k_L a_{O_2}$ using Eq. 18. Thus, $k_L a_{O_2}$ and K_{Conv}^{pH} can be chosen to be optimized, which are actually already determined from calibration experiments. However, their apparent values may be different in cell suspension and can be estimated together with specific rates of ${}^{32}q_{O_2}$ and ${}^{m+i}q_{CO_2}$ from CO_2 and O_2 measurements. Some tips regarding the curve fitting procedure are given in **Note 2**.

6. *Experimental procedure.* Prior to an in vivo experiment, the optimal cell concentration, which gives sufficient dynamic changes in O_2 and CO_2 signals and a decent sensitivity for parameter estimation has to be determined: too low cell concentration can give a poor dynamic response, but too high

concentration can give rise to oxygen limitation during measurements. Yang et al. [7] applied 0.5 g/L cell concentration in the measurement reactors for a culture with a specific growth rate of 0.40 h^{-1}, q_{O_2} 6.9 mmol/g/h, and q_{CO_2} 7.5 mmol/g/h. However, this optimal cell concentration has to be carefully selected by the experimenter depending on biological systems of interest. Once the optimal concentration is determined, the following steps can be implemented for dynamic cell respiration and CO_2 labeling measurements:

(a) Pre-incubate the defined medium which is utilized for dilution in the next step at the cultivation temperature, and saturate with the same CO_2-free synthetic air which is applied to aeration during the dynamic cell respiration measurement. One can choose a slightly higher partial pressure of oxygen than the atmospheric air (20.95 %) to prevent oxygen limitation during the measurements.

(b) Set the measurement system at the stand-by state with continuous aeration with helium as described in Subheading 3.1.

(c) Obtain cell suspension from ^{13}C-cultures with a defined medium and dilute with the pre-incubated medium containing the corresponding ^{13}C-substrates. The final substrate concentration $[S]$ should be well above the critical concentration K_S which can be estimated in terms of the Monod kinetics, i.e., $\mu = \mu_{max}[S]/(K_S + [S])$.

(d) Measure cell concentration, and place cell suspension in measurement reactors, which are aerated with the selected synthetic air (e.g., 10 VVM) at a constant stirring speed (e.g., of 1,000 rpm) and cultivation temperature.

(e) Record time-domain dynamic changes of O_2 and CO_2 ion current signals ($I_{m/z}$) at m/z 32, 44, 45, and 46 for 15–20 min. If more than one reactor is operated, follow the valve operation steps given in Subheading 3.1; otherwise, measure the signals continuously from a reactor.

(f) Collect signals (means and standard deviations at each measurement event), convert signals into concentrations (Subheading 3.2), and conduct nonlinear curve fitting by minimizing the objective function as given in Subheading 3.3 and **Note 2**. The specific growth rate involved in the model functions can be obtained from the main culture.

It has to be noted that the dynamic measurements described above provide instantaneous rates and labeling states of CO_2 due to non-accumulating nature of CO_2 in a system that is continuously aerated with CO_2-free air. In this context, one has to obtain instantaneous rates of substrate uptake, cell growth, and product formation to conduct ^{13}C flux analysis using CO_2 labeling and cell respiration parameters. Also, intracellular metabolite ^{13}C-labeling patterns obtained at the event of dynamic experiments can be combined if those are measureable.

4 Notes

A few theoretical backgrounds relevant to the dynamic cell respiration and CO_2 labeling measurements are supplemented in this section.

1. **Aqueous CO_2 Chemistry and Equilibrium Relations**
For the quantification of dissolved CO_2 in aqueous medium, four different forms of CO_2 have to be considered, where only dissolved CO_2 concentration is directly measurable by the membrane inlet MS. The total concentration C_T of the dissolved species involved in the liquid-phase reaction of CO_2 is

$$C_T = [CO_2]_{aq} + [H_2CO_3] + \left[HCO_3^-\right] + \left[CO_3^{2-}\right]$$

where the concentration of each component can be expressed by its corresponding equilibrium constant as follows.

$$[H_2CO_3] = \frac{[CO_2]_{aq}}{K_0},\ \left[HCO_3^-\right] = K_1 \frac{\left[H_2CO_3^*\right]}{10^{-pH}},\ \text{and}\ \left[CO_3^{2-}\right] = K_1 K_2 \frac{\left[H_2CO_3^*\right]}{10^{-2pH}}$$

$[H_2CO_3^*]$ is the total analytical concentration of dissolved CO_2 and equals the sum of $[CO_2]_{aq}$ and $[H_2CO_3]$, where less than 0.2 % of $[H_2CO_3^*]$ exists as H_2CO_3 at 25 °C. The hydration of CO_2 has a rate constant of 0.025–0.04 s^{-1} at 25 °C and that of the dehydration 10–20 s^{-1} at room temperature. This gives a high uncertainty in K_0. Using the equilibrium relations, the total concentration C_T can be expressed as follows:

$$C_T = \left(1 + \frac{K_1}{10^{-pH}} + \frac{K_1 K_2}{10^{-2pH}}\right)\left(1 + \frac{1}{K_0}\right)[CO_2]_{aq} \qquad (24)$$

Since $1/K_0 \ll 1$, the above equation can be simplified and reformulated such that C_T is directly proportional to the dissolved CO_2 concentration over a pH dependent conversion factor $K_{Conv}{}^{pH}$ at constant temperature and pH. In addition, fermentation media in general contain high concentrations of ions, mainly, inorganic salts. In this case, the so-called mixed acidity constant K' is frequently used to express actual equilibrium between an acid and its corresponding base if pH is measured according to the IUPAC convention ($-\log[H^+] = -\log\{H^+\}$). Thus, Eq. 24 now becomes:

$$\frac{C_T}{[CO_2]_{aq}} = K_{Conv}^{pH} = 1 + \frac{K_1'}{10^{-pH}} + \frac{K_1'K_2'}{10^{-2pH}} \quad (25)$$

Details related to this topic can be found elsewhere [10]. However, using the dynamic method addressed in this chapter, one can determine the conversion factor without measuring pH and knowing the equilibrium constants (mixed acidity constant), as described in Subheading 3.3. One may use Eq. 25 to check if $K_{Conv}{}^{pH}$ numerically estimated is appropriate.

2. **Nonlinear Least-Squares Minimization Using Analytical Gradients and Hessian**

When signals are acquired for O_2 at m/z 32 and CO_2 mass isotopomers at m/z 44, 45, and 46, the objective function Eq. 23 involves the following four ODE functions:

$$f_1: \frac{d[O_2]_{aq}}{dt} = k_L a_{O_2}\left([O_2]_{aq}^* - [O_2]_{aq}\right) - c_{X_0} e^{\mu(t-t_0)} \times q_{O_2} \quad (26)$$

$$f_2, f_3, f_4: \frac{d\,{}^{m+i}[CO_2]_{aq}}{dt} = \left(\frac{D_{CO_2}}{D_{O_2}}\right)^{0.5} \frac{k_L a_{O_2}}{K_{Conv}^{pH}}\left({}^{m+i}[CO_2]_{aq}^* - {}^{m+i}[CO_2]_{aq}\right) + c_{X_0} e^{\mu(t-t_0)} \frac{{}^{m+i}q_{CO_2}}{K_{Conv}^{pH}} \quad (27)$$

Here, $\mathbf{F}(\mathbf{\Theta}, t) = (F_1, F_2, F_3, F_4)^T$ with F denoting the solution of each ODE and f the ODE given for oxygen (f_1) and CO_2 mass isotopomers (f_2, f_3, f_4). Since any ODE problems involve initial value at t_0, we get 10 parameters of

$$\Theta = \begin{pmatrix} {}^{32}q_{O_2}, k_L a_{O_2}, {}^{32}[O_2]_{aq}^0, K_{Conv}^{pH}, {}^{44}q_{CO_2}, \\ {}^{44}[CO_2]_{aq}^0, {}^{45}q_{CO_2}, {}^{45}[CO_2]_{aq}^0, {}^{46}q_{CO_2}, {}^{46}[CO_2]_{aq}^0 \end{pmatrix}^T.$$

Thus, at least 10 measurements ($c_{L,meas}$) are required to get those parameters determined.

The parameters of $k_L a_{O_2}$ and $K_{Conv}{}^{pH}$ are also determined during the dynamic calibrations, and one may set up their reasonable upper and lower bounds. ${}^{32}q_{O_2}$ and ${}^{m+i}q_{CO_2}$ can be defined to be larger than zero. The bounds for the initial values such as ${}^{32}[O_2]_{aq}{}^0$ can also be defined if $c_{L,meas}$ at t_0 and measurement precision are known. Thus, the optimization becomes a constrained nonlinear least-squares minimization problem.

For the minimization, c_{X0} and μ have to be supplied along with $c_{L,meas}$ as the measurements. Thus, statistical qualities of the estimators are determined not only by the measurement errors in $c_{L,meas}$ but also by the errors in c_{X_0} and μ, i.e., the nonlinear model function has errors both in the c_{Lmeas}- and c_X-coordinate, where $c_X = c_{X_0} \times e^{\mu(t-t_0)}$ represents the cell growth. Consequently, statistical analysis by analytical nonlinear regression analysis is not simply applicable to this problem. Due to this, the stochastic Monte Carlo approach is more suitable to investigate to what extent uncertainties of the measured data influence the parameter estimates.

In order to solve the optimization problem efficiently, one may supply analytical gradients ∇ and Hessian (**H**) of the objective function Eq. 23, which are:

$$\nabla = \frac{\partial \mathbf{F}(\Theta,t)^T}{\partial \Theta} \times \left(\mathbf{F}(\Theta,t) - c_{L,\text{meas}}\right) \tag{28}$$

and

$$\mathbf{H} = \frac{\partial \mathbf{F}(\Theta,t)^T}{\partial \Theta} \times \frac{\partial \mathbf{F}(\Theta,t)}{\partial \Theta}, \text{ respectively} \tag{29}$$

Thus, the partial derivatives of the model function **F** with respect to **Θ**, i.e., the Jacobian matrix $\partial\mathbf{F}/\partial\boldsymbol{\Theta}$ is required, which can be also computed while solving the ODE problem Eqs. 26 and 27. For instance, the following holds for a general ODE of $dy/dt=f(t, y, \boldsymbol{\Theta})$:

$$d\left(\frac{\partial y}{\partial \Theta}\right) / dt = \frac{\partial f}{\partial y} \times \frac{\partial y}{\partial \Theta} + \frac{\partial f}{\partial \Theta}. \tag{30}$$

Thus, solving the above ODE gives the desired Jacobian matrix. Note that the initial values are 1 for the partial derivatives of y with respect to its initial value y_0 at t_0; otherwise zeros.

Acknowledgements

The author would like to acknowledge the German Research Foundation DFG for research funding (Project No. 3092/4), Dr. Elmar Heinzle (Biochemical Engineering Institute, Saarland University, Germany), Dr. Christoph Wittmann (Bioengineering Institute, Braunschweig University of Technology, Germany), Nobert Ochs, Peter Skohoutil (Workshop of Physical Chemistry, Saarland University), Genomatica Inc. (San Diego, CA), and Naree Hahn for their contributions as well as supports to this work.

References

1. Heinzle E, Oeggerli A, Dettwiler B (1990) On-line fermentation gas analysis: error analysis and application of mass spectrometry. Anal Chim Acta 238:101–115
2. Yang TH, Coppi MV, Lovley DR, Sun J (2010) Metabolic response of Geobacter sulfurreducens towards electron donor/acceptor variation. Microb Cell Fact 9:90
3. Stephanopoulos G, Aristidou AA, Nielsen JH (1998) Metabolic engineering: principles and methodologies. Academic, San Diego
4. Mahadevan R, Bond DR, Butler JE, Esteve-Nunez A, Coppi MV, Palsson BO, Schilling CH, Lovley DR (2006) Characterization of metabolism in the Fe(III)-reducing organism *Geobacter sulfurreducens* by constraint-based modeling. Appl Environ Microbiol 72:1558–1568
5. Yang TH, Heinzle E, Wittmann C (2005) Theoretical aspects of ^{13}C metabolic flux analysis with sole quantification of carbon dioxide labeling. Comput Biol Chem 29:121–133
6. Yang TH, Wittmann C, Heinzle E (2006) Respirometric ^{13}C flux analysis—part II: in vivo flux estimation of lysine-producing *Corynebacterium glutamicum*. Metab Eng 8:432–446
7. Yang TH, Wittmann C, Heinzle E (2006) Respirometric ^{13}C flux analysis—part I: design, construction and validation of a novel multiple reactor system using on-line membrane inlet mass spectrometry. Metab Eng 8:417–431
8. Buckland B, Brix T, Fastert H, Gbewonyo K, Hunt G, Jain D (1985) Fermentation exhaust gas analysis using mass spectrometry. Nat Biotechnol 3:982–988
9. Heinzle E, Kramer H, Dunn IJ (1985) State analysis of fermentation using a mass spectrometer with membrane probe. Biotechnol Bioeng 27:238–246
10. Yang TH, Wittmann C, Heinzle E (2003) Dynamic calibration and dissolved gas analysis using membrane inlet mass spectrometry for the quantification of cell respiration. Rapid Commun Mass Spectrom 17:2721–2731
11. Oeggerli A, Heinzle E (1992) On-line analysis of volatiles in fermenter exhaust gas using mass spectrometry. IFAC Symposia Series, 295–298
12. Vallino JJ, Stephanopoulos G (1993) Metabolic flux distributions in *Corynebacterium glutamicum* during growth and lysine overproduction. Biotechnol Bioeng 41:633–646
13. Schumpe A, Quicker G, Deckwer W-D (1982) Gas solubilities in microbial culture media. In: Reaction Engineering. Series: Advances in Biochemical Engineering 24:1–38

Part IV

^{13}C Fluxomics Data Processing

Chapter 12

Correction of MS Data for Naturally Occurring Isotopes in Isotope Labelling Experiments

Pierre Millard, Fabien Letisse, Serguei Sokol, and Jean-Charles Portais

Abstract

Mass spectrometry (MS) in combination with isotope labelling experiments is widely used for investigations of metabolism and other biological processes. Quantitative applications—e.g., ^{13}C metabolic flux analysis—require correction of raw MS data (isotopic clusters) for the contribution of all naturally abundant isotopes. This chapter describes how to perform such correction using the software IsoCor. This flexible, user-friendly software can be used to exploit any isotopic tracer, from well-known (^{13}C, ^{15}N, ^{18}O, etc.) to unusual (^{57}Fe, ^{77}Se, etc.) isotopes. It also provides options—e.g., correction for the isotopic purity of the tracer—to improve the accuracy of quantitative isotopic studies, and allows automated correction of large datasets that can be collected with modern MS methods.

Key words Mass spectrometry, Isotope, Labelling experiment, Isotopic cluster, Isotopologue distribution

1 Introduction

Application of mass spectrometry (MS) to stable isotope labelling experiments is extensively used for studies of metabolism. During the last decade, quantitative approaches—such as ^{13}C-metabolic flux analysis—have been increasingly used to provide key information on biological systems in systems biology and metabolic engineering [1, 2]. The reliability of such approaches relies on the accurate quantification of each isotopologue—that refers to molecules of the same molecular entity differing only in the number of isotopes—of metabolites from the fine exploitation of their isotopic cluster [3]. However, stable isotopes exist at natural abundance for most chemical elements [4]. Naturally occurring isotopes contribute to the isotopic clusters, which do not directly reflect the incorporation of the tracer. To extract meaningful, quantitative, labelling information—i.e., isotopologue distribution, which represents the relative fraction of each isotopologue of a metabolite—the contribution of naturally occurring isotopes first

Jens O. Krömer et al. (eds.), *Metabolic Flux Analysis: Methods and Protocols*, Methods in Molecular Biology, vol. 1191, DOI 10.1007/978-1-4939-1170-7_12,

has to be subtracted. Mathematically, a matrix-based correction method was described by van Winden et al. [5]. It requires solving the following equation:

$$IC_{cor} = CM^{-1} \cdot IC_{meas}$$

where IC_{cor} denotes the corrected isotopic cluster (i.e., the isotopologue distribution), CM is a correction matrix (calculated from the elemental formula of the metabolite and the natural abundance of each element), and IC_{meas} is the measured isotopic cluster. This formalism has been implemented in IsoCor [6], an open-source software developed in Python programing language (http://python.org).

Here, we describe how to correct MS data using IsoCor, which includes the following features:

- Correction of any chemical species for any isotopic tracer—from well-known (^{13}C, ^{15}N) to unusual (^{57}Fe, ^{77}Se, etc.) isotopes. Chemical elements with more than two isotopes (e.g., ^{16}O, ^{17}O, ^{18}O) can be considered.
- When metabolite derivatization is required for analytical purposes (e.g., derivatization of amino acids for GC-MS analysis [7]), IsoCor performs appropriate correction by taking into account the contribution of naturally occurring isotopes brought by the derivatization reagent.
- IsoCor provides two options that allow flexible correction of isotopic clusters to account for (or not) the composition of the label input including (1) correction for the isotopic purity of the labeled substrate(s) and (2) correction for natural abundance of the tracer into positions of the labeled substrate that are not isotopically enriched (i.e., for nonuniformly labeled substrates). These options are of particular value for specific applications [8, 9].
- IsoCor calculates the mean isotopic enrichment of molecules, which refers to the molecular content in the tracer isotope. This information is particularly useful for targeted metabolic investigations, such as the quantification of split ratios between two metabolic pathways.
- IsoCor can be applied to large datasets and can deal with the increasing number of data that are generated with modern instruments in a single experiment.

2 Materials

A desktop or a laptop computer under Windows, UNIX/Linux, or Macintosh.

Python and additional python packages (*see* Subheading 3.1).

IsoCor software: The software is freely distributed under OpenSource license and can be downloaded at http://metasys.insa-toulouse.fr/software/.

3 Methods

In this section we provide explicit step-by-step instructions for correcting raw MS isotopic data for naturally occurring isotopes using IsoCor. The correction can be performed for a single metabolite (*see* Subheading 3.3) or for multiple metabolites—taken from a unique biological sample or from different samples—(*see* Subheading 3.4). For each isotopic cluster to correct, the user must provide at least the intensities of peaks originating from the unlabelled to the uniformly labelled molecule, without missing value (e.g., at least peaks M_0 to M_3 for a C_3 compound).

3.1 Installation Instructions

IsoCor requires Python 2.6+ (not 3.0 or higher) and modules wxPython (v2.8.11.0 or higher), NumPy (v1.6.0.2 or higher), and SciPy (v0.9.0.1 or higher). IsoCor was developed on Microsoft Windows but can be used both on Linux (tested on Mandriva 2011 and Ubuntu 12) and Windows (XP and 7) platforms.

To facilitate the installation process under Windows and ensure the compatibility between the different modules, we recommend installing Python and these packages from pythonxy (v2.6.5 or higher), a scientific-oriented Python distribution. To install pythonxy, visit the official download page (http://www.pythonxy.com) and grab the installer. Run it; check that wxPython, NumPy, and SciPy are selected in the "Choose components" step; and finish the installation process with the default settings.

If you have trouble installing system-wide environments like Python, ask some help from your local computer service.

3.2 Input Files

The data required to perform the correction are (1) the natural abundance of isotopes [4], (2) the list of metabolites with their elemental formulas, and (3) if appropriate the list of derivatives with their elemental formulas. These data must be provided in plain text files using syntax rules detailed in **Note 1**. Example files ready to run are provided with the software. They can be easily edited and modified according to the user's needs.

1. File `Isotopes.dat`
   This file contains (1) the list of chemical elements for which isotopic correction can be performed, (2) the list of isotopes of each element, and (3) the natural abundance of each isotope. Chemical elements not included into this file are supposed to have no isotope—i.e., $m_0 = 100$ %. A file containing the natural abundances of isotopes commonly found in biological compounds [4] is provided with the software.

2. File `Metabolites.dat`
   This file contains the list of metabolites, and their elemental formulas, to which the isotopic correction is applied to. Please refer to **Note 2** for detailed definition of elemental formulas. A file containing common metabolites is provided with the software.
3. File `Derivatives.dat`
   This file contains the list of chemical derivatives with their elemental formulas. All isotopes in the derivative (including the tracer isotope if present) are considered to occur at natural abundance. Please refer to **Note 2** for detailed definition of elemental formulas. A file containing some common derivatives is provided with the software.
4. Raw MS data files
   These files contain the raw MS data (measured intensities of isotopic cluster peaks) to be corrected.

3.3 Correcting a Single Metabolite

1. Execute `IsoCor.pyw` to open IsoCor (Fig. 1).
2. Click on the "*Load single meas.*" button and select the input data file containing the raw MS data.
3. Select the chemical element used as isotopic tracer in the isotope-labelling experiment.
4. Select a metabolite (from the `Metabolites.dat` file) in the left panel or enter a formula. If relevant, select a derivative or enter a new derivative in the field "*Derivative formula.*" *See* **Note 2** for more details on the definition of metabolite and derivative formulas.
5. Select options and parameters:
   (a) *Correction for natural abundance of the tracer.* When the label input is not uniformly labelled, it contains "unlabelled" positions in which the tracer isotope is usually occurring at its natural abundance. The MS data can be corrected for the contribution of these naturally occurring tracer isotopes. To perform such correction the option "*nat. ab. of the tracer*" must be enabled. This correction is valid only when the tracer isotope occurs at natural abundance in the "unlabelled" positions of the input substrate(s). That is most often the case but has to be checked from the manufacturers or determined experimentally.
   (b) *Correction for isotopic purity.* Labelled substrates are not isotopically pure; that is, they are not 100 % enriched in the "labelled" position(s). The latter contain small fractions of non-tracer isotopes for which MS data must be corrected. To do so, the fractions of each isotope into the "labelled" positions must be ordered by mass shifts and separated by a semicolon ';' in the field "*Purity.*"

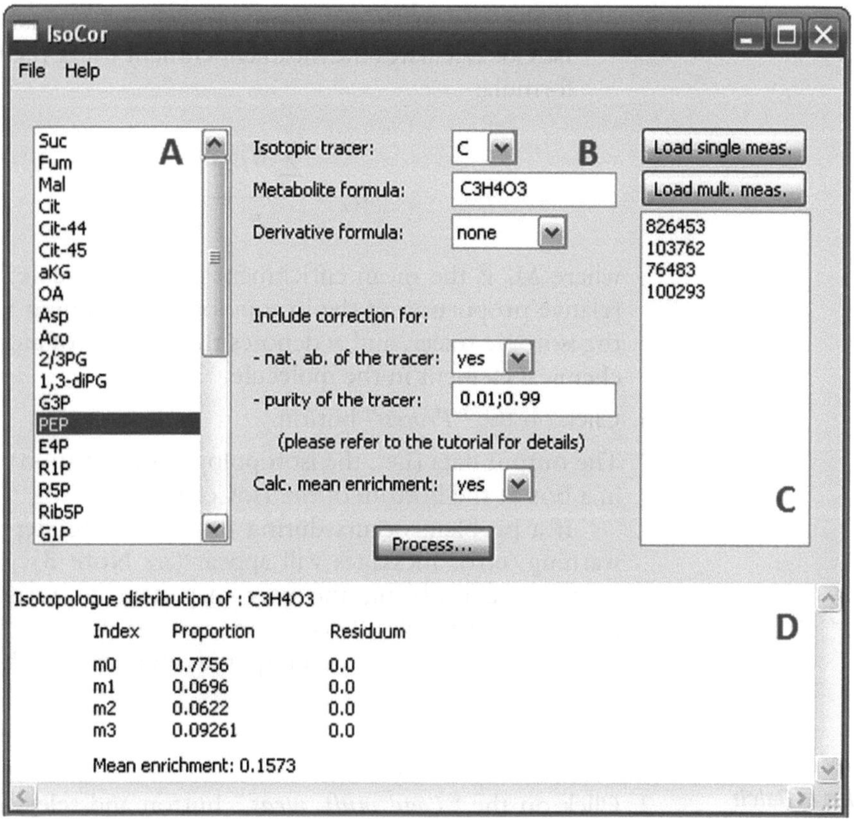

Fig. 1 Screenshot of IsoCor after correction of a single isotopic cluster. (**a**) List of metabolites. (**b**) Correction parameter panel. (**c**) Selection of experimental data files. (**d**) Calculation results. Please refer to **Note 2** for the definition of elemental formula of phosphoenolpyruvate (PEP) shown in this screenshot

For example, if the content in ^{13}C atoms in each position of a U-^{13}C-labeled compound is 99 %, other 1 % being ^{12}C atoms, the purity must be entered as "0.01;0.99". If you do not want to correct isotopic clusters for the isotopic purity of the substrate, just let the default value (purity = 1). This correction is valid only if all the labelled positions of the substrate(s) have the same isotopic purity, which can be checked from the manufacturers or determined experimentally.

(c) *Calculation of the mean isotopic enrichment*. The mean isotopic enrichment of a metabolite refers to the mean content of isotopic tracer in the metabolite, expressed as relative fraction of total atoms of the chemical element in the metabolite. This information is particularly useful for the quantification of split ratios between two metabolic pathways resulting in different contents of the isotopic tracer.

If the option "*Calc. mean enrichment*" is enabled, then IsoCor calculates the mean enrichment using the following formula:

$$M_E = \frac{\sum_{i=0}^{n} M_i \times i}{n}$$

where M_E is the mean enrichment of the molecule, M_i is the relative proportion of the isotopologue containing *i* atoms of the isotopic tracer, and *n* denotes the number of atoms of the chemical element in the molecule.

6. Click on the "*Process*" button.
7. The output data (i.e., the isotopologue distribution) are given in a box at the bottom of the IsoCor window.

   If a problem occurs during one or several steps, explicit warning/error messages will appear (*see* **Note 3**). After correcting the problem, the user may have to rerun IsoCor (to reload data files `Isotopes.dat`, `Metabolites.dat` and `Derivatives.dat` if required) and perform the calculation again.

3.4 Correct Metabolites in Batch

1. Execute `IsoCor.pyw`.
2. Click on the "*Load mult. meas.*" button and select the input data file containing the raw MS data.
3. Select options and parameters (*see* **step 5** of Subheading 3.3 for details).
4. Click on the "*Process*" button. IsoCor proceeds automatically to the correction of all the sets of data that are included into the data file, provided the metabolite formulas, derivative formulas, and MS data have been entered using appropriate formats (*see* **Note 1**).
5. A report containing information such as the correction parameters (e.g., isotopic tracer, purity vector), paths of input and output files, and errors occurred during the correction process is automatically generated by IsoCor. It is displayed in a box at the bottom of the IsoCor window and written in a text file (`InputFile_log.txt`). The corrected isotopic clusters are written in a result file (`InputFile_res.txt`), a plain text files with tabulation as separator. The content of each column is detailed in Table 1. These files are silently overwritten if already existing. So take care to copy your results elsewhere to protect them from overwriting.

Table 1
Content of the result file generated by IsoCor after correction of datasets containing multiple metabolites and/or samples

Column	Name	Description
1	Sample	Name of the sample
2	Metabolite	Name of the metabolite
3	Peak index	This value corresponds to the mass shift of the measured peak for the column "Isotopic cluster," and to the number of tracer atoms in the molecule for the column "Isotopologue distribution"
4	Isotopic cluster	Measured isotopic cluster
5	Isotopologue distribution	Corrected isotopic cluster (normalized to 1)
6	Residuum	$\text{Residuum} = \frac{IC_{meas} - CM \cdot IC_{corr}}{\sum IC_{meas}}$ High values (let us say higher than 10^{-2}) indicate measurement inconsistencies
7	Mean enrichment	Mean content in isotopic tracer in the metabolite (optional)

4 Notes

1. Structure of input files:
All input files are plain text files with tabulation as separator. The structure of these files is given in Tables 2, 3, 4, and 5. Please note that IsoCor is case sensitive; that is, two metabolites or derivatives with the same name but different cases will be considered as two distinct entities.

2. Definition of elemental formulas:
Obviously, errors in elemental formulas will result in erroneous isotopologue distributions; thus special care must be taken when defining these formulas. This section must be considered as guidelines for such definition and provides representative examples to cover the current panel of MS and MS/MS methods dedicated to quantitative isotopic analysis.

   The formula provided in the field "Metabolite formula" or in the file `Metabolites.dat` must correspond to the formula of the molecular form of the analyte that gives rise to the measured isotopic cluster. For example, if pyruvic acid ($C_3H_4O_3$) is analyzed by LC-MS using multiple ion monitoring in the negative mode, and the detected ion is the $[M\text{-}H]^-$ ion, then the formula to enter is $C_3H_3O_3$. When tandem MS/MS is employed, more attention has to be paid to the identification of

Table 2
Structure of the `Isotopes.dat` file

Element_1	proportion_m$_0$	proportion_m$_1$	
Element_2	proportion_m$_0$	proportion_m$_1$	proportion_m$_2$
...			

The first column contains the elements; other columns contain the relative fraction of each isotope (starting from the lowest mass and ordered by mass shifts)

Table 3
Structure of the `Metabolites.dat` and `Derivatives.dat` files, which contain lists of metabolites and derivatives (column 1) with their elemental formulas (column 2)

Succinate	$C_4H_5O_4$
Fumarate	$C_4H_3O_4$
Glucose-6P	$C_6H_{10}O_5$
...	

Table 4
Structure of the input data file for correcting a single isotopic cluster

Intensity_M$_0$
Intensity_M$_1$
Intensity_M$_2$
...

Each line contains the intensity of each peak of the isotopic cluster, starting from the lowest mass and ordered by mass shifts

the molecular species that are actually measured. In the example shown in Fig. 1, phosphoenolpyruvate (PEP) isotopologues were analyzed using the MS/MS method developed by Kiefer et al. [3]. The fragmentation of phosphorylated metabolites results in the efficient release of $[PO_3]^-$ or $[H_2PO_4]^-$ ions, allowing highly sensitive measurement of isotopologue distributions in these compounds in the multiple reaction monitoring (MRM) mode. This is achieved by selecting MRM transitions in which phosphate ions are detected but which encode the isotopic cluster of the complement ion, i.e., the part of the molecule that remains after loss of the phosphate ion that is actually detected. In the case of PEP ($C_3H_5O_6P$), for

Table 5
Structure of the input data file for correcting multiple isotopic clusters (from different metabolites and/or samples)

Sample 1	Metabolite 1		Intensity_M_0
			Intensity_M_1
			Intensity_M_2
	Metabolite 2	Derivative 1	Intensity_M_0
			Intensity_M_1
			Intensity_M_2
Sample 2	Metabolite 3	Derivative 1	Intensity_M_0
			Intensity_M_1
			Intensity_M_2
			Intensity_M_3
...			

Column 1 contains the sample name; columns 2 and 3 contain the name of metabolites and derivatives, respectively; and column 4 contains the isotopic cluster. The names of metabolites and derivatives must be as defined in the `Metabolites.dat` and `Derivatives.dat` files

which the molecular ion that is analyzed is $[C_3H_4O_6P]^-$, the analysis is based on MRM transitions in which $[PO_3]^-$ ions are used, meaning that the isotopic cluster is actually measured for the complement fragment $C_3H_4O_3$. Hence, the formula to enter in IsoCor is $C_3H_4O_3$. In situations where a derivatization step is required before MS analysis, the elemental composition of the derivative (often common to several metabolites) must be declared if it is in the molecular entity that gives rise to the measured isotopic cluster. If this derivative moiety contains atoms corresponding to the isotopic tracer, these atoms are supposed to be at natural abundance for the isotopic tracer. Therefore the derivative moiety must be corrected for natural abundance. This type of correction may not be always requested for metabolite moiety. To allow the correction of all elements of the derivative moiety, its formula must be declared separately from that of the metabolite moiety. For example, alanine ($C_3H_7O_2N$) can be analyzed by GC-MS after t-butyl-dimethyl-silyl (TBDMS) derivatization. A fragment that is classically used for ^{13}C-metabolic flux analysis is the $[M\text{-}57]^-$ fragment that contains the compound of interest and two TBDMS groups, one of which lost the fragment $[C_4H_9]$. The elemental composition of the two TBDMS groups excluding that of the fragment—i.e., $Si_2C_8H_{21}$—must be declared in the "Derivative formula" field; meanwhile the elemental composition of the alanine moiety of the detected ion ($C_3H_5O_2N$) must be declared in the "Metabolite formula" field.

3. Error messages:
   - Errors during software launching:

     "*Isotopes.dat not found.*" Copy this file in the IsoCor's folder.

     "*Isotopes.dat is empty or its structure is incorrect.*" Edit the file to check its structure and correct the errors.
   - Errors during software launching:

     "*Some peaks of the isotopic cluster have negative intensities.*" Since amounts cannot be negative, negative intensities found in input files are set to 0.

     "*Incorrect file structure (line x).*" Edit the data file to correct the error found in line number *x*.
   - Errors during software utilization:

     "*Please load a data file.*" No data file has been loaded.

     "*The isotopic tracer must be present in the metabolite.*" The elemental formula of the metabolite does not contain any atom of the tracer element; please check the formula.

     "*No correction formula.*" Enter an elemental formula in the "Metabolite formula" field.

     "*The length of purity vector must be x for* `'element'`." The purity vector is too short or too long for the considered tracer element; please check it.

     "*The sum of purity vector must be 1.*" The sum of the proportions of each isotopes defined in the "Purity" field is not 1 (i.e., 100 %); please normalize the purity vector to 1.

     "*Check purity vector.*" There is a formal error in the purity vector (e.g., unauthorized characters such as letters); please check it.

     "*The length of the measured isotopic cluster is too short.*" For each isotopic cluster to correct, you must provide (at least) the intensities of peaks originating from the unlabelled to the uniformly labelled molecule, without missing value (e.g., peaks M_0 to M_3 for a C_3 compound).

     "*Problem in matrix size. Fragment does not contain enough atoms to generate this isotopic cluster.*" The length of the isotopic cluster is too long, or there is an error in the elemental formula of the metabolite and/or derivative.

Acknowledgements

The authors are thankful to members of MetaSys team (LISBP, Toulouse) for fruitful discussions. The PhD fellowship of P.M. was funded by the Institut National de la Recherche Agronomique

(INRA) [Program CJS]. The continuous support of Région Midi-Pyrénées, the European Regional Development Fund (ERDF), the French Ministry for Higher Education and Research, the SICOVAL, and INRA is gratefully acknowledged.

References

1. Nicolas C, Kiefer P et al (2007) Response of the central metabolism of Escherichia coli to modified expression of the gene encoding the glucose-6-phosphate dehydrogenase. FEBS Lett 581(20): 3771–3776
2. Perrenoud A, Sauer U (2005) Impact of global transcriptional regulation by ArcA, ArcB, Cra, Crp, Cya, Fnr, and Mlc on glucose catabolism in Escherichia coli. J Bacteriol 187(9):3171–3179
3. Kiefer P, Nicolas C et al (2007) Determination of carbon labeling distribution of intracellular metabolites from single fragment ions by ion chromatography tandem mass spectrometry. Anal Biochem 360(2):182–188
4. Rosman KJR, Taylor PDP (1998) Isotopic compositions of the elements 1997. Pure Appl Chem 70(1):217–235
5. van Winden WA, Wittmann C et al (2002) Correcting mass isotopomer distributions for naturally occurring isotopes. Biotechnol Bioeng 80(4):477–479
6. Millard P, Letisse F et al (2012) IsoCor: correcting MS data in isotope labeling experiments. Bioinformatics 28(9):1294–1296
7. Wittmann C (2007) Fluxome analysis using GC-MS. Microb Cell Fact 6:6
8. Wittmann C, Heinzle E (1999) Mass spectrometry for metabolic flux analysis. Biotechnol Bioeng 62(6):739–750
9. Rodriguez-Castrillon JA, Moldovan M et al (2008) Isotope pattern deconvolution as a tool to study iron metabolism in plants. Anal Bioanal Chem 390(2):579–590

Chapter 13

Steady-State ^{13}C Fluxomics Using OpenFLUX

Lake-Ee Quek and Lars K. Nielsen

Abstract

Metabolic flux estimation using ^{13}C isotopic tracers (^{13}C-MFA) provides a greater resolution of intracellular fluxes than using only cell growth and consumption/production rates. However, ^{13}C-MFA is computationally more demanding. A nonlinear least-square optimization process is employed to constrain metabolic fluxes using atom balance models and experimentally measured ^{13}C labelling pattern of intracellular or proteinogenic metabolites. OpenFLUX was therefore developed for the purpose of streamlining the computational workflow. Here, we describe in detail the computational procedure for performing ^{13}C-MFA using OpenFLUX. We also provide some helpful information on model reconstruction and GC-MS data treatment.

Key words Metabolic flux, ^{13}C tracer, Steady-state, Nonlinear optimization, OpenFLUX, Modeling, MATLAB

1 Introduction

Flux determination using ^{13}C-labelled substrates is an approach often adopted when conventional flux analysis is inadequate [1]. Inevitably, a subset of intracellular fluxes are poorly constrained by experimentally measured growth rate, substrate consumption rates, and product formation rates, particularly when cyclic pathways or large metabolic models are involved. The activity of these under-determined fluxes can be characterized by linear optimization, but the simulated flux ranges may not be physiologically realistic. By providing ^{13}C-labelled substrates, and subsequently measuring and de-convoluting the ^{13}C-labelling pattern, or the mass isotopomer distribution (MID), of metabolic products, one can quantify how these labelled metabolites have flowed and mixed intracellularly during the steady-state growth. This potentially provides experimental evidence on the flux of the intracellular reactions of interest (Fig. 1). Isotopic ^{13}C-MFA becomes

The online version of this chapter (doi:10.1007/978-1-4939-1170-7_13) contains supplementary material, which is available to authorized users.

Jens O. Krömer et al. (eds.), *Metabolic Flux Analysis: Methods and Protocols*, Methods in Molecular Biology, vol. 1191, DOI 10.1007/978-1-4939-1170-7_13, © Springer Science+Business Media New York 2014

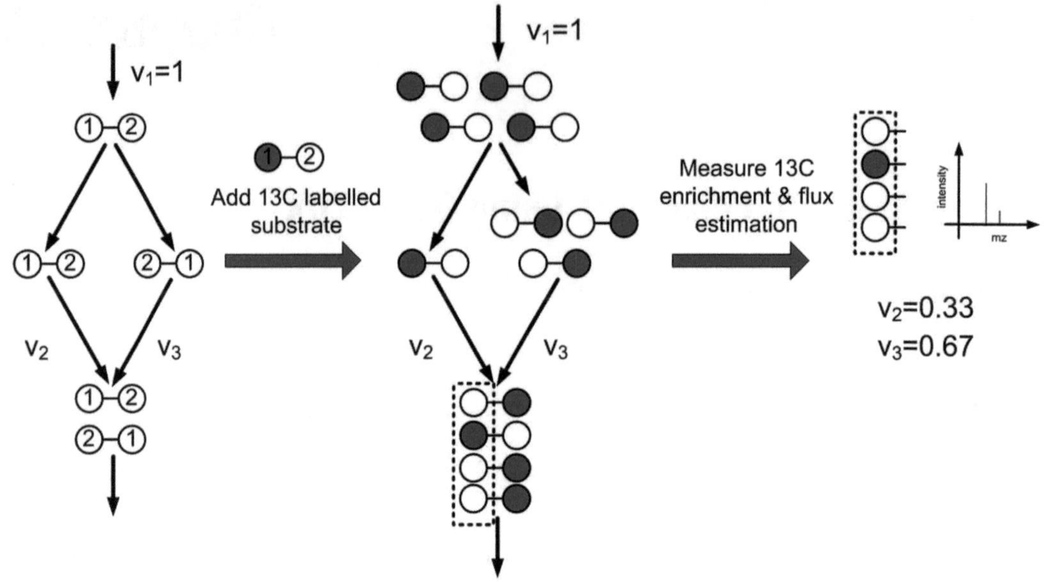

Fig. 1 An illustration on the use of labelled substrates for flux estimation. The backbone carbon of the input substrate is inverted when the right-hand side reaction pathway is used (v_3). The flux through v_2 and v_3 can be determined using ^{13}C enrichment pattern of the first carbon atom of the product

a computational or a modeling challenge once the labelling patterns of metabolites are made available.

In this chapter, we provide instruction on how to use OpenFLUX [2], a software that generates a MATLAB-readable atom balance model and performs flux estimation by nonlinear least-square optimization. The atom balance model describes the relevant carbon atom transitions from substrate to product across the entire metabolic network using elementary metabolite unit (EMU) [3]. The optimization then uses the atom balance model and the experimental MIDs as constraints to estimate fluxes (Fig. 2). We also provide advices on the metabolic model setup, and on the management of the ^{13}C-labelling data and data error. This chapter emphasizes on the OpenFLUX's methodology, while further information about OpenFLUX can be obtained from the original paper [2].

2 Materials

1. Personal computer including the following software: Spreadsheet (e.g., Microsoft Office Excel), MATLAB R2010a and Optimization Toolbox, OpenFLUX (version 2.1), and Java Runtime Environment (Oracle Corporation) (*see* **Note 1**).
2. Metabolic model (*see* **Note 2**).

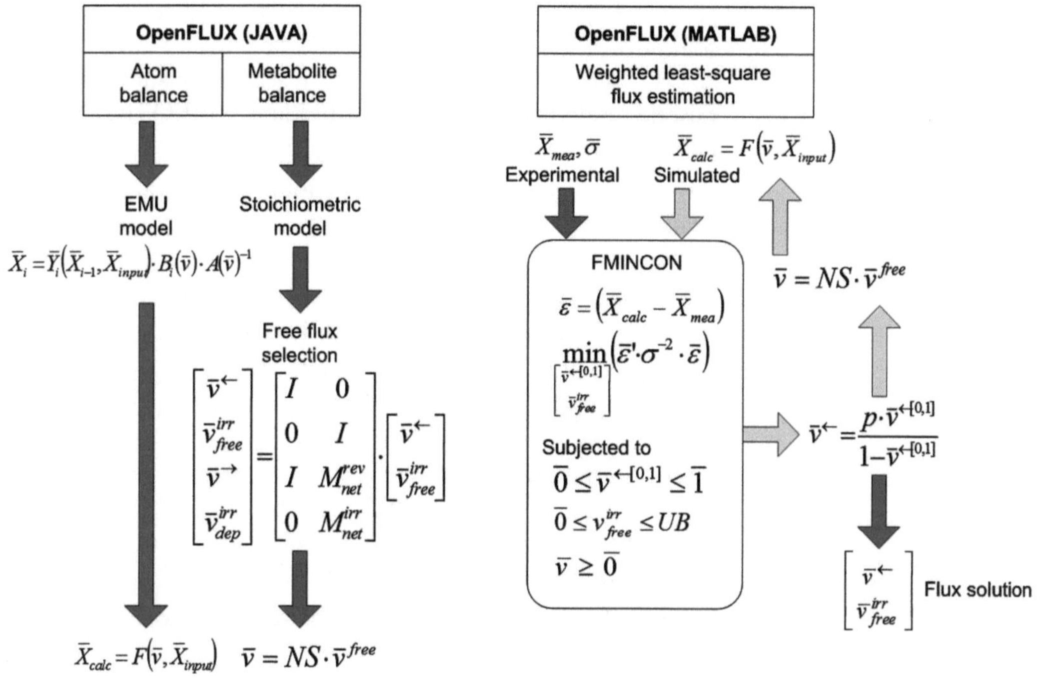

Fig. 2 A summary of OpenFLUX's two core components and its overall workflow

3. ^{13}C labelling intensity measurements and error from GC-MS (*see* **Note 3**).
4. Experimentally measured cell growth rate, substrate consumption rates, and product formation rates (*see* **Note 4**).

3 Methods

We will describe the steps required to estimate fluxes using ^{13}C labelling of metabolites and measured rates. The described protocol will be specific for GC-MS-derived ^{13}C enrichment data. We assume that the reader has a working metabolic network and has performed preliminary flux analyses, and that the network under investigation has under-determined fluxes (*see* **Note 5**). This method section will be split into five subsections, separately detailing the process of (a) converting the conventional flux model to ^{13}C-MFA model, (b) using OpenFLUX to generate a MATLAB-readable metabolic model and to iteratively check the model, (c) running optimization and interpreting optimization results, (d) running sensitivity analysis and interpreting results, and (e) supplementary information to improve OpenFLUX's efficiency. We have provided an *E. coli* metabolic model as working example as the user follows the instructions (*see* **Note 1**).

3.1 ^{13}C-MFA Model Reconstruction

1. Starting from the top left cell in a new spreadsheet, create seven column headers: "rxnID", "rxnEQ", "cTrans", "rates", "type", "basis", and "deviation". Copy the reaction equations from the existing model and paste under the heading "rxnEQ" (*see* **Note 6**).
2. Distinguish whether a reaction participates in the transfer of carbon label from substrate to product. Put a tag "B" under heading "type" for reactions that do not participate in the carbon transfer (*see* **Note 7**).
3. Using flux variability analysis, identify the directionality of a reaction: forward, reverse, or both (*see* **Note 5**).
4. For a reversible reaction (i.e., has both positive and negative fluxes) that participates in the transfer of carbon label, insert a new row immediately below the reaction and add the same reaction equation but written in the reverse direction. In the "type" column, put a tag "FR" for the original forward reaction, and "R" for the reverse reaction. Also put "X" in the "basis" column for reactions with "R" tag.
5. Rewrite the reaction equation in the reverse direction if a reaction has only negative flux.
6. Put a tag "F" for all the remaining reactions that are still empty in column "type".
7. For the "B"-type reactions that have both positive and negative fluxes, replace the original tag with "BR" (*see* **Note 8**).
8. Number the reactions in the "rxnID" column.
9. Provide the atom transition equations for reactions "F", "FR", and "R" types. Use alphabet notations, except capital "X". "X" is reserved for cofactors or metabolites deliberately ignored/skipped in the atom transition equation (*see* **Note 9**).
10. Add measured rates in the "basis" column, and optionally the corresponding measurement errors in the "deviation" column (*see* **Note 10**).
11. Put "X" in the "basis" column corresponding to reactions that are preferably assigned as free fluxes. This step is optional.
12. Identify the boundary metabolites, and put the list under the heading "excludedMetabolites" (*see* **Note 11**). Optionally, energy and redox cofactors (e.g., ATP, NADH, NAD^+, NADPH, $NADP^+$) can be put in this list to omit cofactor balances.
13. Add names of the metabolites being quantified by GC-MS using the same naming convention as in the reaction table above under the heading "simulatedMDVs" (*see* **Note 12**). Binary integers [0, 1] are used in the suffixes (after "#") to distinguish the different ion clusters of the same metabolite; for example, Alanine_232 and Alanine_260 are shown as "M_ala_L_c#011" and "M_ala_L_c#111", respectively. In order to minimize the

size of the atom balance model, the list should only contain ion clusters whose data are used in the optimization.

14. List the names of the input substrates with known MID under the heading "inputSubstrates" (*see* **Note 13**).
15. Provide the MID data under the heading "measurements", and the errors of the MID data under the heading "error" (*see* **Note 14**). A minimum absolute error of 0.003 is used (*see* **Note 15**). MIDs can be provided without any correction for interference originating from non-backbone carbon (*see* **Note 16**).
16. Save spreadsheet.

3.2 Spreadsheet Model Parsing and Iterative Checking

1. Create project folder.
2. Save spreadsheet as a text-delimited text file in the project folder.
3. Run OpenFLUX (JAVA). Point path to the project folder and subsequently to the text model file. Click "Overwrite", and then "Generate Model".
4. Wait for the message "MODEL GENERATION SUCCESSFUL" to appear. Otherwise read status report generated by OpenFLUX (*see* **Note 17**). Fix shown errors, save text (model) file, and click "Generate model" again. Repeat until no more error is shown. Preferably, close OpenFLUX and regenerate model again (*see* **Note 18**).
5. Run MATLAB. Change current folder to project directory.
6. In MATLAB command line, type "start13OF" and enter. Check MATLAB path if OpenFLUX did not start (*see* **Note 1**).
7. Run Task 7. Choose "Yes" when asked to "incorporate flux basis and deviation into measurement and error vector". Choose "Yes" when asked to "check flux range". Proceed if no error is reported. Otherwise revise and regenerate model (*see* **Notes 19** and **20**).
8. Use default upper flux boundary values. Check flux range for all reactions. Make sure that all reactions can carry nonzero fluxes, and have an exitflag of 1. Revise and regenerate metabolic model if necessary.

3.3 Flux Estimation

1. Edit file "x_sim.m" contained in the project folder such that the simulated MIDs are consistent with the measured MIDs in terms of normalization and dimension (*see* **Note 21**). Follow Fig. 3 as an example on how to process signals from Alanine_232 and Alanine_260.
2. Run MATLAB. Change current folder to project directory.
3. In MATLAB command line, type "start13OF" and enter. Check MATLAB path if OpenFLUX did not start (*see* **Note 1**).

GC-MS spectrum and SIM peaks for Ala 232 and 260

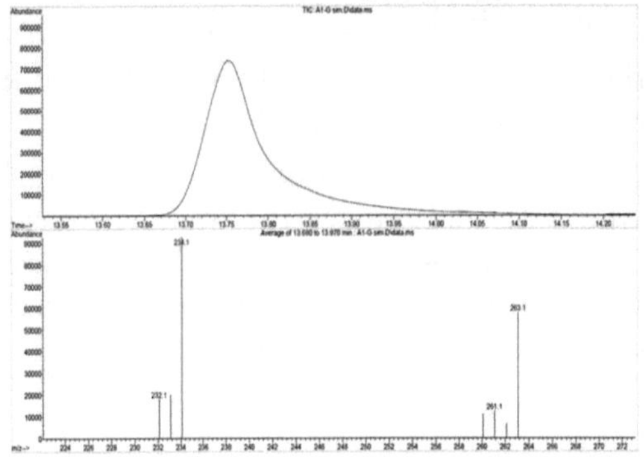

MID data table

ion		mz	signal	normalized signal	
Ala_232	m	232	18109	0.138	M_ala_L_c_011_measured
	m+1	233	20092	0.153	
	m+2	234	92711	~~0.708~~	Largest MID removed
Ala_260	m	260	11117	0.127	M_ala_L_c_111_measured
	m+1	261	12409	0.141	
	m+2	262	6525	0.074	
	m+3	263	57777	~~0.658~~	Largest MID removed

Incorporating natural enrichment and normalizing ALA_232 and ALA_260 EMUs

$$\text{M\_ala\_L\_c\_011\_calculated} = \frac{\text{corrMatGen}([1,2], 2, \text{'C8H26N1O1Si2'}) \cdot \text{M\_ala\_L\_c\_011'}}{\sum \text{corrMatGen}([1,2,3], 2, \text{'C8H26N1O1Si2'}) \cdot \text{M\_ala\_L\_c\_011'}}$$

$$\text{M\_ala\_L\_c\_111\_calculated} = \frac{\text{corrMatGen}([1,2,3], 2, \text{'C8H26N1O2Si2'}) \cdot \text{M\_ala\_L\_c\_111'}}{\sum \text{corrMatGen}([1,2,3,4], 2, \text{'C8H26N1O2Si2'}) \cdot \text{M\_ala\_L\_c\_111'}}$$

Modifications in x_sim.m

```
x_calc = [
    M_ala_L_c_011_calculated
    M_ala_L_c_111_calculated
    .
    .
];
```

Fig. 3 A demonstration on how to modify OpenFLUX's simulated MIDs to match the normalized GC-MS signals obtained for alanine 232 and 260. For the *E. coli* MID data, we chose to remove the largest MID from every ion cluster such that the remaining MIDs are independent after normalization using all $n+1$ signals (n being the number of backbone carbon atoms)

4. Run Task 1. OpenFLUX will check for optimization conditions and parameter saved from the previous runs. Skip to **step 14** if the saved data is reused. For most of the user prompts, default values (in brackets) can be used by immediately pressing "Enter".
5. Choose "Yes" to indicate that measurement error file is available. Otherwise specify a constant relative error for all measurements.
6. Set "100" to indicate the order of magnitude of fluxes (*see* **Note 22**).
7. Use default values for the lower and upper boundary values of the forward and reversible fluxes.
8. Set the number of optimization iterations to "10" (*see* **Note 23**).
9. Set "1" to allow optimization to start with feasible random fluxes (*see* **Note 24**).
10. Set optimality and constraint violation tolerance to default values of 1e−4 and 1e−3, respectively.
11. Set solution clusters to default value of 4. Choose the first "1" cluster to use as output solution, which is the best solution cluster (*see* **Note 25**). Choose cutoff of 1e6 to remove optimizations that have terminated poorly.
12. Choose whether to incorporate measured rates with standard error provided as part of the least-square optimization, or just use the measured rates deterministically to constrain the fluxes (*see* **Note 26**).

	A	B	C	D	E	
1	cluster					
2		1	2	3	4	
3	no of soln in cluster					
4		7	1	1	1	Number of solutions in the cluster
5	best free fluxes					
6	v216	0.995 (0)	0.602 (0)	0.823 (0)	0.995 (0)	
7	v220	0.72 (0.177)	0.391 (0)	0.524 (0)	0.995 (0)	
8	v222	0.938 (0.00687)	0.995 (0)	0.995 (0)	0.995 (0)	
9	v224	0.000551 (0.42)	0.97 (0)	0.995 (0)	0.995 (0)	
10	v226	0.995 (0.496)	0.974 (0)	0.995 (0)	0.995 (0)	Best free flux (std. dev.)
11	v228	0.995 (0.398)	0.468 (0)	0.961 (0)	0.99 (0)	
12	v230	2.87e-018 (0.511)	0.62 (0)	0.231 (0)	0.99 (0)	
40	v368	27.6 (0.182)	32.2 (0)	1.13e+003 (0)	1.48e+003 (0)	
41	v180	779 (454)	561 (0)	249 (0)	655 (0)	
42	v181	1.32e+003 (288)	1.41e+003 (0)	389 (0)	505 (0)	
43	fval					
44		44.8 (0.509)	156 (0)	1.05e+003 (0)	1.34e+004 (0)	Smallest residual error (std. dev.)
45	estimated fluxes					
46	v1	1.46E+03	1.56E+03	531	647	
47	v2	1.46E+03	1.56E+03	531	647	
48	v3	661	707	195	252	
49	v4	-501	-279	28.6	-378	Flux distribution of the best solution in the cluster
50	v5	114	114	114	114	
51	v6	16.7	16.7	16.7	16.7	
411	v366	1.01	1.01	1.01	1.01	
412	v367	67.8	68.1	68.5	74	
413	v368	27.6	32.2	1.13E+03	1.48E+03	
414	chi-square	71				
415	minimum r	44.8				

Fig. 4 Contents in a flux estimation result file "results_PE.txt"

13. Add equality constraints if required (*see* **Note 27**).
14. Accept and save (optional) the current configuration for the optimization.
15. Choose whether to specify input substrates in the MATLAB environment or by using the file "substrate_EMU.m" (*see* **Note 28**).
16. Provide the positional enrichment of the input substrates when prompted. If a different mixture of the same substrate is used, then provide another positional enrichment of the same substrate, as well as the mixing ratio when prompted at the end. Continue until all input substrates have been configured. Accept and save configurations for the input substrates. If input substrates have been provided beforehand, then the user can choose whether to load an existing configuration file.
17. Choose "Yes" to run Task 1 now.
18. Inspect the result files "results_PE.txt", "x_solution.txt", and "MID_solution.txt" once the optimization is completed (Fig. 4 and *see* **Note 29**).

3.4 Sensitivity Analysis (Nonlinear)

1. Run MATLAB. Change current folder to project directory.
2. In MATLAB command line, type "start13OF" and enter. Check MATLAB path if OpenFLUX did not start (*see* **Note 1**).
3. Run Task 1 (*see* **Note 30**). OpenFLUX will check for optimization conditions and parameter saved from the previous runs.
4. Choose "Yes" to indicate that measurement error file is available. Otherwise specify a constant relative error for all measurements.
5. Set "100" to indicate the order of magnitude of fluxes (*see* **Note 22**).
6. Use default values for the lower and upper boundary values of the forward and reversible fluxes.
7. Set optimality and constraint violation tolerance to default values of 1e−4 and 1e−3, respectively.
8. Set the flux parameter confidence interval to 95 %.
9. Set the minimum sensitivity cutoff to 100 % (*see* **Note 31**).
10. Set the maximum difference in convergence cutoff to 5 % (*see* **Note 32**).
11. Choose "No" such that OpenFLUX does not perform fine iteration, particularly for the initial run (*see* **Note 33**).
12. Choose whether to incorporate measured rates with standard error provided as part of the least-square optimization, or just use the measured rates deterministically to constrain the fluxes (*see* **Note 26**).
13. List the flux parameters whose confidence intervals need to be determined. The flux parameters can be a nonlinear function (*see* **Note 34**).
14. Add equality constraints if required (*see* **Note 27**).
15. Accept and save (optional) the current configuration for the optimization.
16. Use the same input substrate configurations specified in Task 1.
17. Accept and save (optional) input substrate configuration.
18. Choose "Yes" to run Task 2 now.
19. Inspect the result files "results_CI.txt" once the optimization is completed. Observe the pair of quality status for each of the flux parameters. Seek alternative approaches if "CHOKE" is observed (Fig. 5 and *see* **Note 35**).

3.5 Supporting Capabilities in OpenFLUX

3.5.1 "S"-Type Reactions

"S"-type reactions do not appear in the stoichiometric matrix. This makes it possible to map ^{13}C enrichment of precursor metabolites in the central metabolism to the final products without invoking metabolite balances. Users can replace full biosynthetic pathways with "S"-type reactions, and at the same time only specify drains of precursors in the central metabolism (e.g., pyruvate, acetyl-CoA, oxo-glutarate, and oxaloacetate).

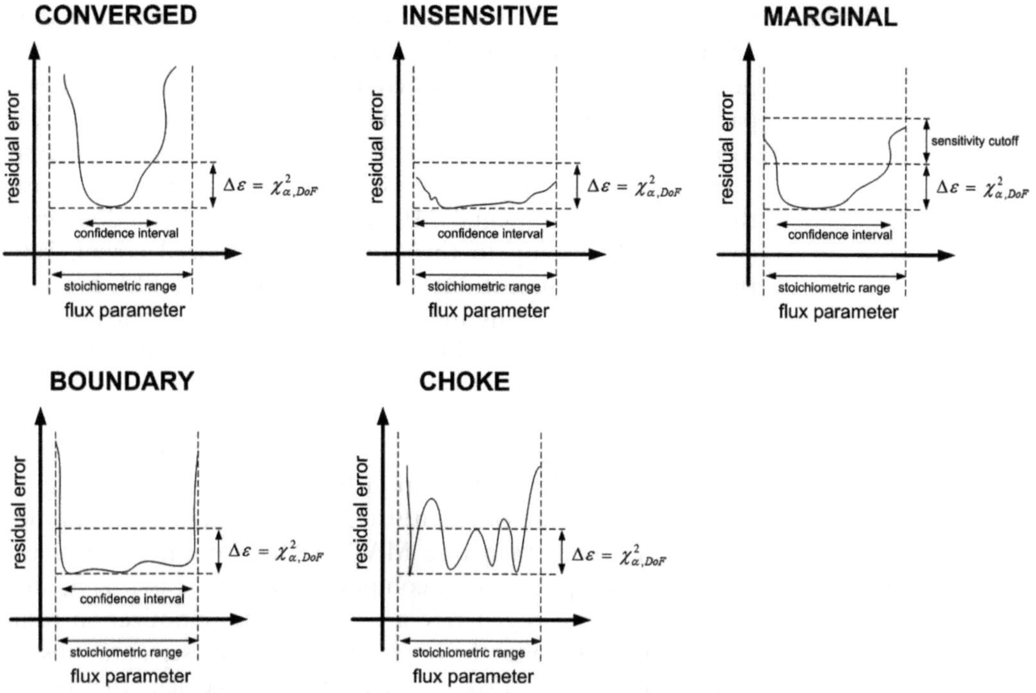

Fig. 5 A graphical illustration of the five statuses describing the quality of the sensitivity analysis results with respect to the residual error landscape

3.5.2 Simulations of MIDs and First-Order Derivatives

Using Task 8, users can generate hypothetical MIDs by providing a flux distribution ($\overline{v}$) and by using the function "mdvGenerator($\overline{v}$)". The MIDs to be generated can be modified in "x_sim.m". Similarly, "[mid dee_mid] = mdvGenerator($\overline{v}$)" can be used to generate first-order derivatives of the MIDs with respect to the free fluxes. Unlike the optimization in Task 1, the reverse fluxes are not compactified. The file "dee_x_sim.m" must be modified accordingly to reflect the existing content in "x_sim.m".

3.5.3 Avoid Repeated Use of corrMatGen1()

Instead of generating the same correction matrices whenever "x_sim.m" is called by MATLAB, it is possible to generate and save to disk the required correction matrices prior to running Task 1. In "preSolverScript.m", these saved matrices can be reloaded during the initialization phase of OpenFLUX and converted into global variables such that they are made available to "x_sim.m".

3.5.4 Manually Changing Configuration Files

The files "setting_PE.mat" and "setting_CI.mat" contain the saved configurations for Task 1 and 2, respectively. The user can directly change the content of these files in MATLAB. Particularly, the user can bypass OpenFLUX's prompts and provide the full list of flux parameters used in sensitivity analysis. Similarly, the user can change positional enrichment of input substrates contained in "inputSubConfig.mat".

3.5.5 Sensitivity Analysis by Monte-Carlo

Confidence interval of flux parameters can also be determined by a Monte-Carlo approach. This is an alternative approach when the nonlinear optimization approach does not return successful results, particularly for numerically challenging problems. OpenFLUX's Monte-Carlo approach is essentially a repeated process of running Task 1 for a large number of iterations, with each iteration using the original experimental MIDs corrupted with random errors (weighted according to their respective standard error). When the series of optimizations are completed, OpenFLUX collates all of the best solutions and generates a confidence interval using the minimum and maximum value found for the given flux parameter.

3.5.6 Running Simultaneous Optimizations

A limitation of OpenFLUX is that it relies on multiple optimizations such that the best solution found is most likely the global minimum. A feature of OpenFLUX is the ability to run optimizations using more than one MATLAB environment, such as in multi-core computers or in computing clusters. Tasks 4, 5, and 6 were designed to regenerate a report of the optimization results once the user compiles the results obtained from separate runs. For flux estimation (i.e., Task 1), the user needs to collate the contents in all the separate "results_flux.txt", "results_fval.txt", and "results_x.txt" files back into a single file, respectively. For sensitivity analysis (i.e., Task 2), the user needs to collate the contents in all the separate "resultsCI_raw.mat" files back into a single file.

3.5.7 Multiple Labelling Experiments in a Single Optimization

It is possible to run a single optimization that utilizes MIDs obtained from more than one labelling experiments of the same growth condition. In "postXSimScript.m", the user can sequentially load different labelling of input substrates and calculate the resulting MIDs using the same flux distribution vector. The subroutine required is shown (Fig. 6). Beforehand, the user must provide the various versions of "inputSubEMU.mat" to reflect the different labelling experiments.

4 Notes

1. OpenFLUX can be downloaded from SourceForge (http://openflux.sourceforge.net/). Unzip application. In MATLAB, add path to the "mFiles" folder.
2. The main purpose of ^{13}C-MFA is to resolve under-determined fluxes in central metabolism, as well as the relative activity of pathways producing the same metabolite [1]. We used a relatively large ^{13}C-MFA *E. coli* model in this demonstration for the purpose of (a) providing detailed atom mappings of precursors in the central carbon metabolism to the amino acids,

```
x_calc_1 = x_calc;                         x_calc = [x_calc_1
                                                     x_calc_2
[ load inputSubEMU_2 ]                               x_calc_3
[ EMUModel           ]                               .
[ x_sim              ]                               .
[ x_calc_2 = x_calc; ]                               x_calc_i
                                                     ];
[ load inputSubEMU_3 ]
[ EMUModel           ]
[ x_sim              ]
[ x_calc_3 = x_calc; ]

[         .          ]
[         .          ]

[ load inputSubEMU_i ]
[ EMUModel           ]
[ x_sim              ]
[ x_calc_i = x_calc; ] i^th experiment
```

Fig. 6 MATLAB subroutine to be added into "postXSimScript.m" in order to perform optimization of multiple labelling experiments in a single optimization

(b) demonstrating the use of biosynthetic drains to constrain fluxes in the central metabolism, and (c) minimizing the use of reaction lumping. The model used is a reduced version of the *E. coli* genome-scale model iAF1260 [4].

3. We limited the scope of this chapter to GC-MS-derived ^{13}C enrichment data, and did not describe how to use LC-MS/MS- or NMR-derived data [5, 6]. Nevertheless, OpenFLUX's simulated outputs can be modified such that they are consistent with the experimental data. We do not cover how to extract signal intensities from the MS spectra [7].
4. These measurements can be in the form of yields, preferably with the denominator being the substrate uptake (e.g., Y_{ps}, Y_{xs}). This is because ^{13}C-MFA often uses a normalized flux distribution, e.g., normalized to 100 % of glucose uptake.
5. Flux variability analysis should be performed on the network beforehand [8], with the flux model constrained using the measured rates (or yield). This will help the user characterize the directionality of a reaction, and troubleshoot if a reaction has zero flux or is overly constrained by the measured rates. The model must have under-determined reactions; otherwise OpenFLUX will fail to run. OpenFLUX has built-in algorithms to perform flux variability analysis. All reactions are kept in the forward direction. An exception is that reactions that do not participate in carbon transfer can be kept as reversible.

6. The spreadsheet interface is designed to contain all information and data required to perform ^{13}C-MFA such that user inputs are minimized in the MATLAB environment. The first seven columns are the "active zones"; all other annotations can go elsewhere. Further information and data can be provided at the bottom of the reaction table. They are marked with "##" for the header, and "#" for the accompanying data/information. The naming of the headers must not be modified.
7. OpenFLUX creates the atom balance model, i.e., elementary metabolite unit (EMU), via a backward tracing algorithm, starting from a product, whose enrichment pattern has been quantified by GC-MS, tracing to the input substrates of known labelling state [3]. It is not necessary to provide atom transitions for reactions that do not participate in the carbon transfer from the input substrates to the measured products. For example, if only the labellings of proteinogenic amino acids are measured, then it is not necessary to provide atom transition equations for reactions in the nucleotide, lipid, and carbohydrate synthesis pathways.
8. There are three types of reactions under OpenFLUX's consideration, namely reactions that involve (a) only carbon atom transfer ("S" tag), (b) only metabolite balance ("B" and "BR" tag), or (c) both ("F", "FR", and "R" tag). This convention keeps the size of the atom balance model to a minimum, and hence improves computational speed. Reactions that involve only carbon atom transfer are artificial reactions that are added for the sole purpose of directly mapping labelling of precursor metabolite to end products without involving the flux vector. These reactions have "S" tag in the "type" column. The flux of reactions that involves carbon atom transfer is strictly positive, whereas the flux of reversible reactions that participate only in metabolite balance can be negative.
9. Take care when writing the atom transition equations. Mistakes come from using different conventions to count and label carbon atoms in sequence with numbers, particularly for branched or cyclic molecules. Do not directly copy atom transition equations for various sources without any further processing. To be consistent, first establish a convention to count the carbon atom sequence (e.g., IUPAC). Then refer to literature sources when interpreting how carbon atoms transfer from substrate to product. Reproduce these atom transitions with respect to the counting convention that has been established. KEGG'S RPAIR provides graphical notation on atom transfer that can be easily interpreted.
10. Make sure that there is no gross error contained in the measured rates. If the measured rates form a redundant set, then use gross-measurement error analysis to adjust the measured

rates prior to ^{13}C-MFA. When measurement errors are provided in the "deviation' column, OpenFLUX will use the measured rates in the least-square fitting; otherwise the measured rates will be used deterministically. To avoid numerical problems during optimization, it is important to scale all the (adjusted) measured rates such that the basis of the problem, e.g., glucose uptake fluxes, is sufficiently large (100–1,000).

11. Boundary metabolites are external metabolites that are excluded from the stoichiometric matrix when performing flux analysis.
12. In the demonstration model, we used labelling of alanine, glycine, valine, leucine, isoleucine, serine, threonine, phenylalanine, aspartate, glutamate, lysine, histidine, and tyrosine. The signal quality for the other amino acids was poor. Refer to the Excel spreadsheet to inspect the complete list of ion clusters (*see* **Supplementary File 1**).
13. We used a 1:3 mix of 1-^{13}C glucose and U-^{13}C glucose in this *E. coli* demonstration. However, in the model, they are separately represented as two different glucose sources, "M_glc_DS_b" and "M_glc_DU_b", respectively. This is because we were attempting to back-calculate the actual mixing ratio. Alternatively, the user can provide one glucose source, but provide the mixing ratio later when prompted in the MATLAB environment.
14. MIDs are often represented as fractional intensities, whereby the sum of intensities from a given ion cluster is one. The number of GC-MS signals per ion cluster quantified in an MS spectrum is often $n+1$, with n being the number of backbone carbon. However, depending on equipment sensitivity and configuration, data quality, and user preference, the actual signals used in the fitting exercise, as well as the method used to normalize these signals, may vary. OpenFLUX can account for these variations, but requires the user to provide a MATLAB subroutine/script such that the optimization can reproduce the normalization process performed on the raw GC-MS intensities.
15. A minimum normalized intensity error cutoff of 0.003 was derived from the average difference found comparing experimental and theoretical MIDs of naturally labelled proteinogenic amino acids. We believe that the high accuracy and high reproducibility of GC-MS data often underestimate the actual error contained in the measured MIDs.
16. OpenFLUX can incorporate the natural interference prior to performing the least-square fitting. See supplementary material for the full setup of simulated MIDs.

17. OpenFLUX has some (but limited) capability in reporting inconsistencies found in the metabolic model. It can check for obvious and not-so-obvious inconsistencies in the naming of metabolites, the number of backbone carbon, the number of reactants and products, and the presence of singleton metabolites.
18. All residual variables kept in memory are cleared by reinitializing OpenFLUX. This is to avoid accidental carryover of errors when parsing the model.
19. ^{13}C-MFA models tend to be small or moderate in scale. For large-scale models, however, MATLAB's "linprog" function may not be as effective compared to other commercial solvers, for example, MOSEK, CPLEX, or GUROBI. For CPLEX and GUROBI, the user will need to separately provide a wrapper "linprog" function.
20. Gross-measurement error analysis does not adjust fluxes such that calculated fluxes satisfy reversibility constraints. It is possible that some of the adjusted fluxes will result in infeasible flux solution. If this is the case, then a subset of the adjusted fluxes must be removed, or alternatively used in the least-square fitting. To help troubleshoot the problem, choose "Yes" when asked to "incorporate flux basis and deviation into measurement and error vector".
21. The original variables in "x_calc" vector are the EMUs describing the labelling distribution of only the backbone carbon atoms [3]. The naturally enriched non-backbone carbon atoms are incorporated into these variables such that the final distributions are analogous to the normalized experimental data. Use OpenFLUX's function "corrMatGen1(x,y,molform)", where "x" is a vector of the indices of the measured signals used for fitting, "y" is the length of the backbone carbon, and "molform" is the molecular formula of the ion cluster minus the number of backbone carbon. Signal truncation is also introduced here by removing unwanted indices in vector "x" (Fig. 3). Refer to the provided Excel spreadsheet (2nd tab) on how to edit the file "x_sim.m" (*see* **Supplementary File 1**).
22. This "flux order-of-magnitude" parameter is used to compactify reversible fluxes, and to set the default boundary values for forward fluxes [2].
23. OpenFLUX relies on multiple optimization runs to find the minimum residual error. Particularly for numerically difficult problems, it is important to provide enough optimization iterations such that the smallest residual error found is more likely the global minimum.
24. Change to "0" if optimization is not starting when running Task 1. This allows OpenFLUX to start optimization with any flux values, instead of being limited to feasible fluxes.

25. OpenFLUX produces a report that clusters solutions of similar residual error from multiple runs. The user can then inspect the solution range of free fluxes within a cluster, and the best flux distribution from the given cluster.
26. By choosing "No", the user is putting greater confidence in the measured rates over the labelling data. By choosing "Yes", the measure rates, weighted using the respective standard errors, will be used in the least-square calculations. The latter option increases the optimization time due to the increase in optimized flux parameters. Choose "Yes" if the measured rates are over-constraining the fluxes and OpenFLUX is unable to produce any feasible flux solutions (*see* **Note 20**).
27. The equality constraints are expressed as a function of fluxes, with the left-hand side of the equation being zero "$0 = f(\overline{v})$". The right-hand side function can be nonlinear.
28. In the MATLAB environment, the user can specify positional enrichment pattern (or atom activity vector, AAV) of input substrates [9]. Alternatively, the user can specify the MIDs of the input substrate EMUs listed in substrate_EMU.m.
29. The file "results_PE.txt" contains the summary of the optimization results from all iterations, while "x_solution.txt" and "MID_solution.txt" contain the best set of free fluxes and simulated MIDs. Results from the individual optimization iterations are saved in "results_x0.txt" (initial random free flux), "results_x.txt" (free flux solutions), "results_mid.txt" (simulated MIDs), "results_fval.txt" (residual errors), "results_flux.txt" (flux distributions), and "results_exitflag.txt" (optimization termination condition).
30. Sensitivity analysis is best performed when the residual error is sufficiently low, typically below the $\chi^2_{\alpha,\mathrm{DoF}}$ cutoff, with α significance (e.g., 0.05) and DoF being the number of independent measurements minus the number of free fluxes. Sensitivity analysis tends to underestimate the actual flux confidence intervals if only a small increase is allowed in the background of a large residual error.
31. Sensitivity analysis involves moving the boundary (i.e., confidence interval) of a given flux parameter with $\Delta\epsilon \leq \chi^2_{\alpha,\mathrm{DoF}}$ [2, 10]. A flux parameter is considered insensitive if moving the flux parameter to the upper or the lower stoichiometric boundary values, respectively, still results in $\Delta\epsilon \leq \chi^2_{\alpha,\mathrm{DoF}}$. If $\Delta\epsilon \leq \chi^2_{\alpha,\mathrm{DoF}} \cdot (1 + \text{minimum sensitivity cutoff \%})$, then the flux parameter is considered as marginally sensitive (Fig. 5).
32. It is difficult for the solver to satisfy the constraint $\Delta\epsilon = \chi^2_{\alpha,\mathrm{DoF}}$. Therefore, as long as $\Delta\epsilon \geq \chi^2_{\alpha,\mathrm{DoF}} \cdot (1 - \text{maximum convergence difference})$, OpenFLUX considers that the optimized flux parameter is adequately close to the true boundary value.

33. OpenFLUX can sub-split the problem and progressively move the flux parameter to the true boundary value. Choose the "Yes" option only if OpenFLUX has difficulty finding the flux confidence interval.
34. At any point in providing the flux parameters, the user can delete, modify, or review the listed flux parameters.
35. There are five different statuses: CONVERGED, INSENSITIVE, MARGINAL, BOUNDARY, and CHOKE (Fig. 5). CHOKE indicates that OpenFLUX failed to find an optimum solution, i.e., boundary value, that satisfies the constraint $\Delta\epsilon \approx \chi^2_{\alpha,\mathrm{DoF}}$. Alternative approaches include a rescale or a transform of the flux parameter, use of sensitivity analysis by Monte-Carlo, or use of a fine-iteration approach.

References

1. Wittmann C (2007) Fluxome analysis using GC-MS. Microb Cell Fact 6:6
2. Quek LE, Wittmann C, Nielsen LK, Kromer JO (2009) OpenFLUX: efficient modelling software for ^{13}C-based metabolic flux analysis. Microb Cell Fact 8:25
3. Antoniewicz MR, Kelleher JK, Stephanopoulos G (2007) Elementary metabolite units (EMU): a novel framework for modeling isotopic distributions. Metab Eng 9:68–86
4. Feist AM, Henry CS, Reed JL, Krummenacker M, Joyce AR, Karp PD, Broadbelt LJ, Hatzimanikatis V, Palsson BO (2007) A genome-scale metabolic reconstruction for *Escherichia coli* K-12 MG1655 that accounts for 1260 ORFs and thermodynamic information. Mol Syst Biol 3:121
5. Wittmann C (2002) Metabolic flux analysis using mass spectrometry. Adv Biochem Eng Biotechnol 74:39–64
6. Zupke C, Stephanopoulos G (1995) Intracellular flux analysis in hybridomas using mass balances and in vitro ^{13}C nmr. Biotechnol Bioeng 45:292–303
7. Antoniewicz MR, Kelleher JK, Stephanopoulos G (2007) Accurate assessment of amino acid mass isotopomer distributions for metabolic flux analysis. Anal Chem 79:7554–7559
8. Mahadevan R, Schilling CH (2003) The effects of alternate optimal solutions in constraint-based genome-scale metabolic models. Metab Eng 5:264–276
9. Zupke C, Stephanopoulos G (1994) Modeling of isotope distributions and intracellular fluxes in metabolic networks using atom mapping matrixes. Biotechnol Prog 10:489–498
10. Antoniewicz MR, Kelleher JK, Stephanopoulos G (2006) Determination of confidence intervals of metabolic fluxes estimated from stable isotope measurements. Metab Eng 8:324–337

Chapter 14

Flux Visualization Using VANTED/FluxMap

Christian Krach, Astrid Junker, Hendrik Rohn, Falk Schreiber, and Björn H. Junker

Abstract

The calculation of metabolic fluxes has been shown to be a valuable asset in systems biology. Several procedures are commonly used to achieve this. Flux balance analyses or metabolic flux analyses usually result in a list of reaction rates (fluxes) provided in a spreadsheet format. This makes it difficult to quickly assess general characteristics of the solution. A fast and easy mapping of these results to a graphical map template facilitates an easy visual data inspection. Here, we describe a protocol that helps in setting up user-specific network templates, mapping flux results to it, and creating multiple exportable flux maps at one time.

Key words Visualization, VANTED, FluxMap, Java, Automation

1 Introduction

The systemic analysis of different cellular components such as DNA, RNA, proteins, and metabolites is represented by the corresponding -omics approaches such as genomics, transcriptomics, proteomics, and metabolomics, respectively. An overall aim of systems biology is the integrated analysis of different -omics domains, which is able to provide a holistic understanding of a biological system [1].

Fluxomics, another -omics field, has gained importance as an efficient tool in metabolic engineering. Metabolic fluxes are defined as the turnover of molecules through a specific metabolic route. In accordance with a systems biology approach, a specific flux value, given in mole per time and mass/weight unit, integrates multiple levels of information connected to a specific metabolic reaction. It is determined by metabolite concentrations and enzyme activities which in turn can be under the influence of different regulatory events such as transcriptional regulation of gene expression, mRNA stability/degradation, posttranslational protein modifications, or feedback loops between substrate/product concentrations and enzyme activities.

Jens O. Krömer et al. (eds.), *Metabolic Flux Analysis: Methods and Protocols*, Methods in Molecular Biology, vol. 1191, DOI 10.1007/978-1-4939-1170-7_14, © Springer Science+Business Media New York 2014

Therefore, on the one hand, a flux value represents a multimodal type of data as it integrates various types of biological information [2]. On the other hand, flux analysis is mostly performed on the level of metabolic networks, which describe the entire set of metabolic reactions in a given biological system such as an organism, an organ/tissue, or a cell type and allow for a detailed analysis of specific metabolic functions. Changing a flux through a specific metabolic reaction (as achieved by gene deletion, mutations) is supposed to affect the flux through other reactions even in distant parts of the metabolic network. Thus, on a network scale, it is possible to monitor global responses of the metabolic network, the state of which is known to largely influence the performance of a biological system.

Metabolic fluxes can be measured using wet lab methods such as metabolic flux analysis (MFA) by feeding ^{13}C-labeled substrates and determining labeling patterns of intracellular metabolites, which are then used to estimate flux distributions in the metabolic network [3]. Alternatively, stoichiometric constraints in metabolic networks are used as a basis for optimization-based simulation of flux distributions (such as flux balance analysis (FBA); [4]). MFA and FBA are based on different approaches, but the formats of their results resemble each other. Results of both methods basically consist a of two-columned TXT or XLS sheet assigning flux values to the respective metabolic reactions. This tabular format makes it very difficult to evaluate flux distributions. Especially for a large number of fluxes it is inevitable to employ visualization approaches, which integrate the topology of the metabolic network and flux measurements. Network-based visualization of flux data substantially facilitates data interpretation and the gain of knowledge from large flux data sets as it sets into context experimental/simulation-derived flux values and metabolic pathways. Thereby significant changes are highlighted and, at a glance, can be assigned by the user to specific pathways in order to draw biological conclusions. In principle flux values can be mapped to any part of the metabolic network, nodes or edges, the latter of which are commonly used. In a network the nodes usually represent metabolites and/or enzymes depending on the chosen way of representation. Edges are represented as arrows which connect substrates and products of a biochemical conversion. Encoding the flux values through edge thickness and the flux direction by the arrow tip represents an intuitive way of flux visualization.

VANTED ("*v*isualization and *a*nalysis of *n*e*t*works with related *e*xperimental *d*ata," [5–7]) was developed to handle large data sets from biochemical experiments in an easy spreadsheet-based way. It allows to map the data to either self-drawn networks or downloaded maps from different sources for a better visual inspection of the data. Several statistical tools are already integrated by default; correlation networks can be automatically generated.

Over time additional functionalities have been added to the core program to adapt it to different applications.

The FluxMap add-on to the VANTED software provides a convenient workflow for flux data visualization: (1) It uses a template-based flux data import together with metadata. (2) Its network import function provides interfaces to metabolic network databases such as KEGG [8] and MetaCrop [9, 10], but also supports common network exchange formats such as SBML. (3) The input data sets are validated automatically for the flux balance of each reaction. (4) During visualization of metabolic fluxes in the context of a metabolic network, FluxMap functionalities enable the user to additionally visualize statistical data such as confidence intervals and provide a series of visualization options for individual user requirements. (5) An interactive exploration of visualized flux distributions by comparative analysis of different conditions, genotypes, and time points is possible and (6) the export of flux visualization graphs is implemented.

This step-by-step tutorial will help to set up a custom metabolic network, incorporate flux data from a steady-state metabolic flux experiment into the model, and generate a graphical flux map using VANTED.

2 Materials

2.1 Soft- and Hardware Requirements

1. Java runtime version 6 or version 7, compatible with Windows XP, Windows Vista, Windows 7, Mac OS X (V10.5.x, V10.6.x), Ubuntu Linux, and Sun Solaris.
2. Microsoft Excel, Open Office Calc, or other spreadsheet software.
3. Screen resolution of at least 1,024 × 768, mouse, 2 GB RAM (32 bit) or 4 GB RAM (64 bit) recommended.

2.2 Data

1. A set of flux values obtained through available flux software.
2. The corresponding standard deviations (SD) or confidence intervals (CI) of the flux values.

3 Methods

3.1 Software Installation

1. For a local installation go to "http://www.vanted.org" and follow the instructions given under "Installation." Alternatively the latest installer can be downloaded directly from "http://sourceforge.net/projects/VANTED."
2. Start the program and click on Help/Preferences on the right-hand side of the screen. Click on "Install/Configure Add-Ons."
3. Make sure that the "Flux Map" add-on is installed and activated (*see* **Note 1**).

3.2 Set Up of the Data Input File

The calculated fluxes from a metabolic flux experiment (MFA) are loaded into the software via a specific input template spreadsheet file.

1. Start VANTED.
2. Click on right-side menu Experiments/Data Input Templates/ Flux Data, name the file, select a location to save it, and click "Create Flux Data Template."
3. The file contains the four sections "Experiment," "Conditions," "Measurements," and "Substance weights."
4. Section "Experiments" can be used to file particular information of the experiments.
5. Section "Conditions" allows for the input of several data sets of different experiments. Different sets have to be increasingly numbered (filled into the green fields; extend manually for bigger data sets). Visualization of different data sets (i.e., numbered columns) can be easily performed (see below).
6. Section "Measurements" contains the reaction details of the flux map. Each reaction in the model network has to be given in the same syntax/stoichiometry as the example reactions (e.g., PYRm → ACCm + CO_2). The naming of the metabolites/ reactions has to be identical to the naming in the network map (*see* **Note 2**)! Each single data set which has been listed in the "Conditions" section has to be represented by two columns in the "Measurements" section. The flux values are pasted in the first. The second "quality" column can be used to type in data for evaluating the fluxes (e.g., confidence intervals) (*see* **Note 3**). If the corresponding data are available, the rows "Time" and "Unit (Time)" permit a further subdivision of the data set.
7. In section "Substance weights" the number of carbon atoms of each metabolite is listed. Take care of consistent metabolite naming (*see* **Note 4**). Via this weighing molecular fluxes can be transformed into carbon fluxes.
8. Save the completed file for later use (*see* Fig. 1).

3.3 Set Up the Model Network

1. Start VANTED.
2. Open an empty file (either via the icon on top or via "File/New").
3. On the very left side there are two tool buttons: "Create Node and Edges" and "Select Node and Edges." Attention: Only one of them is active at a time!
4. To create a node click on "Create," and then click on the background of the graph editor window.
5. To represent an enzymatic reaction correctly, the nodes represent both metabolites and reactions and have to be labeled

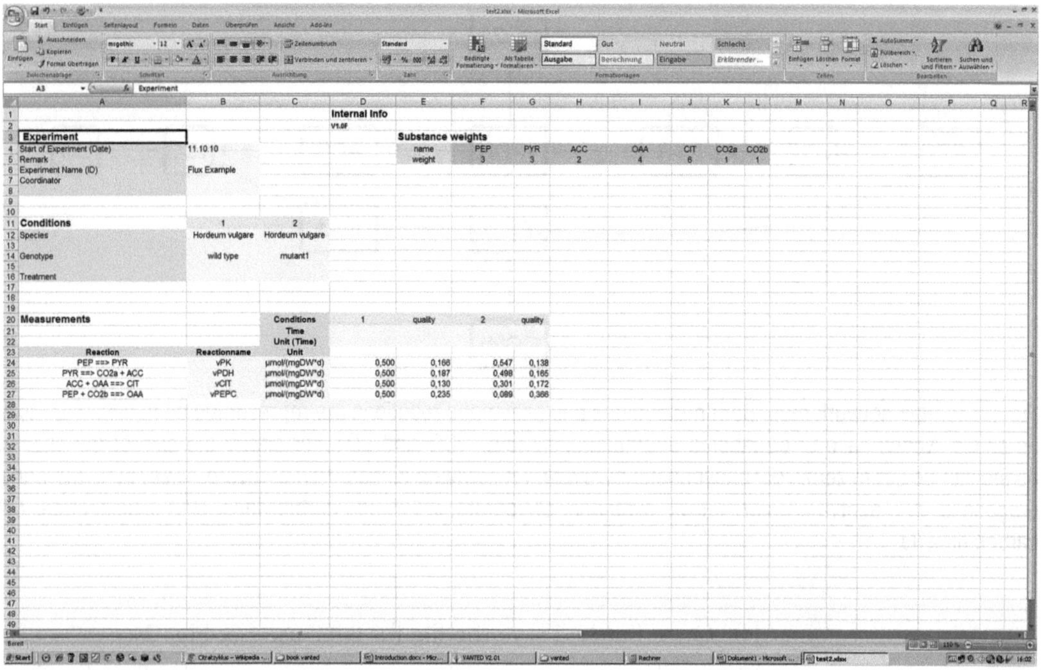

Fig. 1 Screenshot of the input template. The four different input blocks are (1) experiment, (2) condition, (3) measurements, and (4) substance weights

unambiguously, for example, PYRm (mitochondrial pyruvate) or vPDH (pyruvate dehydrogenase). Metabolites occurring in different reactions (e.g., CO2) should be labeled individually (CO2-a, CO2-b, …) (*see* **Note 4**). This will ensure that the nodes are treated separately and will not be merged automatically during the mapping process.

6. To create an edge, which connects two existing nodes, click on the source node, and then on the target node. A single or double click on the canvas prior to selecting the target node will create a smoothly curved or a sharply curved edge, respectively (*see* **Note 5**).
7. To change the name or to apply any further modification switch to "Select Nodes and Edges." Double-click the respective node, fill in its unique identifier in the label field, and then press "OK."
8. To edit the layout select either a node or an edge, and then click on Network/Node (or Edge) on the right-side menu (*see* **Note 6**).
9. For reasons of visual clarity the size of the enzyme nodes can be set to zero; first activate the node by clicking it, then change the size (right-side menu Network/Node/Node Attributes), and click "Apply Changes."

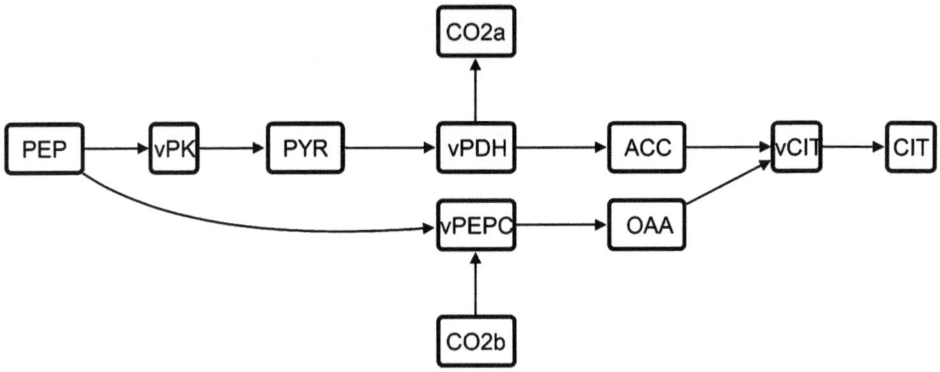

Fig. 2 Example network map. Both metabolites and enzymatic reactions are represented as nodes, and fluxes as edges. Nodes have to be labeled unambiguously (e.g., vCIT and CIT). For clarity the size of the enzyme nodes can be set to zero. *PEP* phosphoenolpyruvate, *PK* pyruvate kinase, *PYR* pyruvate, *vPDH* pyruvate dehydrogenase, *ACC* acetylCoA, *vCIT* citrate synthase, *CIT* citrate, *vPEPC* PEPcarboxylase, *OAA* oxaloacetate, CO_2 carbon dioxide

10. Once finished save the file for later use (*see* Fig. 2).
11. Ready-to-use networks can be imported directly from MetaCrop. On the right-side menu click on "Pathways/ MetaCrop pathways." Choose the network of interest and double click to import it.
12. A detailed step-by-step protocol to set up a model using the KEGG database is described elsewhere [11].

3.4 Data Mapping

1. Start VANTED.
2. Open the model network.
3. On the right-side menu click Experiments/Load Input File/ Load Data Set. Open the prepared input file.
4. Click on "Load all as Flux Data Set," and click "OK." On the right-side menu details about the loaded data (genotypes, conditions, and time points) are shown.
5. Click on "Perform Data Mapping" in the right-side menu.
6. Choose "Initial Charting Style" to be "Hidden" from the newly opened window. Click "OK" (*see* **Note 7**).
7. The newly opened window "Data Mapping Results" indicates potential problems during the mapping process. The following (error) messages occur:
   (a) "x new nodes and y edges nodes have been created": The input file contains more metabolites/reactions than depicted in the model network.

   Solution: Adjust the layout of the newly created edges to fit into the model network or remove the surplus reactions from the input file.

(b) "x nodes and y edges contain data mappings from this operation": Every edge should contain a data mapping.

Solution: Adjust the number of reactions in the input file to the number of reactions in the network map; make sure that all edges are drawn in the right direction.

(c) "At least x and at most y substances map to a single graph element": These numbers should be equal to the number of conditions/genotypes.

Solution: If not, the data has been mapped to a graph that already contains some data. Click on Mapping/Remove Mapped Data and repeat the mapping. Press "OK."

3.5 Visualization

1. To visualize the mapped data click on Mapping/Flux Visualization. The newly opened options help to modify the actual picture. The most common options are as follows:
   (a) "Multiplicator" and "Head/Tail ratio" scale all edges by the same factor, used for esthetics.
   (b) "Colors" let you choose two colors to color-code the quality of the fluxes. Usually a black-to-light grey or a black-to-red scale is chosen.
   (c) Set Prettify Graph/Reaction node style to "Rounded" to hide the reaction nodes. Uncheck the "remove" options to keep bended edges in the model.
   (d) The "Conditions" drop-down box lets you choose between the different data sets while the overall layout of the model is kept constant. Different time points can then be chosen via the slider underneath (*see* **Note 8**).
   (e) "Validate Reactions" can be used to check the balance of the mapped fluxes. An erroneous template (e.g., spelling, weighting, missing reactions) results in an error message.
2. Click "Create Snapshot" to save a png picture of the model in its momentary layout (*see* Fig. 3 and **Note 9**).

4 Notes

1. VANTED can be modified modularly to match one's personal needs and preferences. To do so several different add-ons are available. Click on the right-side menu Help/Settings/"Install/Configure Add-Ons." After clicking "Find Add-ons/Updates" the user can choose from a list of available programs to be automatically integrated into the VANTED environment; a restart of VANTED is required after installation of add-ons. A more detailed list of add-ons can be found at www.vante.ipk-gatersleben.de. The following add-ons are available (summer 2013): FBA-SimVis (an FBA tool), DBE2 (a database system),

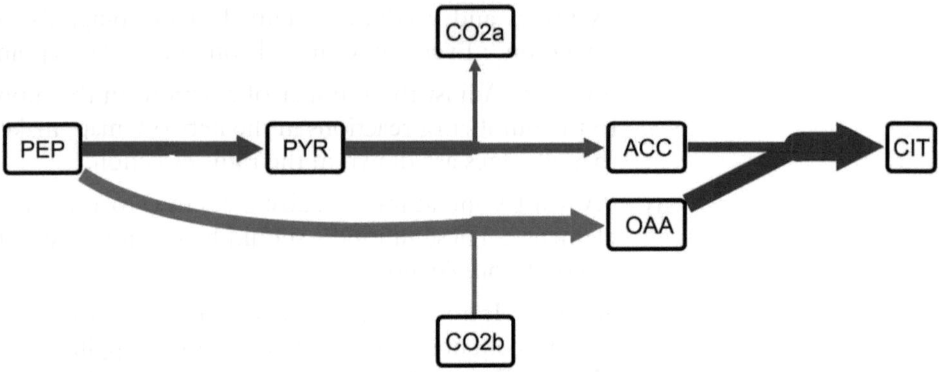

Fig. 3 Final map. The size of the reaction nodes has automatically been set to zero. The color gradient represents the different fluxes' confidence intervals

HIVE (a visualization tool), SBGN-ED (an SBGN editor), CentiLib (a network-analyzing tool), MetaCrop (a database downloader), PetriNet (a petri net editor), and GLIEP (another visualization tool).

2. Inconsistencies in naming will lead to problems when mapping the data. Keep the naming simple and unambiguous; avoid complicated/uncommon names and numberings.
3. Sometimes the flux calculation results in very large confidence intervals (CI). As the quality of fluxes is represented by a linearly encoded color scaling (black to red or black to light grey) over the whole range of the CIs such an outlier would dominate over smaller differences and result in more subtle differences being indistinguishable visually. To avoid these problems during visualization caused by too wide a range of confidence intervals it is preferable to use a normalized representation (i.e., within the interval [0, 1]).
4. Watch out for metabolites occurring in several reactions; they all should be listed individually (e.g., CO2-a/1; CO2-b/1; ...). Otherwise they would be treated as identical and would intervene with the mapping process. Following the CO_2 example the mapping would result in the creation of several new edges connecting all single CO_2 source nodes to one single CO_2 sink node. For reasons of a clear arrangement this node then would have to be split manually.
5. Attention: The edges DO have a head and a tail, direction matters! If there are no arrow heads visible to indicate the direction, check "Directed Edges" under Network/Graph/Network Attributes on the right-side menu, and then click "Apply Changes."

6. The properties of both edges and nodes such as size, shape, color, (gradient) filling, arrow head, or general edge layout can be changed individually. Make sure to click "Apply Changes" at the end. Otherwise the changes do not become visible in the graph.
7. The other charting styles (line chart, bar chart, and pie chart) would generate a graph of this data on top of the corresponding edges. For sake of clarity these options should not be used for flux visualization.
8. This key function has proven to be very helpful when comparing several different data sets from high-throughput experiments. Once the final layout has been chosen it is very easy to switch between the different conditions and see the differences in the flux distribution at one single glance.
9. Other export formats (jpeg, pdf, ppt, or svg) are available under File/Create Image File. When planning to use the pictures for a publication/poster it may be a good idea to export the pictures already at a higher resolution to handle them later more easily. When using them in a PowerPoint presentation, it has given good results to insert the single pictures always at the same position/at the same size. While the overall layout stays the same, the differences in the fluxes can be easily seen by switching between the slides back and forth.

References

1. Kitano H (2002) Systems biology: a brief overview. Science 295(5560):1662–1664
2. Allen DK, Libourel IGL, Shachar-Hill Y (2009) Plant, Cell Environ 32:1241–1257
3. Masakapalli SK, Le Lay P, Huddleston JE, Pollock NL, Kruger NJ, Ratcliffe RG (2010) Subcellular flux analysis of central metabolism in a heterotrophic Arabidopsis cell suspension using steady-state stable isotope labeling. Plant Physiol 152:602–619
4. Grafahrend-Belau E, Schreiber F, Koschützki D, Junker BH (2009) Flux balance analysis of barley seeds: a computational approach to study systemic properties of central metabolism. Plant Physiol 149:585–598
5. Rohn H, Junker A, Hartmann A, Grafahrend-Belau E, Treutler H, Klapperstück M, Czauderna T, Klukas C, Schreiber F (2012) VANTED v2: a framework for systems biology applications. BMC Syst Biol 6:139
6. Rohn H, Hartmann A, Junker A, Junker BH, Schreiber F (2012) FluxMap: a VANTED add-on for the visual exploration of flux distributions in biological networks. BMC Syst Biol 6:33
7. Junker BH, Klukas C, Schreiber F (2006) VANTED: a system for advanced data analysis and visualization in the context of biological networks. BMC Bioinformatics 7(109):1–13
8. Kanehisa M, Goto S, Kawashima S, Okuno Y, Hattori M (2004) The KEGG resource for deciphering the genome. Nucleic Acids Res 32(Database issue):D277–D280
9. Grafahrend-Belau E, Weise S, Koschützki D, Scholz U, Junker BH, Schreiber F (2008) MetaCrop: a detailed database of crop plant metabolism. Nucleic Acids Res 36(Database issue):D954–D958
10. Schreiber F, Colmsee C, Czauderna T, Grafahrend-Belau E, Hartmann A, Junker A, Junker BH, Klapperstück M, Scholz U, Weise S (2012) MetaCrop 2.0: managing and exploring information about crop plant metabolism. Nucleic Acids Res 40(Database issue): D1173–D1177
11. Junker A, Rohn H, Czauderna T, Klukas C, Hartmann A, Schreiber F (2012) Creating interactive, web-based and data-enriched maps with the Systems Biology Graphical Notation. Nat Protoc 7:579–593

Part V

Examples

Chapter 15

Metabolic Flux Analysis for *Escherichia coli* by Flux Balance Analysis

Yu Matsuoka and Kazuyuki Shimizu

Abstract

Conventional metabolic flux analysis (MFA) of *Escherichia coli* wild type and of pathway gene knockout mutants cultivated under anaerobic condition is explained in detail in this chapter. To place the MFA results into the context of the literature, the regulation of central carbon metabolism in terms of catabolite regulation by the phosphotransferase system (PTS) and the response to oxygen limitations via global regulators is reviewed. The effects of gene deletions such as *pflA*, *pta*, *ppc*, *pykF*, *adhE*, and *ldhA* on the metabolic network are presented. Moreover, for the *pflA* mutant the effects of various carbon sources were quantified. The chapter thereby contributes to the discussion of metabolic network function and the design of microbial cell factories.

Key words Metabolic flux analysis, *E. coli*, Metabolic regulation analysis, Catabolite regulation, Oxygen limitation, Anaerobic cultivation

1 Introduction

Escherichia coli is a Gram negative, facultative anaerobic and non-sporulating bacterium. *E. coli* cells are typically rod shaped with a length of about 1–2 μm having a cell volume of 0.6–0.7 μm^3 [1]. This organism has been identified by Theodor Escherich in 1885, and is now classified as part of the *Enterobacteriaceae* family of gamma-proteobacteria [2]. This bacterium grows optimally at 37 °C. *E. coli* is commonly found in the intestine of warm-blooded organisms. Most *E. coli* strains are harmless to host organisms and form part of the normal gut flora that benefits the host organisms by providing some vitamins, and by preventing the establishment of pathogenic bacteria within the intestine. Some strains such as O157:H7, however, cause serious disease to humans [3, 4].

E. coli is the most popular bacterial model organism due to its well-known characteristics, fast growth, and easy genetic manipulation. It is a useful model not only for basic science but also for the medical, pharmaceutical, and fermentation applications.

Jens O. Krömer et al. (eds.), *Metabolic Flux Analysis: Methods and Protocols*, Methods in Molecular Biology, vol. 1191, DOI 10.1007/978-1-4939-1170-7_15, © Springer Science+Business Media New York 2014

The most typical wild-type strain *E. coli* K-12 has about 4,605 open reading frames (ORFs) in a 4.6 Mb genome [2]. Of the total genes, about 300 are predicted for encoding transcription factors (TFs) and 7 for σ factors [5, 6]. The detailed information on the roles of TFs is constantly updated as documented in RegulonDB [7] and EcoCyc [8]. The hierarchical structure of such TFs and the regulated genes has been identified to some extent [1], and some of the metabolic regulation mechanisms have been investigated [9, 10]. Here, the metabolic flux analysis (MFA) together with the metabolic regulation is explained for the typical wild-type *E. coli* and its specific pathway gene knockout mutants under microaerobic or anaerobic conditions.

2 Carbon Catabolite Regulation

Before MFA and other results are presented, we briefly review transcriptional regulation by TFs for catabolite regulation. The central players in carbon catabolite regulation in *E. coli* are the transcriptional activator Crp (cyclic AMP (cAMP) receptor protein; also called catabolite activator protein (CAP)), the signal metabolite cAMP, adenylate cyclase (Cya), and the phosphoenolpyruvate (PEP): pyruvate (PYR) phosphotransferase systems (PTSs). These systems are involved in both transport and phosphorylation of carbohydrates. The PTS in *E. coli* consists of two common cytoplasmic proteins, EI (enzyme I) encoded by *ptsI* and HPr (histidine-phosphorylatable protein) encoded by *ptsH*, as well as carbohydrate-specific EII (enzyme II) complexes. The glucose-specific PTS in *E. coli* consists of the cytoplasmic protein $EIIA^{Glc}$ encoded by *crr* and the membrane-bound protein $EIICB^{Glc}$ encoded by *ptsG*, which transport and concomitantly phosphorylate glucose [11, 12]. The phosphoryl groups are transferred from PEP via successive phosphorelay reactions in turn by EI, HPr, $EIIA^{Glc}$, and $EIICB^{Glc}$ to glucose (Fig. 1). The cAMP-Crp complex and the repressor Mlc are involved in the regulation of the *ptsG* gene and in *pts* operon expression. It has been demonstrated that unphosphorylated $EIICB^{Glc}$ can relieve the *ptsG* gene expression by sequestering Mlc from its binding sites through a direct protein-protein interaction in response to glucose concentration. In contrast to Mlc, which represses the expression of *ptsG*, *ptsHI*, and *crr*, the cAMP-Crp complex activates *ptsG* gene expression (Table 1). Since intracellular cAMP levels are low during growth on glucose, these two antagonistic regulatory mechanisms guarantee a precise adjustment of *ptsG* gene expression levels under various conditions [13]. The unphosphorylated $EIIA^{Glc}$ inhibits the uptake of other non-PTS carbohydrates by the so-called inducer exclusion [14], while phosphorylated $EIIA^{Glc}$ ($EIIA^{Glc}$-P)

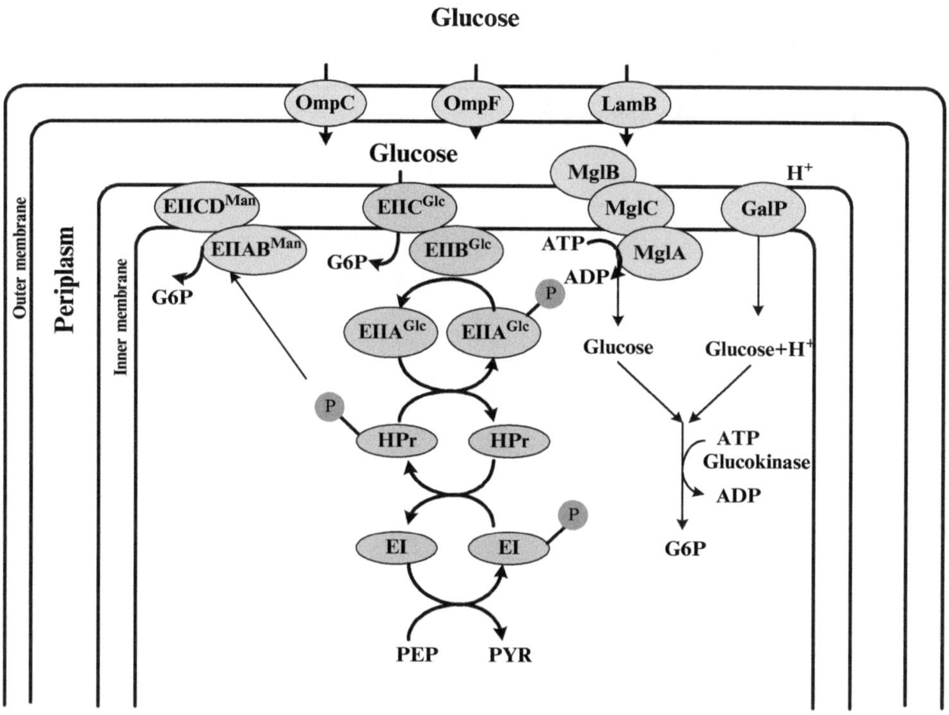

Fig. 1 Glucose transport in *E. coli*: Phosphotransferase system (PTS) and other transporters

activates Cya, which generates cAMP from ATP and increases the intracellular cAMP level [15]. If the concentration ratio between PEP and PYR (PEP/PYR) is high, $EIIA^{Glc}$ is predominantly phosphorylated, whereas if this ratio is low, then $EIIA^{Glc}$ tends to be dephosphorylated [16, 17].

In addition to cAMP-Crp, which acts depending on the level of glucose concentration, the catabolite repressor/activator protein (Cra) originally characterized as the fructose repressor (FruR) plays an important role in the control of carbon flow in *E. coli* [18–20]. The carbon uptake and glycolytic genes such as *ptsHI*, *pfkA*, *pykF*, *zwf*, and *edd–eda* are repressed, while gluconeogenic pathway genes such as *ppsA*, *fbp*, *pckA*, as well as TCA cycle gene such as *icdA* and glyoxylate pathway genes such as *aceA* and *aceB* are activated by Cra (Table 1) (Fig. 2).

The gluconeogenic pathway is deactivated by the knockout of *cra* gene, and the carbon flow toward catabolism and the glucose consumption rate are expected to increase since glycolytic genes such as *ptsHI*, *pfkA*, and *pykF* are activated by the *cra* gene knockout. It has been shown that *cra* gene knockout indeed enables the increase of the glucose consumption rate, and thus it is expected to improve the rate of metabolite production under certain culture conditions [21].

Table 1
The effects of global regulators on the metabolic pathway genes

Global regulator	Metabolic pathway gene
Cra	+: *aceBAK*, *cydB*, *fbp*, *icdA*, *pckA*, *pgk*, *ppsA* −: *acnB*, *adhE*, *eda*, *edd*, *pfkA*, *ptsHI*, *pykF*, *zwf*
Crp	+: *acnAB*, *aceEF*, *focA*, *fumA*, *gltA*, *malT*, *manXYZ*, *mdh*, *mlc*, *pckA*, *pdhR*, *pflB*, *pgk*, *ptsG*, *sdhCDAB*, *sucABCD*, *ugpABCEQ* −: *cyaA*, *lpdA*, *rpoS*
ArcA/B	+: *cydAB*, *focA*, *pflAB*, −: *aceBAK*, *aceEF*, *acnAB*, *cyoABCDE*, *fumAC*, *gltA*, *icdA*, *lpdA*, *mdh*, *nuoABCDEFGHIJKLMN*, *pdhR*, *sdhCDAB*, *sodA*, *sucABCD*, *glpD*, *LctPRD*
IclR	−: *aceBAK*, *acs*
FadR	+: *iclR*
Mlc	−: *crr*, *manXYZ*, *malT*, *ptsG*, *ptsHI*
PdhR	−: *aceEF*, *lpdA*
Fnr	+: *acs*, *focA*, *frdABCD*, *pflB*, *yfiD* −: *acnA*, *cyoABCDE*, *cydAB*, *fumA*, *fnr*, *icdA*, *ndh*, *nuoA-N*, *sdhCDAB*, *sucABCD*
RpoS	+: *acnA*, *acs*, *adhE*, *asmC*, *fumC*, *gadAB*, *talA*, *tktB*, *poxB* −: *ompF*
SoxR/S	+: *acnA*, *fumC*, *sodA*, *zwf*

3 Metabolic Consequences of Oxygen Limitation

Global regulators such as Fnr (fumarate and nitrate reduction) and Arc (anoxic respiration control) system are mainly responsible for the regulation to the availability of oxygen and other electron acceptors in the culture environment, where Fnr regulates the expression of metabolic pathway genes under anaerobic condition [22], while ArcA/B regulates under both anaerobic and microaerobic conditions [23, 24].

Reoxidation of the reducing equivalents such as NADH generated by the oxidation of the energy source occurs in the respiratory chain under aerobic or microaerobic conditions. In *E. coli*, NADH is oxidized in the respiratory chain via a coupled NADH dehydrogenase NDH-1 encoded by *nuo* or an uncoupled dehydrogenase NDH-2 encoded by *ndh*. $FADH_2$ is oxidized by succinate dehydrogenase (SDH) encoded by *sdh* gene. The electrons are transported into the quinone and quinol pool. With the availability of oxygen, quinol is then oxidized by either the cytochrome bo (Cyo) or the cytochrome bd (Cyd) terminal oxidase complex, which in turn passes the electrons to oxygen with concomitant production

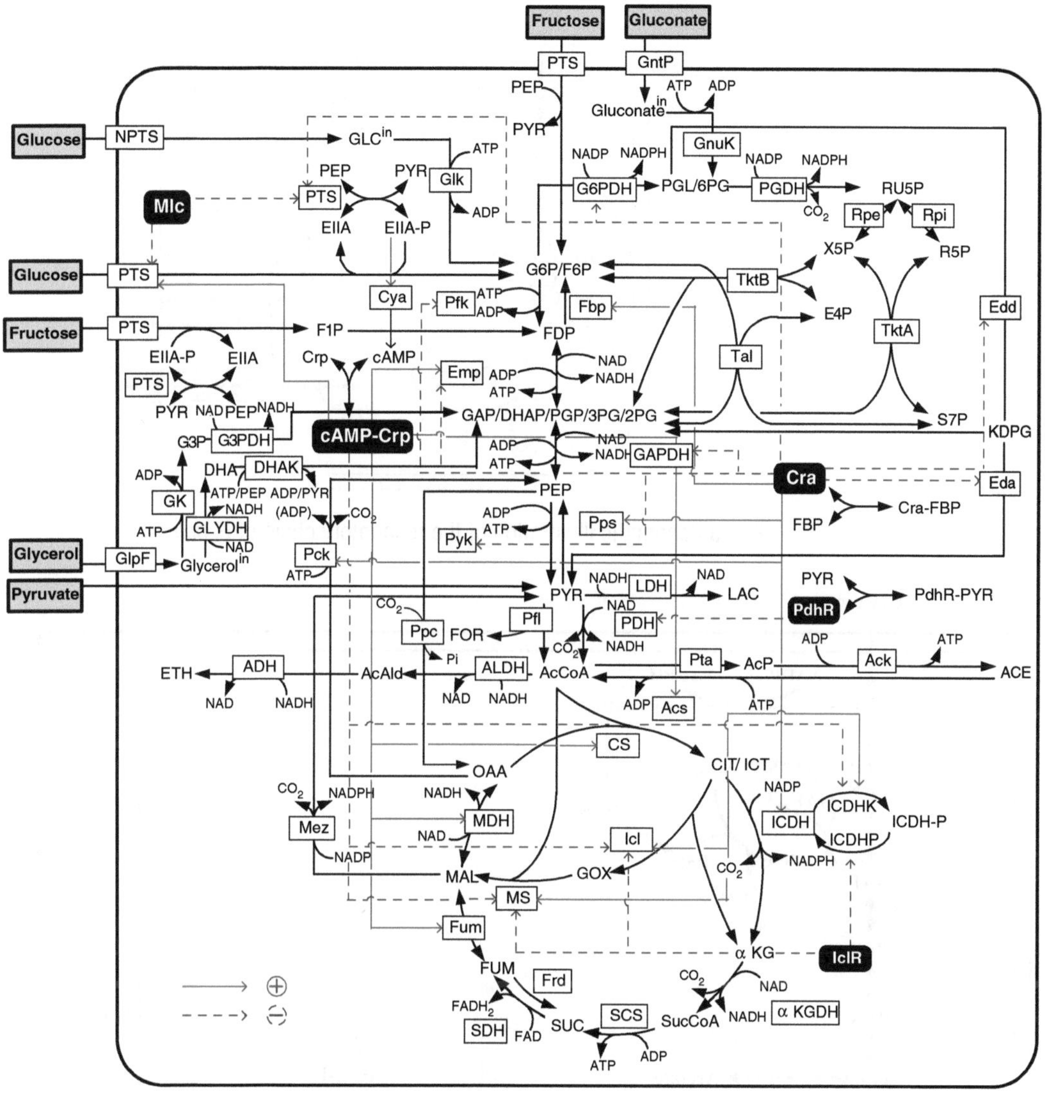

Fig. 2 *E. coli's* main metabolic pathways and their regulation by global regulators

of water (Fig. 3). The minor contribution is made by cytochrome bd-II oxidase (AppCB) encoded by *cyxAB* (or *appCB*). The *cyoABCDE* operon encoding the cytochrome bo is repressed by both ArcA and Fnr, while *cydAB* operon encoding the cytochrome bd is activated by ArcA and repressed by Fnr (Fig. 4) (Table 1) [25].

The reducing equivalents such as NADH and $FADH_2$ are reoxidized in the respiratory chain, where oxygen, nitrate, fumarate, dimethyl sulfoxide, and others are electron acceptors. This process is coupled to the formation of a proton motive force (PMF), which is utilized for ATP generation from ADP and P_i (Fig. 3). In the absence of oxygen, or other electron accepters,

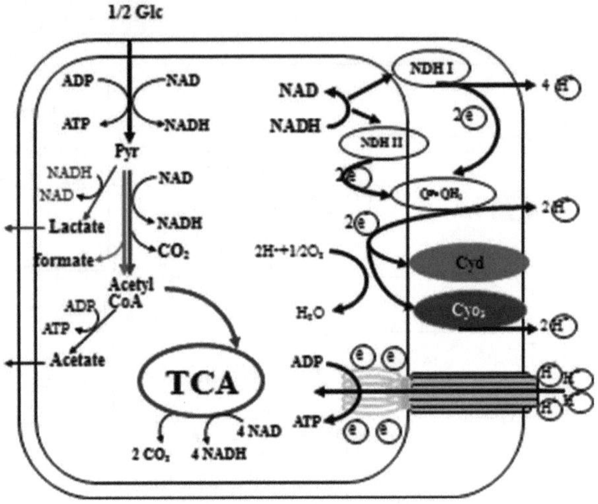

Fig. 3 Schematic illustration of the respiratory chain of *E. coli*

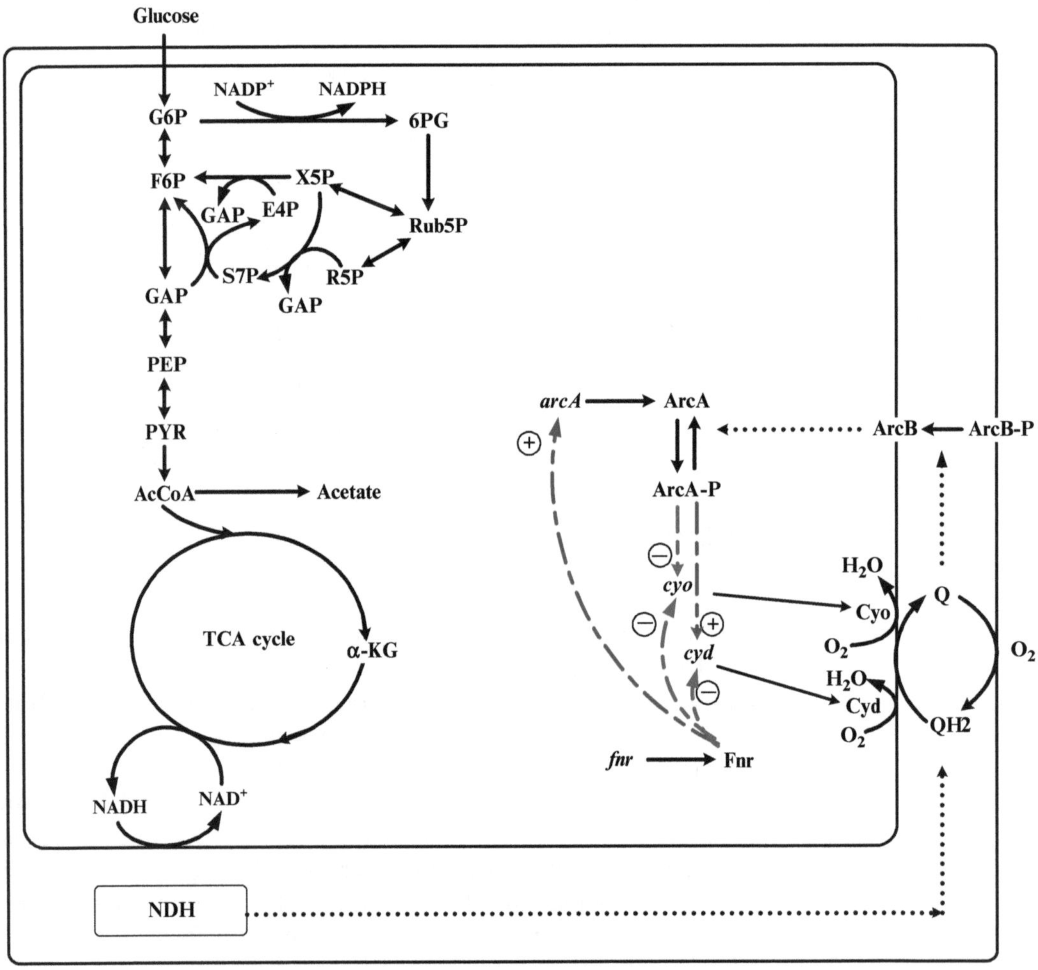

Fig. 4 Regulation of ArcA/B and Fnr and their effects on cytochromes in relation to oxygen availability

ATP is generated via substrate-level phosphorylation through the process of degradation of carbon source in the metabolic pathways. Under anaerobic condition, pyruvate is converted into formate (FOR) and acetyl-coenzyme A (AcCoA) by pyruvate formate lyase (Pfl) reaction. The expression of *pfl* gene which encodes Pfl is activated by ArcA and Fnr, and it becomes higher at lower oxygen concentrations, whereas *aceE* and *F*, which encode α and β subunits of pyruvate dehydrogenase complex (PDHc), are repressed by ArcA under oxygen-limited condition (Table 1). The product of both Pfl and PDHc reactions is AcCoA, which is further converted to acetate and ethanol or subsequently undergoes further oxidation in the TCA cycle.

The Arc system, composed of the cytosolic response regulator ArcA, and the membrane-bound sensor kinase ArcB, regulates the TCA cycle genes depending on the oxygen level or the redox state. Upon stimulation by the redox state or the oxygen limitation, ArcB undergoes autophosphorylation, and the phosphoryl group is transferred to ArcA (Fig. 4). The phosphorylated ArcA represses the expression of the genes involved in the TCA cycle and the glyoxylate shunt such as *gltA*, *acnAB*, *icdA*, *sucABCD*, *sdhCDAB*, *fumA*, *mdh*, and *aceA* and *B* (Table 1). Moreover, the genes which encode for the primary dehydrogenases such as *glpD*, *lctPRD*, *aceE* and *F*, and *lpdA* are also repressed by ArcA (Table 1).

Respiration is a fundamental cellular process utilizing different terminal electron acceptors. The ability to sense these electron acceptors is a key for the cells to survive. *E. coli* is a metabolically versatile chemoheterotroph that can grow on a variety of substrates under various oxygen concentrations, where fumarate or nitrate can be a terminal electron acceptor under anaerobic condition [26]. In *E. coli*, this switch from aerobic to anaerobic respiration is controlled by Fnr. Under anaerobic condition, Fnr binds to a $[4Fe\text{-}4S]^{2+}$ cluster, and becomes a transcriptionally active dimeric form.

The pyridine nucleotides such as NADH and NAD^+ function as important redox carriers in metabolism. These coenzymes not only serve as electron acceptors in the breakdown of substrates but also provide the reducing power for the redox reactions in anaerobic and aerobic respiration. A balance for oxidation and reduction of these nucleotides is regulated for catabolism and anabolism, since the turnover of the nucleotides is very high compared to their concentrations [27]. Under fermentation conditions, the reoxidation of NADH is coupled to the formation of reduced compounds, whereas NADH oxidation is coupled to electron transfer in the respiration, either aerobic with oxygen or anaerobic with nitrate etc.

4 Metabolic Flux Analysis by Flux Balance Analysis (FBA)

Metabolic flux is defined as the intracellular reaction rate per cell and per time (mmol $g_{DCW}^{-1} h^{-1}$). The information on the metabolic flux distribution is particularly useful for metabolic engineering [28, 29]. In principle, MFA is based on mass conservation of key metabolites. Intracellular fluxes can be calculated from the measured specific rates by applying mass balances to the intracellular metabolites together with stoichiometric equations. The number of measurable extracellular fluxes is limited in practice, and the stoichiometric constraints often lead to an underdetermined algebraic system. Therefore, cofactor balances sometimes need to be introduced into the stoichiometric equations, or the appropriate objective functions have to be introduced to determine the fluxes. The central metabolic pathways have both anabolic and catabolic functions, as they provide cofactors and building blocks for macromolecular synthesis (anabolism), as well as energy (ATP) generation (catabolism). Optimizations may be made by maximizing the cell growth rate and/or the cell yield, or others, under the constraints of the stoichiometric equations [30]. Flux balance analysis (FBA) has been extensively used to predict steady-state metabolic fluxes in order to maximize the cellular growth rate [31–33]. This method has been significantly extended to genome-scale models, which include thousand and more metabolic reactions [34]. *E. coli* utilizes carbon efficiently, and this is a special feature of this organism [35–37]. However, the accuracy of the flux calculation depends on the validity of the cofactor assumptions and the appropriate choice of the objective function(s) employed [30]. There might exist unknown pathways that generate or consume the cofactors, and this may invalidate the assumption that their concentrations remain in balance, and the selected objective functions may not be appropriate, or their validity may be limited to certain states of the cell such as the exponential growth phase. Moreover, this approach cannot essentially compute such fluxes as (1) recycled fluxes, (2) bidirectional fluxes, and (3) parallel fluxes, due to the singularity of the stoichiometric matrix. These fluxes can be computed based on ^{13}C-metabolic flux analysis, which is explained in the next chapter. However, the conventional FBA may be reasonably applied for the cases without recycles, such as anaerobic cultivations or by lumping several pathways together. Here, we present in detail metabolic flux analysis based on mass balances with stoichiometric equations of *E. coli* cultivated under anaerobic conditions.

5 Metabolic Flux Analysis of *E. coli* Wild Type and its Pathway Gene Knockout Mutants

Although conventional MFA has some limitations as stated above, it can be used to analyze anaerobic cultures, since the metabolic pathway network does not include cyclic pathways, and the

Table 2
Comparison of the fermentation results of different mutants cultivated under anaerobic condition, where M9 synthetic medium was used for the batch cultivation with 10 g/l of glucose as a carbon source

Strain	Specific growth rate μ (h^{-1})	Yield on glucose ($g_{biomass\ or\ metabolite\ produced}/g_{glucose\ consumed}$) (%)						
		Biomass	Lactate	Acetate	Formate	Succinate	Ethanol	Pyruvate
Wild	0.85 ± 0.08	30.1 ± 0.	1.2 ± 0.0	24.3 ± 0.1	30.0 ± 0.2	5.0 ± 0.2	14.7 ± 0.3	0.2 ± 0.0
Δ*pfl*A	0.18 ± 0.04	8.1 ± 0.1	72.5 ± 0.0	1.1 ± 0.1	0.0 ± 0.2	0.0 ± 0.0	1.0 ± 0.0	1.1 ± 0.0
Δ*pta*	0.24 ± 0.04	6.9 ± 0.1	51.3 ± 0.0	1.6 ± 0.1	2.1 ± 0.2	4.5 ± 0.1	0.0 ± 0.0	0.0 ± 0.0
Δ*ppc*	0.01 ± 0.02	4.6 ± 0.1	32.2 ± 0.0	13.5 ± 0.1	9.4 ± 0.2	0.3 ± 0.0	6.2 ± 0.1	4.6 ± 0.0
Δ*adh*E	0.03 ± 0.02	11.4 ± 0.1	9.7 ± 0.0	42.4 ± 0.1	12.1 ± 0.2	3.2 ± 0.1	0.2 ± 0.0	0.0 ± 0.0
Δ*pyk*F	0.48 ± 0.05	13.0 ± 0.1	2.1 ± 0.0	34.0 ± 0.1	22.9 ± 0.2	6.9 ± 0.2	12.0 ± 0.2	0.0 ± 0.0

The pH was maintained at 7.0, and the dissolved oxygen concentration was kept less than 5 % of air saturation (microaerobic condition)

reactions mainly occur unidirectionally. Here, we consider MFA of *E. coli* wild type and its several single-gene knockout mutants cultivated under anaerobic condition.

Before performing MFA, the fermentation characteristics have to be determined for the mass balance approach (Table 2). Some enzyme activities were also measured (Table 3) for the *pflA*, *pta*, *ppc*, *adhE*, and *pykF* mutants, as well as the wild-type BW25113 [38]. Table 3 indicates that the *pflA* mutant shows much higher activities of glyceraldehyde 3-phosphate dehydrogenase (GAPDH) and lactate dehydrogenase (LDH), which implies coupling between NADH production and consumption. Similarly to the *pflA* mutant, simultaneous up-regulations of GAPDH and LDH are also observed in the *pta* mutant (Table 3). Interestingly, Pfl activity is repressed in this mutant, which implies a common regulatory mechanism for these two strains. The significant difference between the *pta* and *pflA* mutants is the specific activity of PEP carboxylase (Ppc). In the *pflA* mutant, this enzyme activity changed significantly, but this mutant produced less succinate than the *pta* mutant (Table 2), despite higher Ppc activity. This implies that the pool size of PEP caused this phenomenon, where PEP concentration was higher in the *pta* mutant, while it was lower in the *pflA* mutant (Table 4) [38].

The enzyme activities of acetate kinase (Ack) and alcohol dehydrogenase (ADH), both of which are AcCoA assimilation pathway enzymes, increase in the *ppc* mutant (Table 3). However, the enzyme activity of Pfl, which supplies AcCoA for these two reactions, is lower than that of the parent strain (Table 3), and less

Table 3
Comparison of enzyme activities for different mutants cultivated under anaerobic condition

	Enzyme activities (mmol/min mg protein)								
	G6PDH	**6PGDH**	**GAPDH**	**Pyk**	**LDH**	**Ppc**	**Pfl**	**Ack**	**ADH**
wt	0.152 ± 0.023	0.18 ± 0.002	0.008 ± 0.001	0.657 ± 0.102	0.503 ± 0.120	0.008 ± 0.003	0.071 ± 0.012	0.038 ± 0.006	0.006 ± 0.000
Δ*pfl*A	0.252 ± 0.061	0.079 ± 0.010	0.076 ± 0.011	0.726 ± 0.170	2.093 ± 0.512	0.343 ± 0.122	0.004 ± 0.004	4.088 ± 0.200	0.011 ± 0.002
Δ*pta*	0.187 ± 0.027	0.001 ± 0.001	0.060 ± 0.008	0.512 ± 0.006	1.621 ± 0.306	0.005 ± 0.001	0.013 ± 0.001	3.465 ± 0.195	0.005 ± 0.001
Δ*ppc*	0.017 ± 0.002	0.000 ± 0.000	0.014 ± 0.001	0.183 ± 0.060	0.164 ± 0.020	0.003 ± 0.002	0.025 ± 0.006	0.629 ± 0.035	0.040 ± 0.009
Δ*adh*E	0.013 ± 0.002	0.004 ± 0.001	0.008 ± 0.003	0.108 ± 0.021	0.000 ± 0.000	0.005 ± 0.001	0.043 ± 0.001	0.536 ± 0.035	0.000 ± 0.000
Δ*pyk*F	0.052 ± 0.006	0.017 ± 0.003	0.020 ± 0.005	0.009 ± 0.003	0.069 ± 0.030	0.013 ± 0.001	0.050 ± 0.004	0.293 ± 0.016	0.001 ± 0.001

Table 4
Comparison of the intracellular metabolite concentrations in different mutants cultivated under anaerobic condition

	Metabolite concentration (mM/g_{DCW})											
	G6P	FDP	PEP	PYR	AcCoA	ATP	ADP	AMP	ATP/AXP	NADH	NAD	NADH/NAD
Wild	0.05 ± 0.01	4.59 ± 0.02	0.32 ± 0.12	8.21 ± 0.01	0.07 ± 0.00	3.06 ± 0.17	0.38 ± 0.01	0.21 ± 0.01	0.84	0.018 ± 0.002	0.143 ± 0.001	0.126
Δ*pflA*	1.09 ± 0.01	16.4 ± 0.11	0.12 ± 0.04	25.5 ± 0.01	0.05 ± 0.00	1.69 ± 0.11	0.20 ± 0.01	0.20 ± 0.01	0.81	0.056 ± 0.001	0.060 ± 0.002	0.933
Δ*pta*	0.54 ± 0.03	15.1 ± 0.03	1.27 ± 0.20	21.8 ± 0.55	0.01 ± 0.00	1.82 ± 0.16	1.17 ± 0.08	1.12 ± 0.17	0.44	0.047 ± 0.001	0.129 ± 0.007	0.680
Δ*ppc*	0.84 ± 0.03	24.5 ± 0.04	0.96 ± 0.07	38.7 ± 2.26	0.05 ± 0.02	4.19 ± 0.07	3.71 ± 1.14	1.16 ± 0.43	0.53	0.018 ± 0.004	0.170 ± 0.010	0.106
Δ*adhE*	0.32 ± 0.01	11.4 ± 0.03	0.46 ± 0.08	18.4 ± 0.25	0.01 ± 0.00	1.38 ± 0.04	0.69 ± 0.02	0.46 ± 0.04	0.55	0.010 ± 0.006	0.058 ± 0.009	0.167
Δ*pykF*	1.15 ± 0.02	30.9 ± 0.90	3.67 ± 0.06	47.0 ± 2.0	0.08 ± 0.02	2.65 ± 0.10	2.27 ± 0.05	1.01 ± 0.08	0.47	0.002 ± 0.005	0.177 ± 0.008	0.010

acetate and ethanol are produced in the *ppc* mutant compared to the parent strain (Table 2). Although the activity of LDH is lower, lactate production is higher in the *ppc* mutant (Table 2). The activities of LDH and Pfl are both decreased in the *adhE* mutant (Table 3). The *pykF* mutant shows higher Ppc activity, while the activities of LDH and ADH are both decreased (Table 3) [38].

Let us consider how the intracellular metabolite concentrations changed in response to the specific gene knockout (Table 4). The pool sizes of glucose 6-phosphate (G6P), fructose 1,6-bisphosphate (FDP), and PYR increased, while PEP concentration was lower in the *pflA* mutant than the parent strain. The lower PEP concentration causes less succinate production (Table 2), even with high Ppc activity (Table 3). In the case of the *pykF* mutant, all intracellular metabolites in glycolysis are significantly accumulated (Table 4). Two of the lactate-producing strains, the *pflA* and *pta* mutant, show higher $NADH/NAD^+$ ratios, which indicates that lactate production is promoted by the intracellular redox balance pressure in these two strains (Table 4). Table 4 indicates that the value of ATP/AXP (ATP + ADP + AMP) is less in the mutants as compared to wild-type strain, and in particular, this ratio is low in *pta* and *pykF* mutants [38]. Note that FDP concentration affects the pyruvate kinase (Pyk) activity by allosteric regulation. Moreover, FDP concentration modulates the Cra activity, which in turn regulates the metabolic pathway gene expression (Table 1), and thus the enzyme activities and the fluxes might be explained from this point of view.

Consider the metabolic flux distribution for each strain as given in Table 5, where the metabolic pathways are shown in Fig. 5 [38]. Although one of the catabolic pathways is blocked by the specific gene knockout, the higher glucose uptake rate (v_1) may be seen in such strains as *pflA*, *pta*, and *pykF* mutants (Table 5). The flux through GAPDH (v_3) is regulated by the intracellular $NADH/NAD^+$ ratio [27, 39]. It has been shown that NADH competitively combines with GAPDH and inhibits the reaction through this enzyme. The GAPDH activities in the *pflA*, *pta*, and *pykF* mutants are all higher than that of the parent strain (Table 3). The intracellular $NADH/NAD^+$ ratio is significantly lower in the *pykF* mutant (Table 4), and therefore, the higher glycolytic flux is not restricted by the competitive inhibition of NADH on the reaction through relatively high GAPDH activity.

The flux through Ppc (v_5) in the *pykF* mutant is about 2.3-fold that of the parent strain (Table 5). From enzyme activity data, Ppc activity in the *pykF* mutant is up-regulated by about 60 % (Table 3). The reason for the high Ppc flux may be due to high PEP concentration and synergistic activation by FDP (Table 4) [40, 41]. Since the product of the reaction through Ppc is oxaloacetate (OAA), which is the precursor for biosynthesis, the *ppc* gene knockout causes low biomass synthesis flux (v_{12}) (Table 5) [38].

Table 5
Effect of a single-gene knockout on the flux distribution of a cell cultivated under anaerobic condition

	Fluxes (mmol/g_{DCW}/h)											
	v_1	v_2	v_3	v_4	v_5	v_6	v_7	v_8	v_9	v_{10}	v_{11}	v_{12}
Wild	6.30 ± 0.06	5.90 ± 0.07	11.73 ± 0.10	3.85 ± 0.09	1.39 ± 0.09	0.42 ± 0.06	0.09 ± 0.06	7.28 ± 0.35	1.97 ± 0.23	4.47 ± 0.12	3.52 ± 0.11	3.94 ± 0.10
Δ*pflA*	8.15 ± 0.06	7.58 ± 0.07	14.93 ± 0.10	5.71 ± 0.10	0.78 ± 0.10	0.18 ± 0.11	11.69 ± 0.7	0.00 ± 0.00	1.64 ± 0.11	0.46 ± 0.2	0.45 ± 0.10	1.62 ± 0.11
Δ*pta*	6.51 ± 0.06	5.79 ± 0.06	11.27 ± 0.10	3.03 ± 0.10	1.37 ± 0.10	0.77 ± 0.10	7.36 ± 0.83	0.70 ± 0.06	0.92 ± 0.10	0.61 ± 0.3	0.31 ± 0.12	1.26 ± 0.12
Δ*ppc*	1.59 ± 0.02	1.48 ± 0.02	2.90 ± 0.03	1.12 ± 0.02	0.14 ± 0.02	0.00 ± 0.03	1.13 ± 0.1	0.62 ± 0.07	0.66 ± 0.03	0.72 ± 0.08	0.45 ± 0.06	0.20 ± 0.02
Δ*adhE*	1.10 ± 0.02	1.04 ± 0.02	2.05 ± 0.03	0.76 ± 0.04	0.15 ± 0.03	0.07 ± 0.02	0.23 ± 0.03	0.53 ± 0.06	1.03 ± 0.05	1.45 ± 0.05	0.00 ± 0.02	0.27 ± 0.04
Δ*pykF*	9.02 ± 0.07	7.05 ± 0.07	13.21 ± 0.07	0.00 ± 0.00	3.21 ± 0.33	1.54 ± 0.23	0.13 ± 0.2	7.82 ± 0.29	2.56 ± 0.43	7.75 ± 0.4	2.93 ± 0.31	2.48 ± 0.23

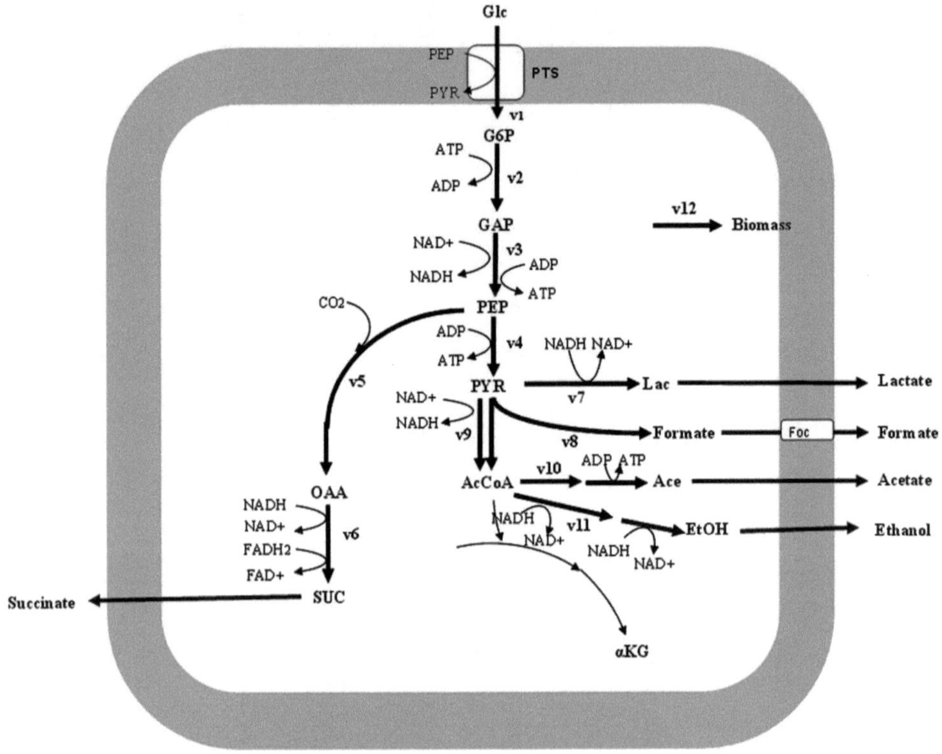

Fig. 5 Metabolic pathways of *E. coli* cultivated under anaerobic condition

Pyruvate is the substrate for the reactions catalyzed by Pfl and LDH under anaerobic condition. In wild-type *E. coli*, the LDH-catalyzed reaction is not as competitive as the reaction through Pfl, and therefore, acetate and formate are the main metabolites instead of lactate (Table 2). The knockout of *pflA* gene blocks the pyruvate assimilation through the Pta–Ack and ADH pathways, which are commonly used for ATP production and NADH reoxidation, respectively, in the parent strain. Since the glycolytic flux is promoted to increase ATP production by substrate-level phosphorylation under anaerobic condition, lactate is produced to satisfy both the stoichiometric and intracellular redox balances [42].

In PTS, PEP is used to transport phosphate for glucose utilization. The reaction through Pyk also uses PEP to produce PYR and ATP. When the *pykF* mutant was cultivated, the specific glucose uptake rate increased (Table 5). The reduced ATP production due to lack of the Pyk reaction might be the reason for the increased specific glucose uptake rate. Namely, the flux through glycolysis is controlled by the ATP requirement in *E. coli* [43]. Since the glucose uptake rate (v_1) affects the phosphorylation of EIIA in PTS, this affects the cAMP-Crp level, and in turn affects the TCA cycle gene expression (Table 1).

Consider next the effect of the *ldhA* gene knockout on the metabolism. Deletion of the LDH pathway blocks carbon flow from pyruvate and results in a higher pyruvate excretion rate. MFA result shows that most of the carbon flux in *ldhA* mutant is forced through formate, acetate, and ethanol production pathways, resulting in a concomitant increase in these fluxes [44]. When PEP is converted to malate (MAL) via OAA by Ppc and malate dehydrogenase (MDH), the free energy is wasted. In contrast, one ATP is produced when PEP is directly converted to pyruvate by Pyk. Therefore, a large fraction of PEP is channeled through the Pyk pathway, and produces more pyruvate. This might be why *ldhA* mutant induces the NADH forming malic enzyme (Sfc) pathway to supply C_4 intermediates such as malate in the TCA cycle for conserving the free energy of PEP, as evidenced by the significant up-regulation of Sfc, where it catalyzes the reaction for converting pyruvate to malate.

Figure 6 shows the brief summary of the effect of the specific pathway gene knockout on the metabolism [45].

6 Metabolic Flux and Regulation Analyses of the *pfl* Mutant Grown on Various Carbon Sources

Consider next the effects of using other carbon sources such as gluconate, pyruvate, fructose, and glycerol on the metabolism of *pfl* mutant under anaerobic condition. The results are shown in Tables 6 and 7 [42]. A large amount of acetate was produced, when gluconate or pyruvate was used as a carbon source (Table 6). In particular, acetate is the major product when pyruvate was used as a carbon source (Table 6). The fermentation patterns, when using fructose or glycerol as a carbon source, are similar to the case of using glucose. However, fructose and glycerol uptake rates are significantly lower than glucose uptake rates (Table 6), and the lactate yields vary from 60 % to about 70 % (Table 7).

Tables 8 and 9 show the enzyme activities and the intracellular metabolite concentrations for *pflA* mutant [42]. Compared to the case of using glucose as a carbon source, Ack activity is about tenfold higher for the case of using gluconate or pyruvate as a carbon source, while this enzyme activity is reduced for the case of using glycerol or fructose as a carbon source (Table 8). Ppc activity is the highest when using glucose as a carbon source, while it is significantly lower for the case of using fructose or glycerol as a carbon source (Table 8). Table 9 shows that NADH/NAD$^+$ ratio is the highest when using glycerol, and this ratio is relatively low when using gluconate, pyruvate, or fructose. In the case of glycerol, NADH is formed in the glycerol dehydrogenase pathway. In *E. coli*, the Pta–Ack pathway is related to the AcCoA pool. Since AcCoA formation through Pfl is deficient in the *pflA* mutant, Pta–Ack

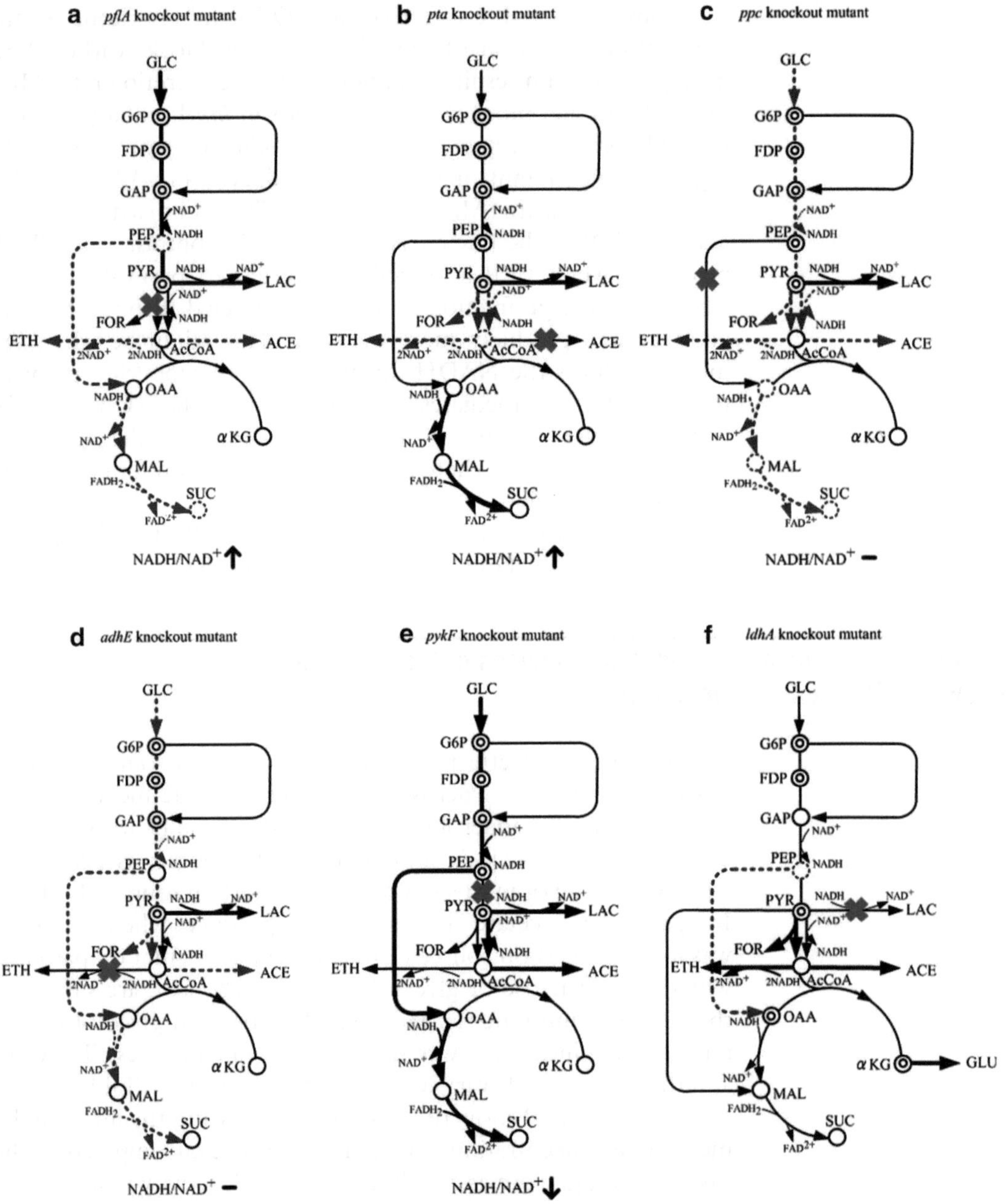

Fig. 6 Schematic illustration of the specific pathway gene knockout on the metabolism under anaerobic condition: (**a**) *pflA*, (**b**) *pta*, (**c**) *ppc*, (**d**) *adhE*, (**e**) *pykF*, and (**f**) *ldhA* knockout mutants

reactions may occur in the direction of forming AcCoA using acetate and ATP. The significant induction of Ack may be related to the activation of pyruvate oxidase (Pox), which produces acetate directly from pyruvate, and the cell utilizes the Ack–Pta and acetyl coenzyme A synthetase (Acs) pathways to supply the AcCoA pool.

The cell produces less NADH when using gluconate or pyruvate as a carbon source. This is confirmed by the lower NADH/NAD^+ ratio as compared to the case of using glucose (Table 9).

Table 6
The specific carbon source uptake rates and the specific product formation rates for the *pflA* mutant using different carbon sources under anaerobic condition

	q (mmol gDCW^{-1} h^{-1})					
Carbon source	**Substrate**	**Lactate**	**Succinate**	**Formate**	**Acetate**	**Ethanol**
Glucose	8.15	11.01	0.0	0.0	0.27	0.2
Gluconate	6.57	8.22	0.21	0.0	5.63	0.32
Pyruvate	6.87	1.78	0.25	0.0	4.58	0.11
Fructose	3.21	4.42	0.0	0.0	0.23	0.0
Glycerol	1.33	0.83	0.0	0.0	0.05	0.0

Table 7
The yields (Y) of cell mass and metabolites for different carbon sources for *pflA* mutant grown under anaerobic condition

	Yields on carbon source (g g^{-1})%						
Carbon source	$Y_{x/s}$	$Y_{lactate/s}$	$Y_{acetate/s}$	$Y_{ethanol/s}$	$Y_{succinate/s}$	$Y_{formate/s}$	$Y_{pyruvate/s}$
Glucose	8.1	72.5	1.1	3.8	0.0	0.0	1.1
Gluconate	8.1	57.4	26.2	1.1	1.9	0.0	1.0
Pyruvate	5.9	26.5	45.4	0.2	1.3	0	–
Fructose	9.7	69.0	3.7	0.0	0.0	0.0	1.2
Glycerol	0.0	61.1	2.6	0.0	0.0	0.0	0.0
Glucose (AN)	6.7	48.2	0.6	–	3.6	0.0	–

AN: The *pflA* mutant grown on glucose under anaerobic condition. The culture vessel was filled with CO_2

The activity of GAPDH is significantly higher when using gluconate as a carbon source (Table 8), which indicates the partial release of NADH inhibition on this enzyme. Correspondingly, the lactate yield when using gluconate is lower (Table 7). The results using gluconate or pyruvate as a carbon source may also be distinguished from those using other carbon sources, because of the significantly lower ATP/AMP ratio (Table 9). One reason may be due to the difference in the metabolic networks. Namely, when using gluconate as a carbon source, some of the carbon atoms go directly to PYR through the Entner-Doudoroff (ED) pathway, which reduces the flux through glycolysis where ATP is generated. It should be noted that the activity of Ack is up-regulated significantly when using gluconate or pyruvate as a carbon source (Table 8).

Table 8
Enzyme activities for *pflA* mutant grown on different carbon sources under anaerobic condition

Carbon source	Enzyme activity (mmol (mg protein)$^{-1}$ min^{-1})							
	G6PDH	6PGDH	GAPDH	Pyk	LDH	Ppc	Ack	ADH
Glucose	0.252 ± 0.019	0.080 ± 0.003	0.076 ± 0.003	0.73 ± 0.04	2.09 ± 0.12	0.343 ± 0.032	4.09 ± 0.77	0.010 ± 0.001
Gluconate	0.449 ± 0.02	0.244 ± 0.009	0.126 ± 0.005	0.90 ± 0.04	4.38 ± 0.42	0.084 ± 0.010	38.15 ± 2.5	ND
Pyruvate	0.264 ± 0.019	ND	ND	ND	0.42 ± 0.04	ND	39.82 ± 2.8	ND
Fructose	0.170 ± 0.019	0.066 ± 0.004	0.030 ± 0.006	0.88 ± 0.09	0.6 ± 0.07	0.013 ± 0.001	0.036 ± 0.006	0.015 ± 0.002
Glycerol	0.141 ± 0.013	0.098 ± 0.005	0.026 ± 0.002	0.44 ± 0.03	0.52 ± 0.10	0.025 ± 0.002	0.072 ± 0.009	0.006 ± 0.001

Table 9
Intracellular metabolite concentrations in the *pflA* mutant grown on different carbon sources under anaerobic condition

	Intracellular concentration (μmol $gDCW^{-1}$)				
Metabolite	**Glucose**	**Gluconate**	**Pyruvate**	**Fructose**	**Glycerol**
G6P	1.093 ± 0.006	0.123 ± 0.001	0.120 ± 0.001	0.240 ± 0.001	0.045 ± 0.003
FDP	16.41 ± 0.11	27.08 ± 0.10	164.9 ± 2.7	28.92 ± 0.04	33.02 ± 0.04
PEP	0.12 ± 0.04	0.22 ± 0.11	0.50 ± 0.13	0.45 ± 0.13	0.10 ± 0.05
PYR	25.49 ± 0.01	41.81 ± 0.70	250.6 ± 0.85	47.22 ± 0.50	52.98 ± 0.30
AcCoA	0.050 ± 0.001	0.003 ± 0.001	0.039 ± 0.004	0.132 ± 0.035	0.037 ± 0.001
ATP	0.052 ± 0.012	0.008 ± 0.004	0.034 ± 0.011	0.027 ± 0.006	0.056 ± 0.026
ADP	0.195 ± 0.010	0.143 ± 0.003	0.835 ± 0.011	0.458 ± 0.092	0.531 ± 0.091
AMP	0.197 ± 0.004	0.226 ± 0.007	0.932 ± 0.049	0.127 ± 0.013	0.499 ± 0.006
NADH	0.056 ± 0.001	0.032 ± 0.002	0.016 ± 0.002	0.092 ± 0.012	0.085 ± 0.020
NAD^+	0.060 ± 0.002	0.046 ± 0.001	0.032 ± 0.001	0.244 ± 0.003	0.037 ± 0.001
ATP/AMP	0.264	0.035	0.036	0.213	0.112
NADH/NAD^+	0.93	0.7	0.5	0.38	2.3

Figure 7 shows the brief illustration for the metabolism of *pflA* mutant cultivated under anaerobic condition using various carbon sources.

7 Concluding Remarks

In this chapter, FBA-based metabolic flux analysis of *E. coli* grown under anaerobic condition is presented in detail. In particular, an attempt was made to clarify the metabolic regulation mechanisms in response to the specific pathway gene knockout based on different levels of information such as fermentation data, enzyme activities, intracellular metabolite concentrations, as well as metabolic fluxes. These results may be useful for multiple gene mutations toward the design of microbial cell factories.

Nomenclature

FBA	Flux balance analysis
MFA	Metabolic flux analysis
ED pathway	Entner-Doudoroff pathway

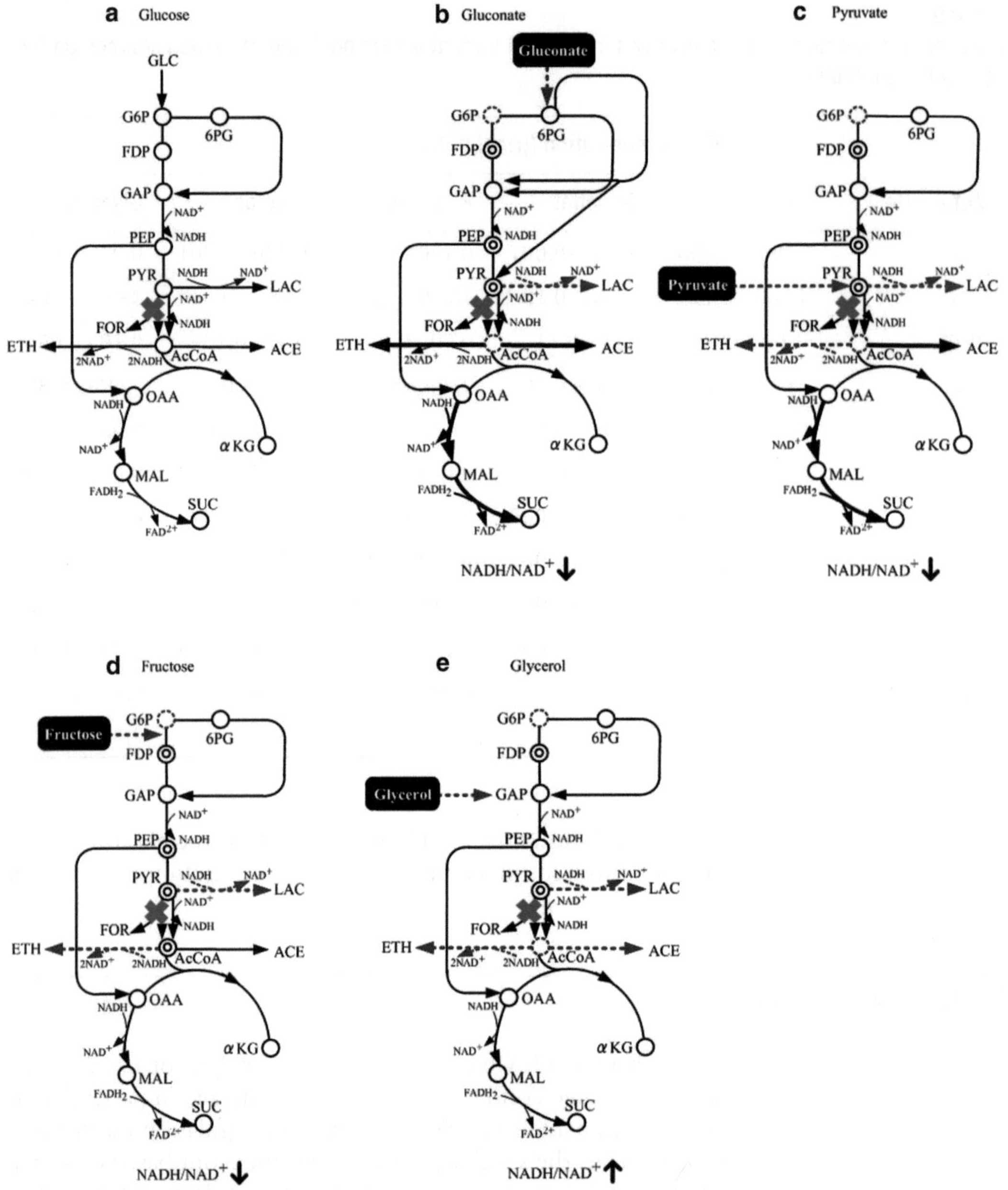

Fig. 7 Brief illustration of the metabolic changes of *pfl* mutant with different carbon sources: (**a**) glucose, (**b**) gluconate, (**c**) pyruvate, (**d**) fructose, and (**e**) glycerol. *Blue dashed lines* denote down-regulation, while *black solid lines* denote up-regulation as compared to wild-type strain (Color figure online)

EMP pathway	Embden-Meyerhof-Parnas pathway
TCA cycle	Tricarboxylic acid cycle
Arc system	Anoxic respiration control system
PTS system	Phosphotransferase system
Crp	cAMP receptor protein

Cya	Adenylate cyclase
DHAK	Dihydroxyacetone kinase
GlpF	Glycerol facilitator protein
GLYDH	Glycerol dehydrogenase
G3PDH	Glycerol-3-phosphate dehydrogenase
EI	Enzyme I
EII	Enzyme II
$EIIA^{Glc}$-P	Phosphorylated $EIIA^{Glc}$
HPr	Histidine-phosphorylatable protein
Cra	Catabolite repressor/activator protein
Fnr	Fumarate and nitrate reduction

Enzyme

Ack	Acetate kinase
Acn	Aconitase
Acs	Acetyl coenzyme A synthetase
ADH	Alcohol dehydrogenase
CS	Citrate synthase
Eno	Enolase
Fbp	Fructose bisphosphatase
Frd	Fumarate reductase
Fum	Fumarase
GAPDH	Glyceraldehyde-3-phosphate dehydrogenase
G6PDH	Glucose-6-phosphate dehydrogenase
Glk	Glucokinase
ICDH	Isocitrate dehydrogenase
Icl	Isocitrate lyase
KGDH	Ketoglutarate dehydrogenase
LDH	Lactate dehydrogenase
LPD	Lipoamide dehydrogenase
MDH	Malate dehydrogenase
Mez	Malic enzyme
MS	Malate synthase
Pck	Phosphoenolpyruvate carboxykinase
PDHc	Pyruvate dehydrogenase complex
Pfk	Phosphofructokinase
Pfl	Pyruvate formate lyase
6PGDH	6-Phosphogluconate dehydrogenase
Pgi	Phosphoglucose isomerase
Pox	Pyruvate oxidase
Ppc	Phosphoenolpyruvate carboxylase
Pps	Phosphoenolpyruvate synthase
Pta	Phosphotransacetylase
Pyk	Ribose phosphate isomerase
SCS	Succinyl-CoA synthetase
SDH	Succinate dehydrogenase
Tal	Transaldolase
TktA	Transketolase I
TktB	Transketolase II

Metabolite	
AcCoA	Acetyl-coenzyme A
ACE	Acetate
AcP	Acetyl phosphate
CIT	Citrate
E4P	Erythrose-4-phosphate
ETH	Ethanol
FDP	Fructose-1,6-bisphosphate
FOR	Formate
F6P	Fructose-6-phosphate
G6P	Glucose-6-phosphate
GAP	Glyceraldehyde-3-phosphate
GLC	Glucose
GLU	Glutamate
GOX	Glyoxylate
ICT	Isocitrate
KDPG	2-Keto-3-deoxy-6-phosphogluconate
αKG	α-Ketoglutarate
LAC	Lactate
MAL	Malate
OAA	Oxaloacetate
PEP	Phosphoenol pyruvate
6PG	6-Phosphogluconate
PYR	Pyruvate
R5P	Ribose-5-phosphate
RU5P	Ribulose-5-phosphate
S7P	Sedoheptulose-7-phosphate
SUC	Succinate
X5P	Xylulose-5-phosphate

References

1. Martinez-Antonio A (2011) *Escherichia coli* transcriptional regulatory network. Network Biol 1:21–33
2. Blattner FR, Plunkett G, Bloch CA et al (1997) The complete genome sequence of *Escherichia coli* K-12. Science 277:1453–1474
3. Ohnishi M, Tanaka C, Kuhara S et al (1999) Chromosome of the enterohemorrhagic *Escherichia coli* O157:H7; comparative analysis with K-12 MG1655 revealed the acquisition of a large amount of foreign DNAs. DNA Res 6:361–368
4. Steed H, Macfarlane GT, Macfarlane S (2008) Prebiotics, synbiotics and inflammatory bowel disease. Mol Nutr Food Res 52:898–905
5. Madan BM, Teichmann SA (2003) Evolution of transcription factors and the gene regulatory network in *Escherichia coli*. Nucleic Acids Res 31:1234–1244
6. Pérez-Rueda E, Collado-Vides J (2000) The repertoire of DNA-binding transcriptional regulators in *Escherichia coli* K-12. Nucleic Acids Res 28:1838–1847
7. Gama-Castro S, Jimenez-Jacinto V, Peralta-Gil M et al (2008) RegulonDB (version 6.0): gene regulation model of *Escherichia coli* K-12 beyond transcription, active (experimental) annotated promoters and Textpresso navigation. Nucleic Acids Res 36:D120–D124
8. Keseler IM, Collado-Vides J, Santos-Zavaleta A et al (2011) EcoCyc: a comprehensive database of *Escherichia coli* biology. Nucl Acids Res 39:D583–D590
9. Shimizu K (2012) Bacterial cellular metabolic systems. Woodhead, Oxford
10. Shimizu K (2013) Metabolic regulation of a bacterial cell system with emphasis on *Escherichia coli* metabolism. International

Scholarly Research Network, ISRN Biochemistry, Article ID 645983
11. Gosset G (2005) Improvement of *Escherichia coli* production strains by modification of the phosphoenolpyruvate:sugar phosphotransferase system. Microb Cell Fact 4:14
12. Deutscher J, Francke C, Postma PW (2006) How phosphotransferase system-related protein phosphorylation regulates carbohydrate metabolism in bacteria. Microbiol Mol Biol Rev 70(4):939–1031
13. Bettenbrock K, Fischer S, Klemling A et al (2006) A quantitative approach to catabolite repression in *Escherichia coli.* J Biol Chem 281: 2578–2584
14. Gorke B, Stulke J (2008) Carbon catabolite repression in bacteria: many ways to make the most out of nutrients. Nat Rev Microbiol 6: 613–624
15. Park YH, Lee BR, Seok YJ et al (2006) In vitro reconstitution of catabolite repression in *Escherichia coli.* J Biol Chem 281:6448–6454
16. Bettenbrock K, Sauter T, Jahreis K et al (2007) Correlation between growth rates, EIIACrr phosphorylation, and intracellular cyclic AMP levels in *Escherichia coli* K-12. J Bacteriol 189: 6891–6900
17. Hogema BM, Arents JC, Bader R et al (1998) Inducer exclusion in *Escherichia coli* by non-PTS substrates: the role of the PEP to pyruvate ratio in determining the phosphorylation state of enzyme IIAGlc. Mol Microbiol 30:487–498
18. Moat AG, Foster JW, Spector MP (2002) Microbiology, 4th edn. Wiley, New York, NY
19. Saier MH Jr, Ramseier TM, Reizer J (1996) Regulation of carbon utilization. In: Neidhardt C, Curtiss R III, Ingraham JL et al (eds) *Escherichia coli* and salmonella: cellular and molecular biology. ASM, Washington, DC, pp 1325–1343
20. Saier MH Jr, Ramseier TM (1996) The catabolite repressor/activator (Cra) protein of enteric bacteria. J Bacteriol 178:3411–3417
21. Sarkar D, Shimizu K (2008) Effect of *cra* gene knockout together with other genes knockouts on the improvement of substrate consumption rate in *Escherichia coli* under microaerobic conditions. Biochem Eng J 42:224–228
22. Kang Y, Weber KD, Qin Y, Kiley PJ, Blattner FR (2005) Genome-wide expression analysis indicates that FNR of *Escherichia coli* K-12 regulates a large number of genes of unknown function. J Bacteriol 187:1135–1160
23. Gunsalus RP (1992) Control of electron flow in *Escherichia coli*: co-ordinated transcription of respiratory pathway genes. J Bacteriol 174: 7069–7074
24. Alexeeva S, Hellingwerf KJ, de Mattos MJT (2003) Requirement of ArcA for redox regulation in *Escherichia coli* under microaerobic but not anaerobic or aerobic conditions. J Bacteriol 185:204–209
25. Alexeeva S, de Kort B, Sawers G et al (2000) Effects of limited aeration and of the ArcAB system on intermediary pyruvate catabolism in *Escherichia coli.* J Bacteriol 182:4934–4940
26. Crack JC, Le Brun NE, Thomson AJ et al (2008) Reactions of nitric oxide and oxygen with the regulator of fumarate and nitrate reduction, a global transcriptional regulator, during anaerobic growth of *Escherichia coli.* Methods Enzymol 437:191–209
27. De Graef MR, Alexeeva S, Snoep JL et al (1999) Steady state internal redox state (NADH/NAD) reflects the external redox state and is correlated with catabolic adaptation in *Escherichia coli.* J Bacteriol 181:2351–2357
28. Bailey JE (1991) Toward a science of metabolic engineering. Science 252:1668–1675
29. Stephanopoulos G, Vallino JJ (1991) Network rigidity and metabolic engineering in metabolite overproduction. Science 252:1675–1681
30. Schuetz R, Kuepfer L, Sauer U (2007) Systematic evaluation of objective functions for predicting intracellular fluxes in *Escherichia coli.* Mol Syst Biol 3:119
31. Schilling CH, Palsson BO (1998) The underlying pathway structure of biochemical reaction networks. Proc Natl Acad Sci U S A 95: 4193–4198
32. Edwards JS, Palsson BO (2000) Metabolic flux balance analysis and the *in silico* analysis of *Escherichia coli* K-12 gene deletions. BMC Bioinformatics 1:1
33. Price ND, Papin JA, Schilling CH, Palsson BO (2003) Genome-scale microbial *in silico* models: the constraints-based approach. Trends Biotechnol 21:162–169
34. Oberhardt MA, Palsson BO, Papin JA (2009) Applications of genome-scale metabolic reconstructions. Mol Syst Biol 5:320
35. Edwards JS, Ibarra RU, Palsson BO (2001) *In silico* predictions of *Escherichia coli* metabolic capabilities are consistent with experimental data. Nat Biotechnol 19:125–130
36. Ibarra RU, Edwards JS, Palsson BO (2002) *Escherichia coli* K-12 undergoes adaptive evolution to achieve *in silico* predicted optimal growth. Nature 420:186–189
37. Yuan LZ, Rouvière PE, LaRossa RA, Suh W (2006) Chromosomal promoter replacement of the isoprenoid pathway for enhancing carotenoid production in *E. coli.* Metab Eng 8:79–90
38. Zhu J, Shimizu K (2005) Effect of a single-gene knockout on the metabolic regulation in *E. coli* for d-lactate production under microaerobic condition. Metab Eng 7:104–115
39. Garrigues C, Loubiere P, Lindle ND et al (1997) Control of the shift from homolactic acid to mixed-acid fermentation in *Lactococcus*

lactis: predominant role of the NADH/NAD+ratio. J Bacteriol 179:5282–5287
40. Smith TE, Balasubramanian KA, Beezley A (1980) *Escherichia coli* phosphoenol-pyruvate carboxylase: studies on the mechanism of synergistic activation by nucleotides. J Biol Chem 255:1635–1642
41. McAlister LE, Evans EL, Smith TE (1981) Properties of a mutant *Escherichia coli* phosphoenolpyruvate carboxylase deficient in coregulation by intermediary metabolites. J Bacteriol 146:200–208
42. Zhu J, Shimizu K (2004) The effect of *pfl* genes knockout on the metabolism for optically pure D-lactate production by *Escherichia coli*. Appl Microbiol Biotechnol 64:367–375
43. Koebmann BJ, Westerhoff HV, Snoep JL et al (2002) The glycolytic flux in *Escherichia coli* is controlled by the demand for ATP. J Bacteriol 184:3909–3916
44. Kabir MM, Ho PY, Shimizu K (2005) Effect of *ldhA* gene deletion on the metabolism of *Escherichia coli* based on gene expression, enzyme activities, intracellular metabolite concentrations, and metabolic flux distribution. Biochem Eng J 26:1–11
45. Matsuoka Y, Shimizu K (2013) Metabolic regulation of *Escherichia coli* cultivation under anaerobic and aerobic conditions in response to the specific pathway gene knockouts. Adv Biosci Biotechnol 4:455–468

Chapter 16

^{13}C-Metabolic Flux Analysis for *Escherichia coli*

Yu Matsuoka and Kazuyuki Shimizu

Abstract

^{13}C-Metabolic flux analysis (^{13}C-MFA) is used here to study the effects of the knockout of such genes as *pgi*, *zwf*, *gnd*, *ppc*, *pck*, *pyk*, and *lpdA* on the metabolic changes in *Escherichia coli* cultivated under aerobic condition. The metabolic regulation mechanisms were clarified by integrating such information as fermentation data, gene expression, enzyme activities, and metabolite concentrations as well the result of ^{13}C-MFA.

Key words ^{13}C-Metabolic flux analysis, *E. coli*, Metabolic regulation analysis, Catabolite regulation, Aerobic cultivation

1 Introduction

The enzymes which form the metabolic pathways are subject to multiple levels of regulation, where the enzyme level regulation and the transcriptional regulation by global regulators play important roles in the metabolic regulation [1–3]. Undoubtedly, it is quite important to understand the regulatory processes which govern the cellular metabolism in order to develop new production strains or to design targeted drugs for pathogens. This is a key focus of microbial systems biology. Although transcriptional regulation is typically analyzed by measurement of mRNA abundance, the intracellular reaction rates or metabolic fluxes must also be quantified to assess metabolic function in the network context [3, 4].

A powerful approach for accurately estimating metabolic fluxes is ^{13}C-metabolic flux analysis (^{13}C-MFA) based on the measurement of isotopomers using either nuclear magnetic resonance (NMR) or mass spectrometry [5, 6]. The detailed analysis of NMR spectra has been proved to be useful for metabolic flux analysis of *Escherichia coli* [5]. The problem of using NMR is that it requires a relatively large amount of sample. Since the amount of sample required for GC/MS (gas chromatography mass spectrometry) is significantly less, GC/MS is becoming a popular tool for labeling

Jens O. Krömer et al. (eds.), *Metabolic Flux Analysis: Methods and Protocols*, Methods in Molecular Biology, vol. 1191, DOI 10.1007/978-1-4939-1170-7_16, © Springer Science+Business Media New York 2014

analysis in ^{13}C-MFA [7, 8]. Moreover, CE/MS (capillary electrophoresis mass spectrometry) or CE/TOF (time of flight)-MS may become more popular due to their high sensitivity and high throughput performance, where these can detect the mass isotopomer distribution of intracellular metabolites instead of proteinogenic amino acids [9–11]. Based on the mass balance of metabolites and isotopomers, the metabolic flux distribution can be estimated by the best fit to the measured data [12–14].

Isotope labeling experiments are usually applied to continuous culture, and are initiated after the culture reached a metabolic steady state, which may be inferred from the stable oxygen and CO_2 concentrations in the off-gas and stable optical density (OD) in the effluent medium for at least three or four times as long as the residence time. The feed medium with unlabeled glucose is then replaced by an identical medium containing, for example, a mixture of uniformly labeled glucose [U-^{13}C], the first carbon labeled glucose [1-^{13}C], and naturally labeled glucose [U-^{12}C]. Other combinations of labeling substrates may be used, but the composition of the substrate mix should be made based on the available measurement apparatus, the cost of labeled substrate, and the accuracy of the specific pathway fluxes. The accuracy of the metabolic flux distribution depends on the mixture of labeled substrate and the measurement apparatus [14–16]. After the establishment of isotopic steady-state (after about five residence times on the tracer substrate by assuming the first order dynamics such that $(1-e^{-t})\times 100$ % is replaced by the labeled substrate, where t is the residence time), samples for labeling analysis can be harvested, followed by centrifugation, and then the cell is hydrolyzed to obtain proteinogenic amino acids.

In the case of using CE/MS or CE/TOF-MS, the isotopomer distribution of intracellular metabolites can be detected, and thus the flux analysis of the time-varying system such as batch culture can be made [9, 11].

In this Chapter, ^{13}C-MFA is explained for *E. coli* and its several pathway gene knockout mutants. Moreover, the metabolic regulation analysis is made focusing on the roles of transcription factors. As mentioned in Chapter 15, the central players in the carbon catabolite regulation in *E. coli* are Crp, cAMP, and Cya, while Cra controls the carbon flow. Figure 1 illustrates how these regulate the main metabolic pathway genes.

2 ^{13}C-MFA of Wild-Type *E. coli* at Different Growth Rate

Consider the metabolic flux analysis of *E. coli* based on GC-MS [13, 17]. This technique has been applied successfully to estimate the intracellular flux distribution for a variety of organisms [8]. Let us consider the metabolic flux analysis for wild-type *E. coli* cultivated using glucose as a carbon source, where the stoichiometric equations are given in Table 1, and the measured specific rates are

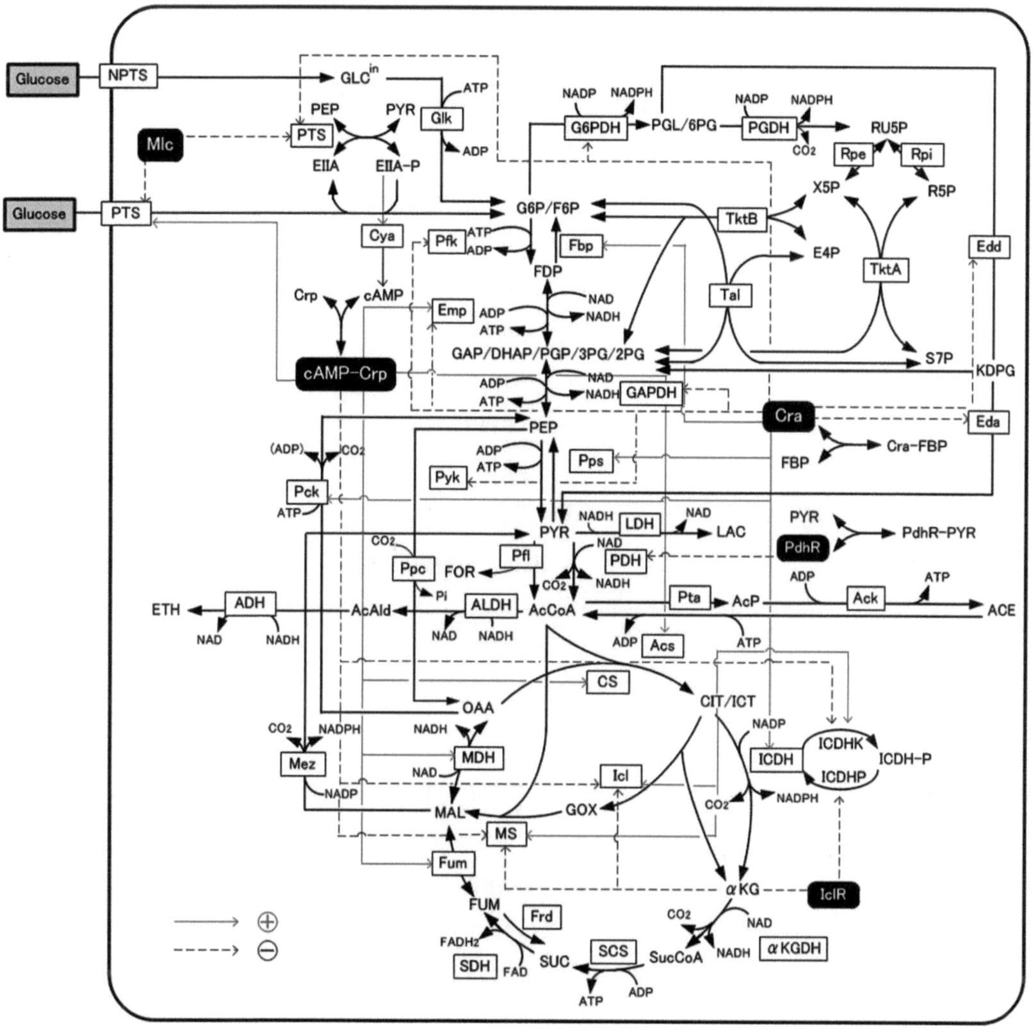

Fig. 1 Main metabolic pathways and the regulation by global regulators

shown in Table 2 [13]. Let the forward fluxes and the backward fluxes be expressed by the net fluxes v^{net} and the exchange fluxes v^{exch} as defined as

$$v^{net} = \vec{v} - \overleftarrow{v} \tag{1a}$$

$$v^{exch} = \min(\vec{v}, \overleftarrow{v}) \tag{1b}$$

Then the following equation is satisfied under quasi-steady state

$$Av^{net} = 0 \tag{2}$$

where A is the stoichiometric matrix as represented in Table 1. Based on the rank of A, the number of degrees of freedom is determined. The labeling data obtained from mass isotopomer

Table 1
Metabolic reactions for glucose metabolism of wild-type *E. coli*

	Reaction	Gene coding for enzymes
1.	GLC ⇒ G6P	*glk (pts)*
2.	G6P ⇔ F6P	*pgi*
3.	F6P ⇔ T3P1 + T3P2	*pfk*
4.	T3P ⇔ PGA	*pgk*
5.	PGA ⇔ PEP	*eno*
6.	PEP ⇔ PYR	*pyk*
7.	PYR ⇔ ACA + CO_2	*aceEF, lpdA*
8.	ACA ⇔ $(Acetate)_{in}$	*acs*
9.	$(Acetate)_{in}$ ⇔ $(Acetate)_{out}$	
10.	G6P ⇒ Ru5P + CO_2	*gnd*
11.	Ru5P ⇔ X5P	*rpe*
12.	Ru5P ⇔ R5P	*rpi*
13.	R5P + X5P ⇔ S7P + T3P (GAP)	*tkl*
14.	S7P + T3P (GAP) ⇔ E4P + F6P	*tal*
15.	X5P + E4P ⇔ F6P + T3P(GAP)	*tk2*
16.	PGA ⇔ SER	*ser*
17.	SER ⇔ GLY + C1	*glyA*
18.	GLY ⇔ C1 + CO_2	*gcvHTP*
19.	AcA + OAA ⇔ CIT	*gltA*
20.	CIT ⇔ ICIT	*acnAB*
21.	ICIT ⇔ OGA + CO_2	*icdA*
22.	OGA ⇔ SUC + CO_2	*sucABCD*
23.	SUC ⇔ FUM	*sdhABCD*
24.	FUM ⇔ MAL	*fumABC*
25.	MAL ⇔ OAA	*mdh*
26.	PEP + CO_2 ⇔ OAA	*ppc-pckA*
27.	$(CO_2)_{in}$ ⇒ $(CO_2)_{out}$	
28.	Precursors ⇒ Biomass	

Table 2
Growth parameters of *E. coli* K12 at the dilution rates of 0.11 and 0.22 h^{-1}, where glucose is used as the sole carbon source

Dilution rate (h^{-1})	Glucose consumption rate (mmol/gDCW/h)	CO_2 formation rate (mmol/gDCW/h)	Acetate formation rate (mmol/gDCW/h)
0.11	2.78	5.25	0.38
0.22	5.92	11.37	0.97

measurements can be used to determine the fluxes. The exchange coefficient is defined as

$$v^{\text{exch}[0,1]} = \frac{v^{\text{exch}}}{\beta + v^{\text{exch}}} \tag{3}$$

where this representation is useful, in particular, for numerical computation.

Table 3 shows the measured mass isotopomer distribution (MID) (Exp), and the simulated MID for the best-fit flux distribution (Cal), where the objective function to be minimized is defined as

$$\varepsilon\left(v,I_{\text{glc}}\right) = \sum_{i=1}^{n}\left(\frac{M_{i,\text{meas}} - M_{i,\text{calc}}\left(v,I_{\text{glc}}\right)}{\delta_i}\right)^2 + \sum_{j=1}^{m}\left(\frac{S_{j,\text{meas}} - S_{j,\text{calc}}}{\delta_j}\right)^2 \tag{4}$$

where v is the flux vector to be optimized, and I_{glc} is the isotopomer distribution of the input glucose. $M_{i,\text{meas}}$ are the n individual labeling data obtained from mass isotopomer measurements, and $M_{i,\text{calc}}$ are the corresponding simulated values which are computed based on the assumed values for v. $S_{j,\text{meas}}$ are the measured values of m extracellular fluxes and $S_{j,\text{calc}}$ are the simulated values of extracellular fluxes. δ_i and δ_j are the corresponding measurement errors.

The best-fit flux distributions in the glucose-limited chemostat cultures at the dilution rates of 0.11 and 0.22 h^{-1} are given in Fig. 2 [13], where the 90 % confidence limits are given in Table 4. Figure 2 indicates that the relative flux for the pentose phosphate (PP) pathway increases, while that of the TCA cycle decreases as the cell growth rate increases. The lower activity of the TCA cycle and the higher activity of PP pathway may be the notable characteristics at the higher dilution rate.

Let us further consider how the specific growth rate (dilution rate in the chemostat culture) affects the global regulators and the metabolic pathway genes of wild-type *E. coli* (BW25113) [18]. It was shown that the specific acetate production rate and the specific

Table 3
Experimental determined (Exp) and calculated (Cal) fragment mass distribution of TBDMS-derivatized amino acids from *Escherichia coli* K12 hydrolysates obtained from the chemostat culture at the dilution rate of 0.22 h^{-1}

			Glucose metabolism		
Amino acids	**Fragment**	**Origin**	***m0***	***m1***	***m2***
Gly	[M-57]$^+$	Exp:	0.9202	0.0154	0.0644
		Cal:	0.9176	0.0274	0.0550
Val	[M-57]$^+$	Exp:	0.7646	0.0944	0.1140
		Cal:	0.7602	0.0952	0.1152
Leu	[M-159]$^+$	Exp:	0.7487	0.1097	0.1103
		Cal:	0.7422	0.1112	0.1121
Ile	[M-159]$^+$	Exp:	0.7689	0.0964	0.1116
		Cal:	0.7632	0.0977	0.1132
Ser	[M-159]$^+$	Exp:	0.9044	0.0260	0.0696
		Cal:	0.8990	0.0274	0.0737
Thr	[M-57]$^+$	Exp:	0.8267	0.0788	0.0565
		Cal:	0.8237	0.0797	0.0573
Phe	[M-159]$^+$	Exp:	0.7683	0.0466	0.1058
		Cal:	0.7553	0.0470	0.1092
Glu	[M-159]$^+$	Exp:	0.7789	0.1112	0.0930
		Cal:	0.7733	0.1129	0.0950

CO_2 evolution rate (CER) increased as the dilution rate was increased (Table 5) [18]. Figure 3 shows the effect of the dilution rate (the specific growth rate) on gene transcript levels and fluxes [18, 19]. As the glucose uptake rate increases, or PTS is enhanced in accordance with the specific growth rate, the absolute glycolysis flux increases, and thus F16BP (FDP) concentration increases. The increased FDP allosterically enhances the activity of Pyk (and also Ppc) by feed-forward control (Fig. 4). The PEP concentration tends to be decreased due to activation of Pyk (and Ppc) as well as the increased flux of PTS, although it is not obvious since the glycolysis flux increases, and tends to give more PEP. It has been known that PEP molecule allosterically inhibits Pfk activity by feed-back regulation (Fig. 4). The decrease in PEP concentration thus causes Pfk activity to be increased, and the glycolysis flux

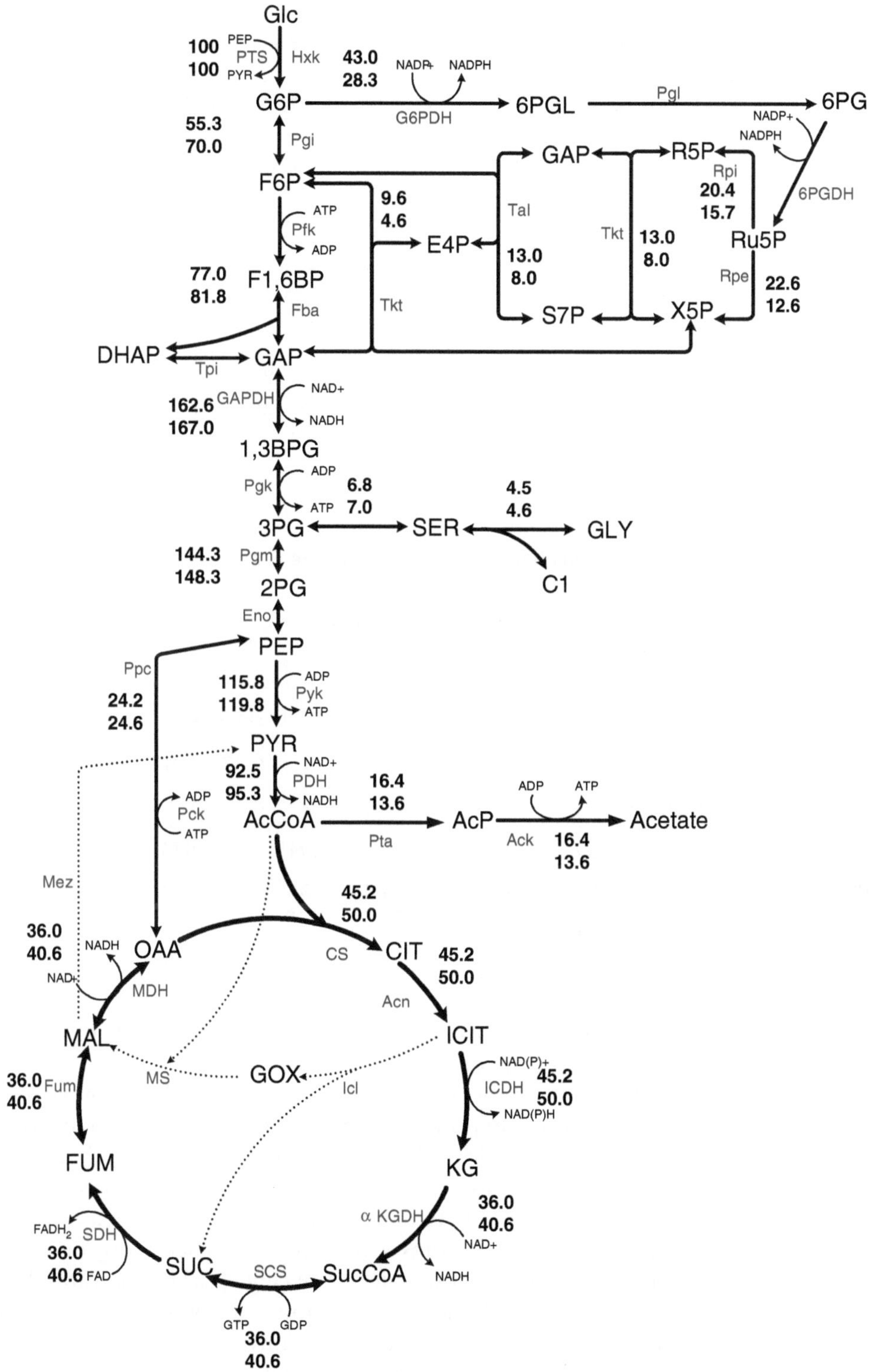

Fig. 2 Effect of the growth rate on the metabolic flux distribution in *E. coli* cultivated at the dilution rates of 0.11 (*lower values*) and 0.22 h^{-1} (*upper values*) of the aerobic continuous culture

Table 4
90 % Confidence limits for estimated net fluxes and exchange coefficients

	$D=0.22\ h^{-1}$		$D=0.11\ h^{-1}$	
	Optimal estimates	90 % confidence region	Optimal estimates	90 % confidence region
tk1	13.00	[12.99, 13.50]	8.00	[7.90, 8.45]
$Exch_{pgi}^{[0,1]}$	0.761	[0.753, 0.768]	0.919	[0.910, 0.925]
$Exch_{eno}^{[0,1]}$	0.402	[0.050, 0.920]	0.251	[0.050, 0.950]
$Exch_{tk1}^{[0,1]}$	0.443	[0.140, 0.520]	0.503	[0.250, 0.630]
$Exch_{tal}^{[0,1]}$	0.125	[0.110, 0.139]	0.100	[0.080, 0.115]

Table 5
Effect of dilution rate on fermentation characteristics of wild-type *E. coli*

Strains	Biomass conc. (g/L)	Glucose conc. (g/L)	Glucose uptake rates (mmol/g/h)	Acetate formation rates (mmol/g/h)	Biomass yield (g/g)	CER (mmol/g/h)	Carbon recovery
BW25113(0.2D)	1.45 ± 0.06	ND	3.07 ± 0.13	ND	0.37 ± 0.015	9.15	94 %
BW25113(0.4D)	1.87 ± 0.09	ND	4.75 ± 0.23	0.01	0.47 ± 0.023	11.61	98 %
BW25113(0.6D)	2.0 ± 0.09	ND	6.67 ± 0.3	0.88 ± 0.04	0.5 ± 0.023	13.17	98 %
BW25113(0.7D)	1.93 ± 0.08	ND	8.05 ± 0.34	1.33 ± 0.06	0.48 ± 0.02	15.83	97 %

ND not detectable, where glucose detectable limit was 0.038 g/l

increases, and in turn FDP concentration increases more. On the other hand, the decrease in PEP and increase in PYR make PEP/PYR ratio to be decreased. This causes phosphorylated EIIA (EIIA-P) concentration to be decreased, and in turn less activates Cya, and thus cAMP is less formed from ATP. As a result, cAMP-Crp level decreases, which in principle decreases the expression of *ptsG*, which codes for EIICB, and this causes the decrease in glucose uptake rate. This forms a negative feed-back loop for the initial increase in the glucose uptake rate (Fig. 4). This indicates that PTS plays an essential role from the robustness point of view in addition to phosphorylation of substrate. Conversely, this robustness is not guaranteed for the case of non-PTS carbohydrates, *pts* mutant, and for the cells without having PTS such as yeast, etc. [20].

The analysis may be further extended to the regulation of the TCA cycle. As mentioned above, the increase in glucose uptake

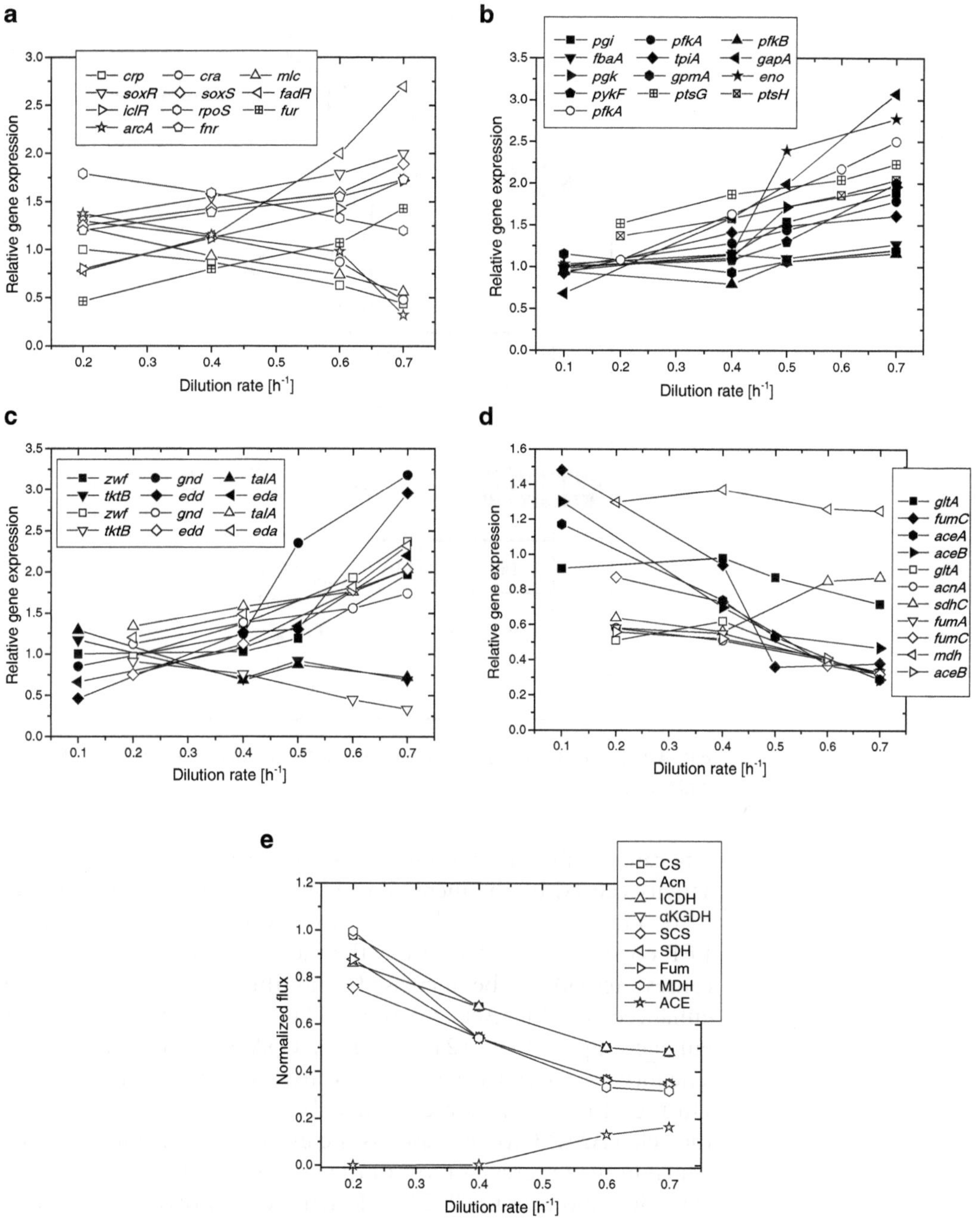

Fig. 3 Effect of the growth rate (dilution rate) on the glycolysis and TCA cycle gene expressions and fluxes in *E. coli*: (**a**) Global regulatory gene expression, (**b**) Glycolysis gene expression, (**c**) PP pathway gene expression, (**d**) TCA cycle gene expressions, (**e**) TCA cycle fluxes and acetate formation rate

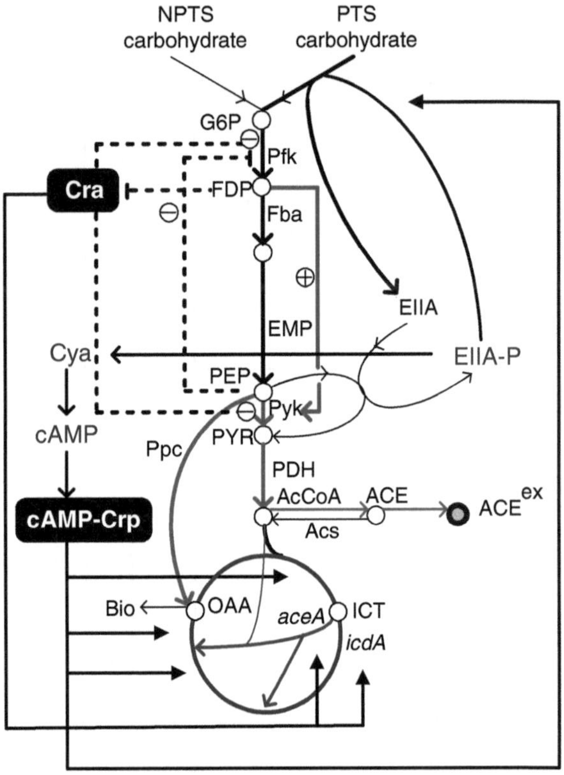

Fig. 4 Enzyme level regulation and the transcriptional regulation of the main metabolism in *E. coli*

rate causes PYR concentration to be increased. The increase in Ppc activity caused by the increase in FDP affects OAA concentration to be increased, although it is not obvious due to the decreased PEP concentration. Moreover, the increase in PYR concentration causes AcCoA to be increased, and thus acetate formation is enhanced by Pta–Ack reaction. The increase in AcCoA also enhances Ppc activity [21], where AcCoA is the activator of Ppc. The increases in AcCoA and OAA cause TCA cycle to be activated. On the other hand, as mentioned above, if PTS was furnished in the cell, cAMP-Crp decreases as the glucose uptake rate increases, which in turn represses most of the expression of TCA cycle genes (Fig. 3d), and thus the TCA cycle activity is repressed by transcriptional regulation (Figs. 3e and 4).

In fact, there is another global regulator Cra (FruR) that plays an important role for the control of carbon flows, where Cra detects FDP concentration, and Cra activity decreases with the increase in FDP concentration. In the above example, the increase in glucose uptake rate causes the increase in FDP concentration, and thus Cra activity decreases. This causes the increases of the expression of glycolysis genes such as *pfkA* and *pykF* genes (Fig. 3b), which may have caused the increased glycolysis fluxes. Moreover, the decrease in Cra activity also affects TCA cycle and glyoxylate pathway genes

(Fig. 3d), and thus TCA cycle is further repressed by this mechanism (Fig. 4). The increase in glycolysis activity and the decrease in TCA cycle activity cause more acetate production (Fig. 3e) [18, 22–24].

Although the TCA cycle activity may be repressed based on gene expression data, the specific glucose consumption rate increases, and thus the absolute TCA cycle flux tends to increase on the mmol basis by taking into account the specific glucose consumption rate [18]. This means that the respiration is rather activated with the increased production of NADH in the glycolysis and the TCA cycle, and more radical oxygen derivatives are generated, where this is reflected in the increase in *soxR/S* gene transcript levels (Fig. 3a), where SoxR/S is the global regulatory system for the oxidative stress regulation. The increased activity of SoxS causes transcriptional activation of *zwf* gene expression (*see* Table 1 of Chapter 15). It has been shown that 6PGDH activity is the growth rate dependent [25], where it has been believed that the activation of the oxidative PP pathway is due to NADPH production for biosynthesis. The present analysis gives additional viewpoint for the activation of PP pathway.

3 Effects of the Specific Pathway Mutation on the Metabolism

The proper understanding for the metabolic regulation of the central carbon metabolism is critical for the efficient metabolic engineering. For this, it may be useful to understand the effect of the specific pathway mutation on the metabolic changes based on ^{13}C-metabolic flux distribution (^{13}C-MFD) together with different levels of information [3, 4, 25, 26]. While the specific-gene knockout mutations in the central metabolism preclude growth on glucose, a majority of such mutations seem to be potentially compensated by the use of alternative enzymes or by rerouting of the carbon fluxes through alternative pathways, which results in a robust phenotype such as little effect on the cell growth rate.

3.1 Effect of pgi Gene Knockout on the Metabolism

If *pgi* gene was knocked out, the primary route for glucose catabolism is the PP pathway together with Entner-Doudoroff (ED) pathway [11, 27–29]. NADPH is over-produced, which inhibits the activity of G6PDH, and thus the glucose consumption rate becomes significantly low in this mutant [11]. This in turn causes glycolysis fluxes as well as the related intermediate concentrations to be decreased [11]. The decreases in FDP and PEP concentrations cause Ppc activity to become lower by enzyme level regulation, which causes OAA concentration to be significantly decreased, which further activates the glyoxylate pathway [11, 29]. The acetate formation becomes significantly low due to the activation of the glyoxylate pathway [11]. Figure 5 shows the metabolic flux distribution of *pgi* mutant as compared to wild-type *E. coli* (W3110) cultivated at the

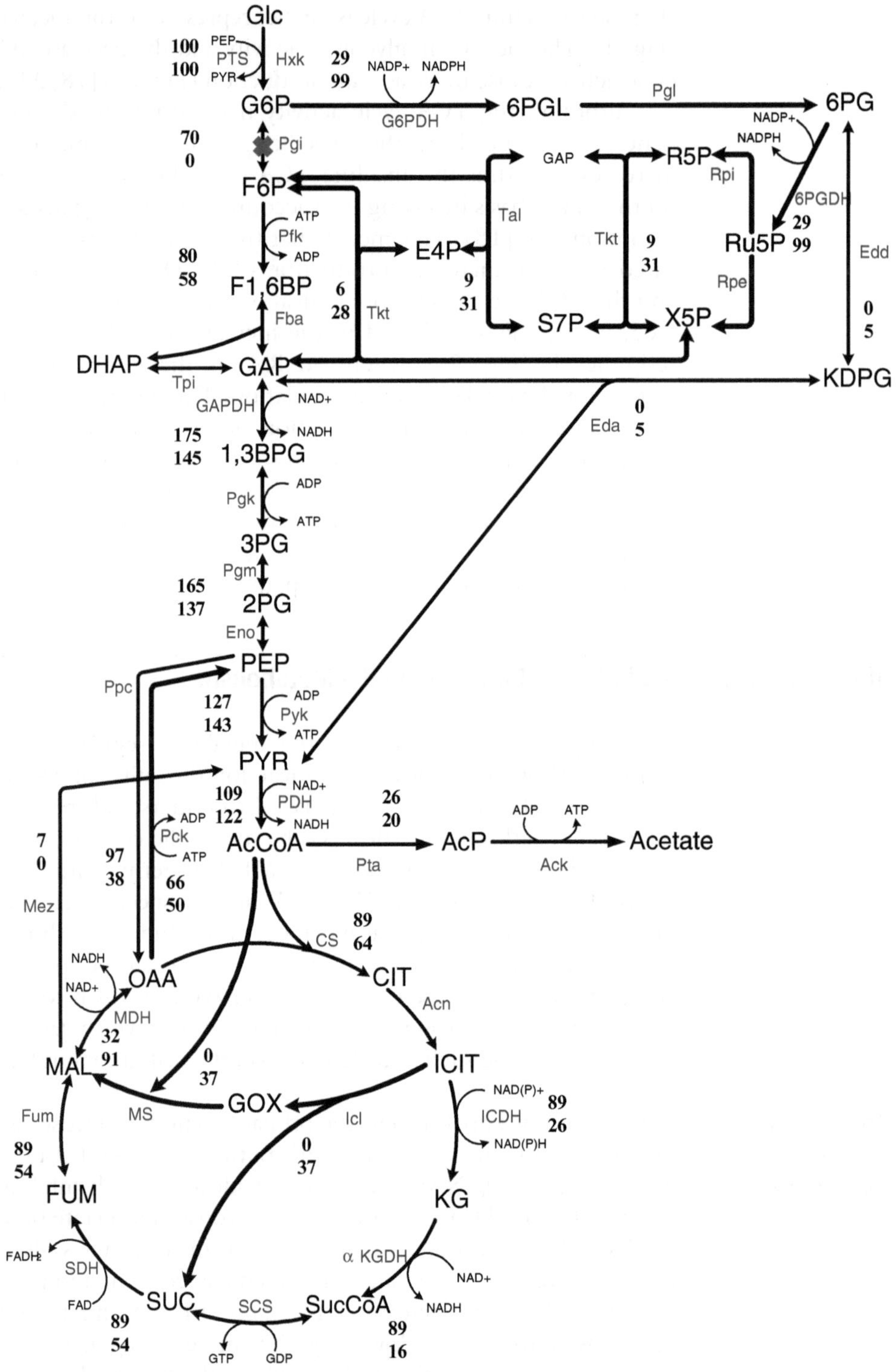

Fig. 5 ^{13}C-metabolic flux distribution of *pgi* mutant (*lower values*) as compared to wild-type W3110 (*upper values*) cultivated at the dilution rate of 0.1 h^{-1} in the aerobic continuous culture

Table 6
Fermentation characteristics of *pgi* mutant as compared to wild type cultivated at the dilution rate of 0.1 h^{-1} in the continuous culture

Chemostat conditions	Strain	$Y_{X/S}$ (g/g)	Glucose conc (mM)	Ammonium conc (mM)	$q_{glucose}$ (mmol/g/h)	q_{O2} (mmol/g/h)	q_{CO2} (mmol/g/h)	$q_{acetate}$ (mmol/g/h)	q_{EPS} (mmol/g/h)	C balance (%)
C limited	W3110	0.40 ± 0.03	<0.1	38.0 ± 3.4	1.4 ± 0.1	4.0 ± 0.7	4.2 ± 0.4	0	0	99 ± 7
	pgi mutant	0.43 ± 0.03	<0.1	37.1 ± 3.3	1.3 ± 0.1	3.9 ± 0.7	3.7 ± 0.4	0	0	97 ± 7
N limited	W3110	0.23 ± 0.02	8.3 ± 0.6	<0.1	2.9 ± 0.3	7.1 ± 1.4	7.2 ± 0.9	1.60 ± 0.14	0.04 ± 0	93 ± 9
	pgi mutant	0.39 ± 0.03	15.4 ± 1.3	<0.1	1.6 ± 0.2	3.8 ± 0.7	4.0 ± 0.4	0.31 ± 0	0	97 ± 7

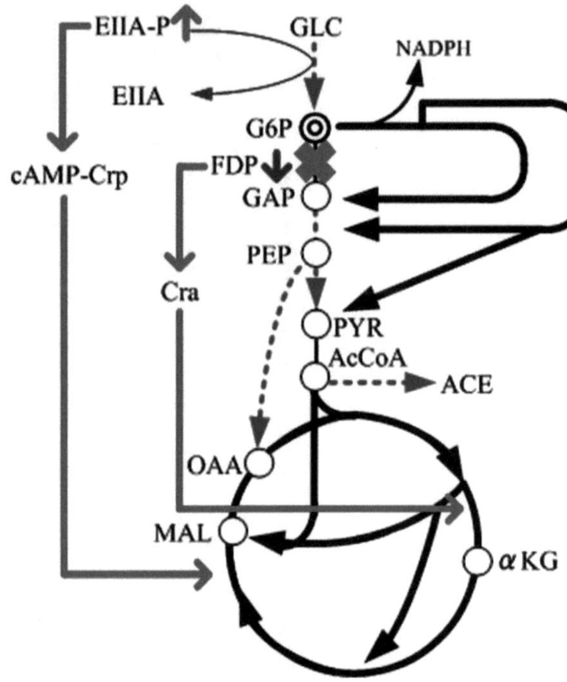

Fig. 6 Brief overall regulation mechanism of *pgi* mutant

dilution rate of 0.1 h^{-1} in the aerobic continuous culture, where the fermentation data are given in Table 6 [29]. The activation of glyoxylate pathway reduces NADPH production at ICDH, and some of the NADPH overproduced at PP pathway may be converted to NADH by transhydrogenase such as Udh [29]. NADPH and $NADP^+$ have been identified as effectors that regulate the reversible phosphorylation or inactivation of ICDH in *E. coli* [30].

As the glucose consumption rate decreases in *pgi* mutant, cAMP-Crp is considered to increase, and this in turn activates TCA cycle gene expressions. Moreover, the decrease in FDP concentration may cause Pyk and Ppc activities to be decreased by allosteric regulation in *pgi* mutant. The decreased FDP concentration activates Cra, which in turn represses glycolysis gene expression, while activates gluconeogenic pathway gene expression as well as *icdA* and *aceA/B* genes. These cause the acetate production to be decreased in *pgi* mutant.

In *pgi* mutant, the ED pathway is used to some extent to avoid further production of NADPH at 6PGDH. A similar observation is seen on other *pgi*-deficient *E. coli* strains [27, 28, 31]. Figure 6 shows the brief overall regulation mechanism for *pgi* mutant in *E. coli*.

3.2 Effects of zwf and gnd Gene Knockout on the Metabolism

In the case of *zwf* gene knockout, the cell growth phenotype is little affected, where the non-oxidative PP pathway flux is reversed [29, 32]. Figure 7 shows ^{13}C-MFD of *zwf* mutant and *gnd* mutant in comparison to that of wild-type *E. coli* cultivated at the dilution

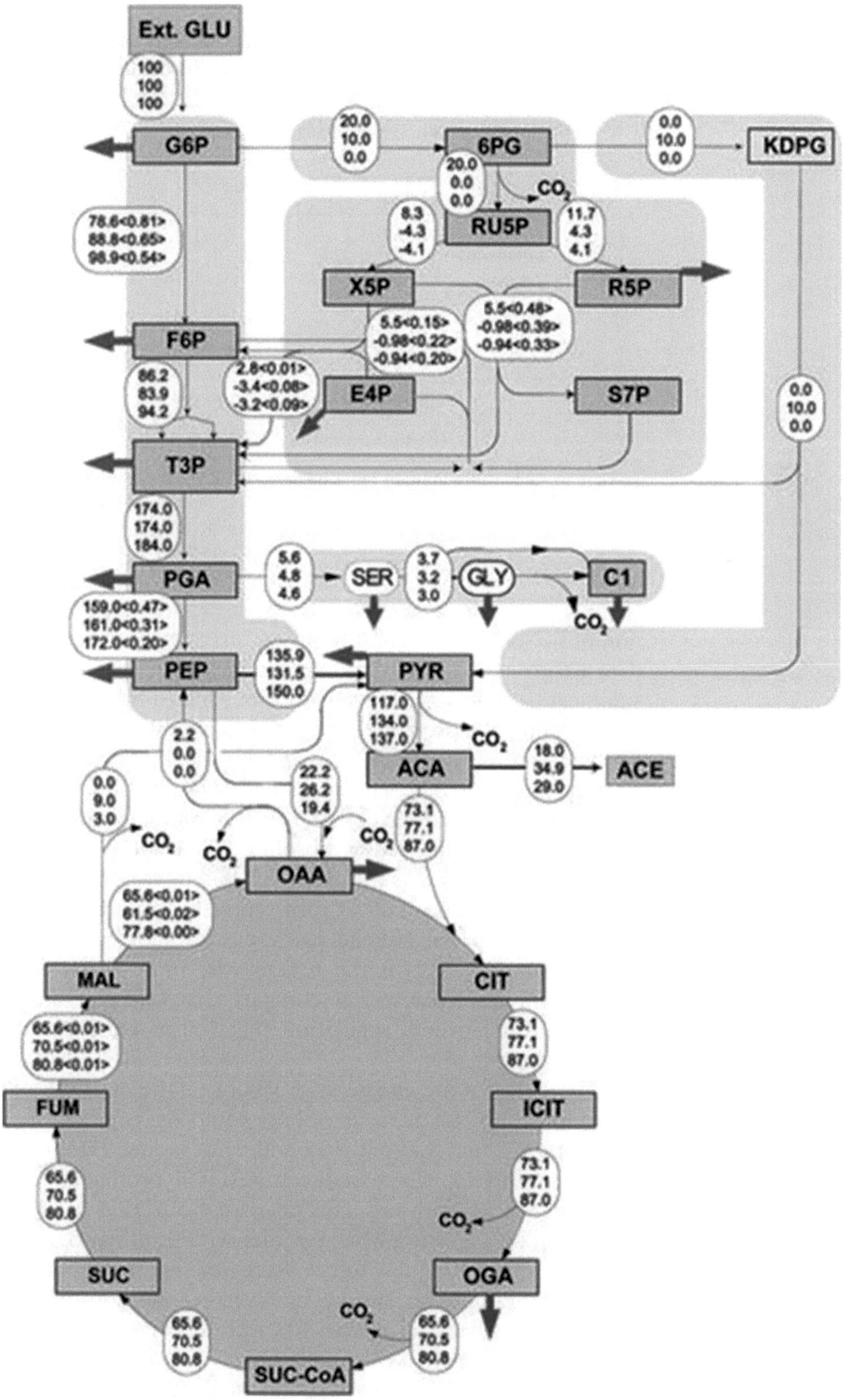

Fig. 7 ^{13}C-metabolic flux distribution of *zwf* (*third values*) and *gnd* mutants (*second values*) as compared to wild-type BW25113 (*first values*) cultivated at the dilution rate of 0.2 h^{-1} in the aerobic continuous culture

Table 7
Fermentation characteristics of *zwf* and *gnd* mutants as compared to wild type cultivated at the dilution rate of 0.2 h^{-1} in the continuous culture

	Glucose culture		
Parameter	**Wild**	***gnd***	***zwf***
q_S	3.20 ± 0.10	3.75 ± 0.15	3.82 ± 0.07
q_{CO2}	8.17 ± 0.30	9.90 ± 0.50	11.00 ± 0.61
q_{ace}	0.58 ± 0.07	1.31 ± 0.10	1.11 ± 0.07
$Y_{X/S}$	0.35 ± 0.01	0.30 ± 0.02	0.29 ± 0.01
$Y_{CO2/S}$	2.55 ± 0.09	2.64 ± 0.03	2.88 ± 0.12
$Y_{ace/S}$	0.18 ± 0.03	0.35 ± 0.04	0.29 ± 0.02

q_S: Carbon source consumption rate (mmol/g/h), q_{CO2}: CO_2 evolution rate (mmol/g/h), q_{ace}: acetate formation rate (mmol/g/h), $Y_{X/S}$: biomass yield on substrate (g/g), $Y_{CO2/S}$: CO_2 yield on substrate (mmol/mmol), $Y_{ace/S}$: acetate yield on substrate (mmol/mmol)

rate of 0.2 h^{-1} for the continuous culture, where the fermentation data are given in Table 7 [32, 33]. In particular, *zwf* mutant shows significant acetate overflow metabolism, and some of the NADPH, which cannot be produced at the oxidative PP pathway, is backed up by activating Mez (Fig. 7), and by the transhydrogenase Pnt from NADH [29, 32]. The *zwf* gene knockout causes the increase in glycolysis flux, and thus EIIA-P decreases and in turn cAMP-Crp decreases. This causes absolute TCA cycle activity to be repressed. Moreover, as the glycolysis flux increases, FDP concentration also increases, and in turn represses Cra activity, and thus TCA cycle is repressed by the down regulation of *icdA* and *aceA/B* gene expression. This is the reason why more acetate overflow metabolism was observed as compared to wild-type strain [29, 32]. The brief overall regulation mechanism for *zwf* mutant is illustrated in Fig. 8a.

The effect of the *gnd* gene knockout on the metabolism is different from the case of *zwf* gene knockout (Fig. 8b). The *gnd* gene knockout activates ED pathway and causes a decrease in the flux through G6PDH, which reduces NADPH production through the oxidative PP pathway. This decreased NADPH production is backed up by activating Mez together with transhydrogenase Pnt. The shortage of R5P, due to the decreased flux through the oxidative PP pathway, is backed up by reversing the non-oxidative PP pathway. This back-up system has little effect on the phenotype such as the cell growth rate.

Further inspection of NADPH generating reactions for the mutants reveals the negative correlation between Mez and ICDH

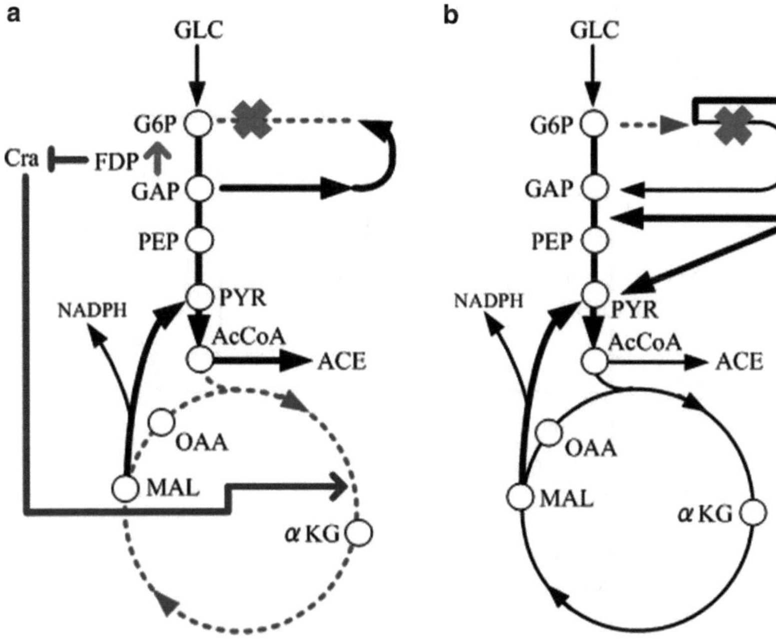

Fig. 8 Brief overall regulation mechanism of *zwf* (**a**) and *gnd* (**b**) mutants

fluxes in the mutants, i.e., higher flux of Mez pathway is accompanied by lower flux through the ICDH pathway. Moreover, there is a positive correlation between Mez and Ppc fluxes.

The flux via G6PDH is reduced but not blocked due to the shutdown of 6PGDH pathway. The activity of the oxidative PP pathway is, therefore, partly maintained in *gnd* mutant. In *E. coli*, the oxidative PP pathway plays a major role in the generation of NADPH, while ICDH is also important in producing NADPH [34]. Although the shortage of NADPH due to disruption or depression of the oxidative PP pathway can be partially compensated by the flux through ICDH, it appears that the flux via ICDH alone cannot satisfy the shortage of NADPH. For this, malate is drawn out of the TCA cycle through Mez to supply NADPH to meet the biosynthesis requirements. When glucose is used as a carbon source, a higher flux through ICDH is observed for *zwf* mutant (Fig. 7). This can generate more NADPH than that in *gnd* mutant, thus rendering the pathway via Mez less active (Fig. 7). The drain of carbon skeletons from the TCA cycle through Mez enables the cell to replenish the OAA pool by regulating the carbon flux through Pps–Ppc. To increase the synthesis of OAA from PEP, Ppc is therefore up-regulated in accordance with the activity of the Mez pathway in the mutants (Fig. 7).

3.3 Effects of ppc and pck Gene Knockout on the Metabolism

Figure 9 shows ^{13}C-MFD of *pckA* mutant in comparison with the wild-type *E. coli* (W3110) cultivated at the dilution rate of 0.1 h^{-1}, where the fermentation data are given in Table 8 [21]. When *pckA*

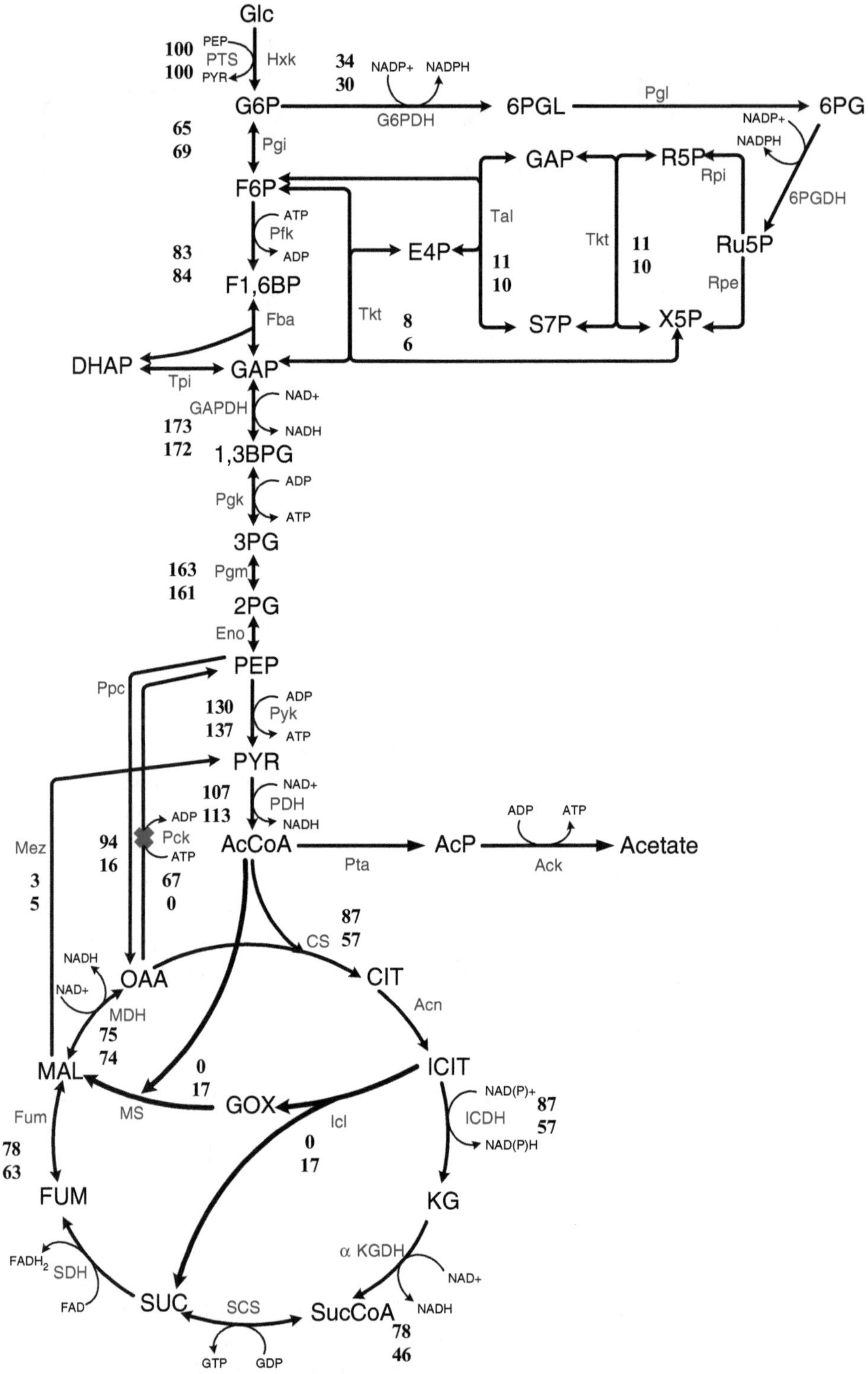

Fig. 9 ^{13}C-metabolic flux distribution of *pckA* mutant (JW3366) (*lower values*) as compared to wild-type (W3110) (*upper values*) cultivated at the dilution rate of 0.1 h^{-1} in the aerobic continuous culture

Table 8
Fermentation characteristics of *pckA* mutant as compared to wild type cultivated at the dilution rate of 0.1 h^{-1} in the continuous culture

	E. coli strain (growth rate)			
Growth parameters	W3110 (0.10 h^{-1})	W3110 (0.32 h^{-1})	W3110 (0.55 h^{-1})	JWK3366 (0.10 h^{-1})
Biomass yield (g/g)	0.40 ± 0.02	0.44 ± 0.02	0.48 ± 0.03	0.46 ± 0.02
Glucose uptake rate (mmol/g/h)	1.4 ± 0.1	4.0 ± 0.2	6.4 ± 0.3	1.2 ± 0.1
O_2 uptake rate (mmol/g/h)	4.0 ± 0.7	10.7 ± 1.6	16.3 ± 2.4	2.7 ± 0.5
CO_2 evolution rate (mmol/g/h)	4.2 ± 0.4	11.1 ± 1.2	16.6 ± 1.8	2.9 ± 0.3
C balance (%)	99 ± 7	103 ± 8	104 ± 8	97 ± 7

gene was knocked out, OAA concentration decreases [21]. This also causes the phosphorylation or inactivation of ICDH, since OAA acts as one of the effectors which inhibit phosphorylation of ICDH. This results in an increase in isocitrate concentration, and the flux through Icl also significantly increases (Fig. 9) [21]. Moreover, the *pckA* gene knockout causes accumulation of PEP, which in turn inhibits Pfk activity. This mutant produces less acetate and CO_2, resulting in higher cell yield than the wild type [21]. It is shown that in vivo regulation of the Pck (and Ppc) flux occurs mainly by modulation of enzyme activity caused by the changes in PEP and OAA concentrations rather than by the ATP/ADP concentrations under aerobic condition. This indicates that the reaction catalyzed by Pck can respond flexibly to the availability of PEP and OAA, which may form the metabolic cycle caused by Pck and Ppc fluxes, in particular, at low glucose concentrations [28]. PEP is an important intermediate in the metabolism, since it alone directly regulates PTS but also affects Pfk and Pyk activities. Since PEP can be formed gluconeogenically through Pck from the TCA cycle, Pck serves to maintain the relative balance between OAA and PEP pools and drain off excess carbon of the TCA cycle to supply PEP for cellular requirements [21].

In relation to *pckA* gene knockout, consider the effect of *ppc* gene knockout on the metabolism. Among the central metabolic pathways of *E. coli*, Ppc plays an anaplerotic role in replenishing OAA for biosynthetic reactions, thus keeping the TCA cycle intermediates from starvation. The specific growth rate and the specific glucose consumption rate of *ppc* mutant are lower than those of the wild type (BW25113) (Table 9) [35]. The CO_2 evolution rate (CER) is also reduced, and little acetate is excreted in the *ppc* mutant (Table 9). As a consequence, an improvement in the

Table 9
Fermentation characteristics of *ppc* mutant as compared to wild type cultivated at the dilution rate of 0.2 h^{-1} in the continuous culture

Parameter	Wild type	*ppc* mutant
μ_{max} (h^{-1})	0.41	0.34
q_{glc} (mmol/g/h)[a]	5.61	3.16
q_{CO2}(mmol/g/h)[b]	3.38	2.47
Q_{ace} (mmol/g)[c]	0.22	0.00
$Y_{biomass/glc}$ (g/g)[d]	0.48	0.55

[a]Specific glucose consumption rate
[b]Specific carbon dioxide evolution rate
[c]Specific acetate production rate
[d]Yield of biomass on glucose

biomass yield on glucose can be observed in *ppc* mutant (Table 9). The activity of Pck is considerably lower in *ppc* mutant than wild type [35]. The activities of the glycolytic enzymes such as Pgi, Pfk, Fba, and GAPDH, and the PP pathway enzymes such as G6PDH and 6PGDH are all significantly decreased in *ppc* mutant [35]. These data correspond to the lower glucose consumption rate and the slower growth rate in *ppc* mutant (Table 9).

The up-regulation of Pyk is expected to channel more carbon flux from PEP to PYR due to blockage of Ppc in *ppc* mutant [35]. The activities of Acn and MDH, but not ICDH, in the TCA cycle, increased in a coordinated manner, and the activity of Icl in the glyoxylate shunt, significantly increased in *ppc* mutant [35]. This regulation pattern clearly demonstrates that the mutant utilizes an alternative anaplerotic pathway, the glyoxylate shunt, to replenish OAA in response to blockage at Ppc. It is shown that the glycolytic intermediates such as G6P, F6P, FDP, and PEP, and the PP pathway intermediate such as 6PG, accumulate in *ppc* mutant [35]. In contrast, the intracellular concentration of AcCoA decreases, implying higher activity of the glyoxylate pathway [35].

Figure 10 shows MFD of the wild-type *E. coli* (BW25113) and its *ppc* mutant. Three notable differences between the two strains may be observed. First, 50.7 % of carbon flux is channeled through Ppc, and the backflow through Pck, is also high, accounting for 23.6 % of the total glucose consumed in the wild-type strain. The flux from OAA to PEP catalyzed by Pck is significantly decreased in *ppc* mutant. Up-regulation of Icl activity in the mutant is consistent with this flux analysis result (Fig. 10) [35]. Second, the remaining carbon flux through ICDH via the TCA cycle is still higher than that of the wild type, implying that more NADPH required for biosynthesis can be generated via the TCA cycle in *ppc* mutant. Third, 26.8 % of the carbon flux is channeled through the oxidative PP pathway in the wild type, while only 15.1 % in *ppc* mutant.

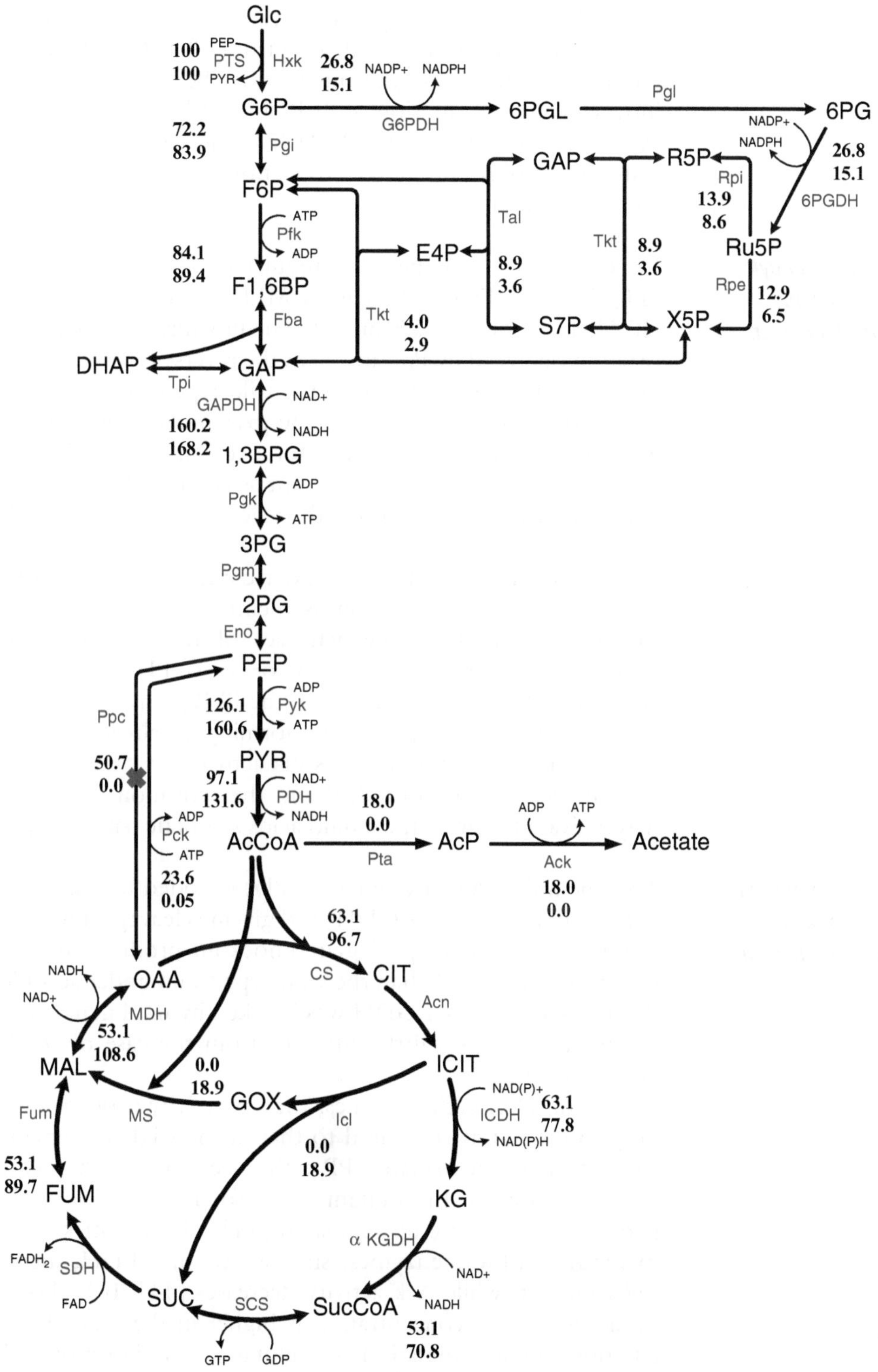

Fig. 10 ^{13}C-metabolic flux distribution of *ppc* mutant (JW3928) (*lower values*) as compared to wild type (BW25113) (*upper values*) cultivated at the dilution rate of 0.2 h^{-1} in the aerobic continuous culture

PEP accumulates in *ppc* mutant, and allosterically inhibits some of the glycolytic enzymes such as Pgi and Pfk [36]. The remarkably reduced Pck activity in the mutant may be caused by the higher intracellular concentration of PEP, since Pck is allosterically inhibited by nucleotide ATP and PEP [37]. The activation of the glyoxylate shunt contributes to the reduction of CO_2 production in *ppc* and *pckA* mutants.

3.4 Effect of pyk Gene Knockout on the Metabolism

Let us consider how the cell is regulated upon *pyk* gene knockout [11, 38, 39]. Figure 11 shows MFDs of the wild-type *E. coli* (BW25113) and its *pykF* mutant. The flux through the anaplerotic reaction catalyzed by Ppc as well as the reverse flux of Pck is higher in *pykF* mutant as compared to the wild-type strain (Fig. 11). The flux through Mez is higher for the *pykF* mutant, while it is much lower for the wild-type strain. Moreover, the glycolytic flux from G6P to F6P is lower for the mutant as compared to the wild-type strain, while the fluxes through the oxidative PP pathway are higher (Fig. 11) [40, 41].

The TCA cycle genes are under control of cAMP-Crp which mediates catabolic repression as mentioned before [42]. In the case of *pykF* mutant, the activities of the glycolytic pathway decrease, while the activities of the oxidative PP pathways increase (Fig. 11). The blockage of PEP to the PYR pathway activates both Ppc and Mez to supply PYR by forming an alternative pathway, and increased PEP pool inhibits Pfk activity.

The accumulation of PEP and the activation of PP pathway may activate the aromatic amino acids synthetic pathways [43].

3.5 Effect of lpdA Gene Knockout on the Metabolism

The lipoamide dehydrogenase (LPD) encoded by *lpdA* gene is a component of PDHc, KGDH, and glycine cleavage multi-enzyme (GCV) system. This *lpdA* gene knockout produces more pyruvate and L-glutamate than the wild-type strain under aerobic conditions [44]. If only KGDH was blocked by *sucA* gene knockout, the metabolism is a little different from the case of *lpdA* gene knockout [45].

As for TCA cycle enzymes, the CS activity is down-regulated in *lpdA* mutant as compared to that of the wild-type strain [44]. The activity of the oxidative PP pathway enzyme, such as G6PDH, is up-regulated in the mutant and also the activities of the ED pathway enzymes are up-regulated [44]. The activities of the fermentative pathway enzymes, such as Acs and LDH, increase in *lpdA* mutant, while Ack activity decreases [44]. It is shown that intracellular PYR concentration is higher in the mutant [44]. In addition, the accumulation of L-glutamate is higher than that in the wild-type strain [44]. The concentration of αKG also increases in the mutant [44]. It is shown that the biomass yield of the mutant is higher than that of the wild-type strain, while glucose uptake rate is lower for the mutant, and the specific CO_2 evolution rate is much reduced in the mutant (Table 10) [44].

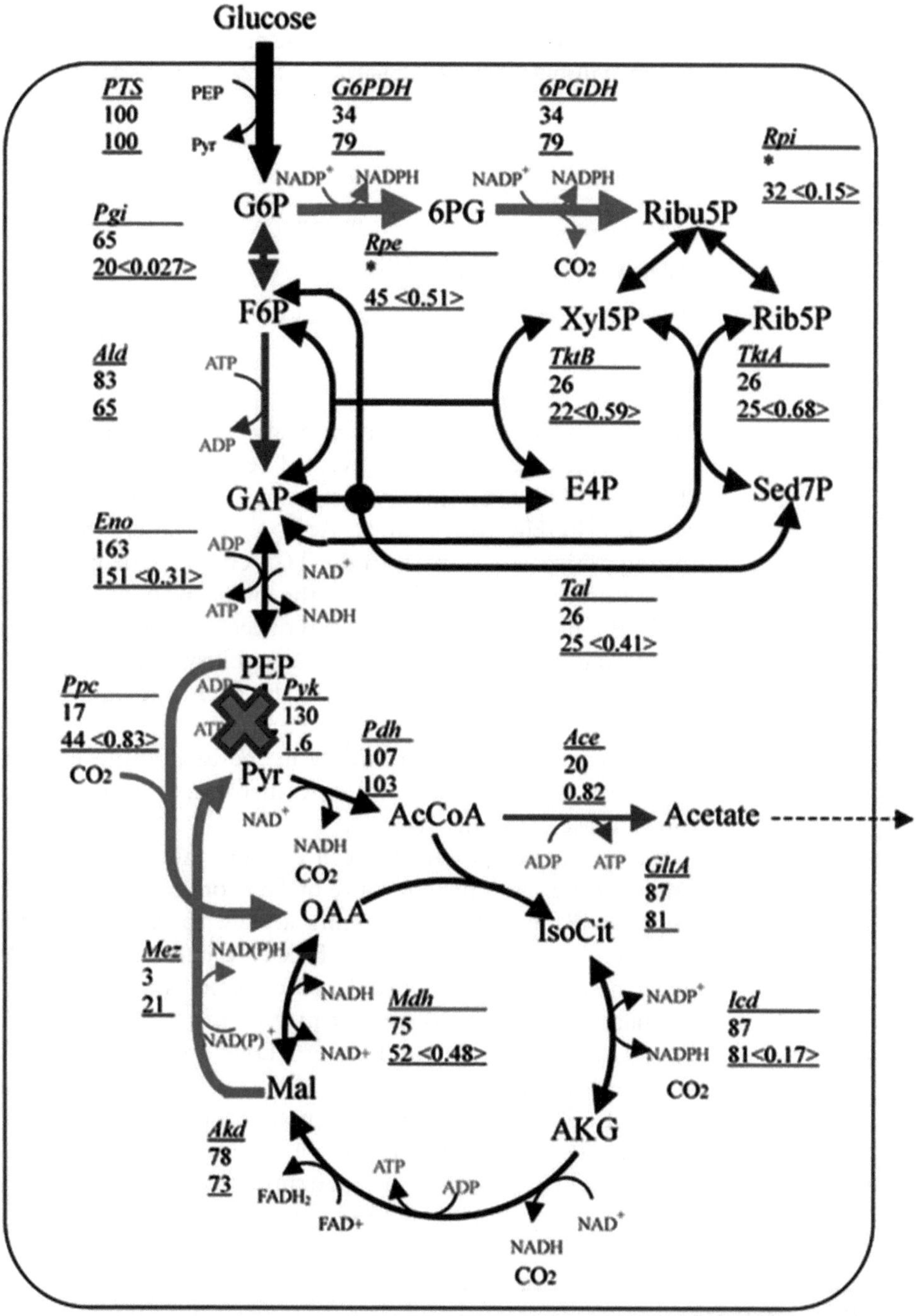

Fig. 11 ^{13}C-metabolic flux distribution of *pyk* mutant (*lower values*) as compared to wild-type BW25113 (*upper values*) cultivated at the dilution rate of 0.2 h^{-1} in the aerobic continuous culture

Figure 12 shows the comparison of MFDs between *lpdA* mutant and the wild-type strain (BW25113) [44], where it shows that with blockage of the fluxes through the reactions of PDHc and KGDH, the ED pathway is activated in the mutant.

Table 10
Growth characteristics of *E. coli* BW25113 at the dilution rate of 0.2 h^{-1} and its *lpdA* mutant at the dilution rate of 0.22 h^{-1} in the continuous culture

Variables	BW25113	*lpdA* mutant
Biomass yield (g/g)	0.40	0.49
Glucose uptake rate (mmol/g/h)	3.04	2.48
Specific CO_2 evolution rate (mmol/g/h)	8.17	4.08

The blockage of the PDHc reaction also induced the PEP synthase (Pps), which catalyzes the reversed reaction from PYR to PEP (Fig. 12) [44]. Since the *lpdA* mutant cannot utilize the pathway through PDHc to produce AcCoA, the alternative pathway through PoxB is activated. In the TCA cycle, all the net flux values are reduced. With blockage of the pathway between αKG and SUC, the flux through ICDH is reduced. The glyoxylate shunt is activated in order to replenish OAA for biosynthesis (Fig. 12). For the reactions through Fum and MDH, exchange fluxes are higher in the case of *lpdA* mutant (Fig. 12). The fluxes through anaplerotic reactions increase drastically in the case of the mutant, i.e., the fluxes through Pck and Ppc increase. Moreover, the flux through Mez also increases.

To maintain growth, the cell can generate AcCoA from PYR using another less efficient route by first converting PYR to acetate using PoxB, and then converting acetate to AcCoA via Acs and Ack–Pta. The activation of PoxB can be detected in the PDHc-deficient mutant by in vitro PoxB enzyme assay [46]. The PYR assimilation in the *lpdA* mutant may be made via PoxB and Pps–Ppc. The double mutants lacking PDHc and Pps, or the PDHc and Ppc, could not metabolize PYR [47]. Thus, the direct utilization of PYR is made by the combining reactions catalyzed by PoxB and Pps–Ppc. The inactivation of KGDH in the *lpdA* mutant causes the TCA cycle to be interrupted, which may activate the glyoxylate shunt to supply OAA for cell growth (Fig. 12).

4 Concluding Remarks

In this Chapter, ^{13}C-MFA of *E. coli* was considered for the case of aerobic cultivation. In particular, an attempt was made to clarify the metabolic regulation mechanisms in response to the specific pathway gene knockout based on different levels of information. Among these, the metabolic fluxes are located on top level, and the flux information is of primary interest for metabolic engineering.

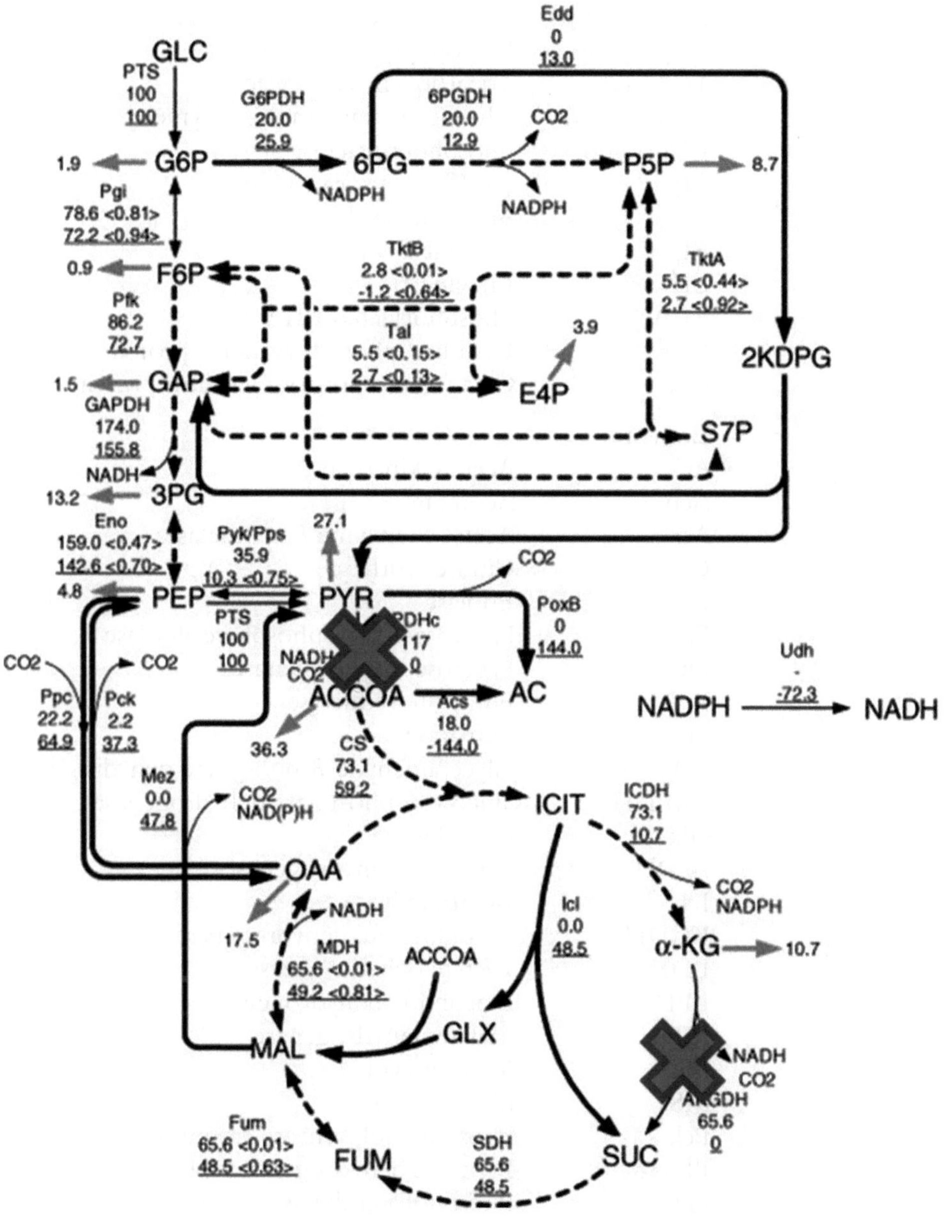

Fig. 12 ^{13}C-metabolic flux distribution of *lpdA* mutant (*lower values*) as compared to wild-type BW25113 (*upper values*) cultivated at the dilution rate of 0.2 h^{-1} in the aerobic continuous culture

Nomenclature

^{13}C-MFA	^{13}C-metabolic flux analysis
^{13}C-MFD	^{13}C-metabolic flux distribution
MID	Mass isotopomer distribution
ED pathway	Entner-Doudoroff pathway

EMP pathway	Embden-Meyerhof-Parnas pathway
PP pathway	Pentose phosphate pathway
TCA cycle	Tricarboxylic acid cycle
GCV system	Glycine cleavage multi-enzyme system
PTS system	Phosphotransferase system
Crp	cAMP receptor protein
Cya	Adenylate cyclase
EI	Enzyme I
EII	Enzyme II
$EIIA^{Glc}$-P	Phosphorylated $EIIA^{Glc}$
HPr	Histidine-phosphorylatable protein
Cra	Catabolite repressor/activator protein

Enzyme

Ack	Acetate kinase
Acn	Aconitase
Acs	Acetyl coenzyme A synthetase
CS	Citrate synthase
Eno	Enolase
Fba	Fructose-1,6-bisphosphate aldolase
Fbp	Fructose bisphosphatase
Frd	Fumarate reductase
Fum	Fumarase
GAPDH	Glyceraldehyde-3-phosphate dehydrogenase
G6PDH	Glucose-6-phosphate dehydrogenase
Glk	Glucokinase
ICDH	Isocitrate dehydrogenase
Icl	Isocitrate lyase
KGDH	Ketoglutarate dehydrogenase
LDH	Lactate dehydrogenase
LPD	Lipoamide dehydrogenase
MDH	Malate dehydrogenase
Mez	Malic enzyme
MS	Malate synthase
Pck	Phosphoenolpyruvate carboxykinase
PDHc	Pyruvate dehydrogenase complex
Pfk	Phosphofructokinase
Pfl	Pyruvate formate-lyase
6PGDH	6-Phosphogluconate dehydrogenase
Pgi	Phosphoglucose isomerase
Pox	Pyruvate oxidase
Ppc	Phosphoenolpyruvate carboxylase
Pps	Phosphoenolpyruvate synthase
Pta	Phosphotransacetylase
Pyk	Pyruvate kinase
Rpe	Ribulose phosphate epimerase
Rpi	Ribose phosphate isomerase
SCS	Succinyl-CoA synthetase

SDH	Succinate dehydrogenase
Tal	Transaldolase
TktA	Transketolase I
TktB	Transketolase II
Metabolite	
AcCoA	Acetyl-coenzyme A
ACE	Acetate
AcP	Acetyl phosphate
CIT	Citrate
E4P	Erythrose-4-phosphate
ETH	Ethanol
FDP	Fructose-1,6-bisphosphate
FOR	Formate
F6P	Fructose-6-phosphate
G6P	Glucose-6-phosphate
GAP	Glycelaldehyde-3-phosphate
GLC	Glucose
GLU	Glutamate
GOX	Glyoxylate
KDPG	2-Keto-3-deoxy-6-phosphogluconate
αKG	α-Ketoglutarate
LAC	Lactate
MAL	Malate
OAA	Oxaloacetate
PEP	Phosphoenol pyruvate
6PG	6-Phosphogluconate
PYR	Pyruvate
R5P	Ribose-5-phosphate
RU5P	Ribulose-5-phosphate
S7P	Sedoheptulose-7-phosphate
SUC	Succinate
X5P	Xylulose-5-phosphate

References

1. Matsuoka Y, Shimizu K (2011) Metabolic regulation in *Escherichia coli* in response to culture environments via global regulators. Biotechnol J 6:1330–1341
2. Shimizu K (2013a) Metabolic regulation of a bacterial cell system with emphasis on *Escherichia coli* metabolism. ISRN Biochem, Article ID645983, 1–47
3. Shimizu K (2013) Bacterial cellular metabolic systems. Woodhead Publishing Ltd., Oxford
4. Shimizu K (2004) Metabolic flux analysis based on ^{13}C-labeling experiments and integration of the information with gene and protein expression patterns. Adv Biochem Eng Biotechnol 91:1–49
5. Szyperski T (1995) Biosynthetically directed fractional 13C-labeling of proteinogenic amino acids. An efficient analytical tool to investigate intermediary metabolism. Eur J Biochem 232:433–448
6. Marx A, deGraaf AA, Wiechert W, Eggeling L, Sahm H (1996) Determination of the fluxes in the central metabolism of *Corynebacterium glutamicum* by nuclear magnetic resonance spectroscopy combined with metabolite balancing. Biotechnol Bioeng 49:111–129
7. Sauer U (2006) Metabolic networks in motion: ^{13}C-based flux analysis. Mol Syst Biol 2:62
8. Wittmann C (2007) Fluxome analysis using GC-MS. Microb Cell Fact 6:6

9. Toya Y, Nakahigashi K, Tomita M, Shimizu K (2012) Metabolic regulation analysis of wild-type and *arcA* mutant *Escherichia coli* under nitrate conditions using different levels of omics data. Mol Biosyst 8:2593–2604
10. Toya Y, Ishii N, Hirasawa T, Naba M, Hirai K, Sugawara K, Igarashi S, Shimizu K, Tomita M, Soga T (2007) Direct measurement of isotopomer of intracellular metabolites using capillary electrophoresis time-of-flight mass spectrometry for efficient metabolic flux analysis. J Chromatogr A 1159:134–141
11. Toya Y, Ishii N, Nakahigashi K, Tomita M, Shimizu K (2010) ^{13}C-metabolic flux analysis for batch culture of *Escherichia coli* and its Pyk and Pgi gene knockout mutants based on mass isotopomer distribution of intracellular metabolites. Biotechnol Prog 26:975–992
12. Wiechert W, de Graaf AA (1997) Bidirectional reaction steps in metabolic networks: I. Modeling and simulation of carbon isotope labeling experiments. Biotechnol Bioeng 55: 101–117
13. Zhao J, Shimizu K (2003) Metabolic flux analysis of *Escherichia coli* K12 grown on 13 C-labeled acetate and glucose using GC-MS and powerful flux calculation method. J Biotechnol 101:101–117
14. Arauzo-Bravo MJ, Shimizu K (2003) An improved method for statistical analysis of metabolic flux analysis using isotopomer mapping matrices with analytical expressions. J Biotechnol 105:117–133
15. Wiechert W, Mollney M, Isermann N, Wurzel M, de Graaf AA (1999) Bidirectional reaction steps in metabolic networks: III Explicit solution and analysis of isotopomer labeling systems. Biotechnol Bioeng 66:69–85
16. Matsuoka Y, Shimizu K (2010) The relationships between the metabolic fluxes and ^{13}C-labeled isotopomer distribution for the flux analysis of the main metabolic pathways. Biochem Eng J 49:326–336
17. Schmidt K, Carlsen M, Nielsen J, Villdsen J (1997) Modeling isotopomer distributions in biochemical networks using isotopomer mapping matrices. Biotechnol Bioeng 55:831–840
18. Yao R, Hirose Y, Sarkar D, Ye Q, Shimizu K (2011) Catabolic regulation analysis of *Escherichia coli* and its *crp*, *mlc*, *mgsA*, *pgi* and *ptsG* mutants. Microb Cell Fact 10:67
19. Ishii N, Nakahigashi K, Baba T, Robert M, Soga T, Kanai A, Hirasawa T, Naba M, Hirai K, Hoque A, Ho PY, Kakazu Y, Sugawara K, Igarashi S, Harada S, Masuda T, Sugiyama N, Togashi T, Hasegawa M, Takai Y, Yugi K, Arakawa K, Iwata N, Toya Y, Nakayama Y, Nishioka T, Shimizu K, Mori H, Tomita M (2007) Multiple high-throughput analyses monitor the response of *E-coli* to perturbations. Science 316:593–597
20. Matsuoka Y, Shimizu K (2013) Importance of understanding the main metabolic regulation in response to the specific pathway mutation for metabolic engineering of *Escherichia coli*. Comput Struct Biotechnol J 3:e201210018
21. Yang C, Hua Q, Baba T, Mori H, Shimizu K (2003) Analysis of *E. coli* anaplerotic metabolism and its regulation mechanism from the metabolic responses to alter dilution rates and *pck* knockout. Biotechnol Bioeng 84:129–144
22. Valgepea K, Adamberg K, Nahku R, Lahtvee PJ, Arike L, Vilu R (2010) Systems biology approach reveals that overflow metabolism of acetate in *Escherichia coli* is triggered by carbon catabolite repression of acetyl-CoA synthetase. BMC Syst Biol 4:166
23. Valgepea K, Adamberg K, Vilu R (2011) Decrease of energy spilling in *Escherichia coli* continuous cultures with rising specific growth rate and carbon wasting. BMC Syst Biol 5:106
24. Kayser A, Weber J, Hecht V, Rinas U (2005) Metabolic flux analysis of *Escherichia coli* in glucose-limited continuous culture I. Growth-rate-dependent metabolic efficiency at steady state. Microbiology 151:693–706
25. Wolf RE Jr, Prather DM, Shea FM (1979) Growth-rate-dependent alteration of 6-phosphogluconate dehydrogenase and glucose 6-phosphate dehydrogenase levels in *Escherichia coli* K-12. J Bacteriol 139:1093–1096
26. Shimizu K (2009) Toward systematic metabolic engineering based on the analysis of metabolic regulation by the integration of different levels of information. Biochem Eng J 46: 235–251
27. Canonaco F, Hess TA, Heri S, Wang T, Szyperski T, Sauer U (2001) Metabolic flux response to phosphoglucose isomerase knockout *Escherichia coli* and impact of overexpression of the soluble transhydrogenase UdhA. FEMS Microbiol Lett 204:247–252
28. Fischer E, Sauer U (2003) Metabolic flux profiling of *Escherichia coli* mutants in central carbon metabolism using GC-MS. Eur J Biochem 270:880–891
29. Hua Q, Yang C, Baba T, Mori H, Shimizu K (2003) Responses of the central carbon metabolism in *Escherichia coli* to phosphoglucose isomerase and glucose-6-phosphate dehydrogenase knockouts. J Bacteriol 185:7053–7067
30. Holms WH (1986) The central metabolic pathways of *Escherichia coli*: relationship between flux and control at a branch point, efficiency of

conversion to biomass, and excretion of acetate. Curr Top Cell Regul 28:69–105

31. Fraenkel DG, Levisohn SR (1967) Glucose and gluconate metabolism in an *Escherichia coli* mutant lacking phosphoglucose isomerase. J Bacteriol 93:1571–1578
32. Zhao J, Baba T, Mori H, Shimizu K (2004) Effect of *zwf* gene knockout on the metabolism of *Escherichia coli* grown on glucose or acetate. Metab Eng 6:164–174
33. Zhao J, Baba T, Mori H, Shimizu K (2004) Analysis of metabolic and physiological responses to *gnd* knockout in *E. coli* by using C-13 tracer experiment and enzyme activity measurement. FEMS Microbiol Lett 220:295–301
34. Choi IY, Sup KI, Kim HJ, Park JW (2003) Thermosensitive phenotype of *Escherichia coli* mutant lacking NADP(+)-dependent isocitrate dehydrogenase. Redox Rep 8:51–56
35. Peng L, Arauzo M, Shimizu K (2004) Metabolic flux analysis for a *ppc* mutant *Escherichia coli* based on ^{13}C-labelling experiments together with enzyme activity assays and intracellular metabolite measurements. FEMS Microbiol Lett 235:17–23
36. Fraenkel DG, Neidhardt F (eds) (1999) *Escherichia coli* and *Salmonella*: cellular and molecular biology. American Society for Microbiology, Washington, DC
37. Krebs A, Bridger WA (1980) The kinetic properties of phosphoenolpyruvate carboxykinase of *Escherichia coli*. Can J Biochem 58: 309–318
38. Siddiquee KAZ, Arauzo-Bravo M, Shimizu K (2004) Metabolic flux analysis of *pykF* gene knockout *Escherichia coli* based on ^{13}C-labeled experiment together with measurements of enzyme activities and intracellular metabolite concentrations. Appl Microbiol Biotechnol 63:407–417
39. Siddiquee KAZ, Arauzo-Bravo M, Shimizu K (2004) Effect of pyruvate kinase (*pykF* gene) knockout mutation on the control of gene expression and metabolic fluxes in *Escherichia coli*. FEMS Microbiol Lett 235:25–33
40. Ponce E, Martínez A, Bolívar F, Valle F (1998) Stimulation of glucose catabolism through the pentose pathway by the absence of the two pyruvate kinase isoenzymes in *Escherichia coli*. Biotechnol Bioeng 58:292–295
41. Zhu T, Phalakornkule C, Koepsel PR, Domach MM, Ataai MM (2001) Cell growth and by-product formation in a pyruvate kinase mutant of *E.coli*. Biotechnol Prog 17:624–628
42. Cronan JE, LaPorte DC (1996) Tricarboxylic acid cycle and glyoxylate bypass. In: Neidhardt FC, Curtiss R III, Ingraham JL, Lin ECC, Low KB, Magasanik B Jr, Reznikoff WS, Riley M, Schaechter M, Umbarger HE (eds) *Escherichia coli* and *Salmonella*: cellular and molecular biology, 2nd edn. American Society for Microbiology, Washington, DC, pp 206–216
43. Kedar P, Colah R, Shimizu K (2007) Proteomic investigation on the *pyk-F* gene knockout *Escherichia coli* for aromatic amino acid production. Enz Microb Technol 41:455–465
44. Li M, Yao S, Shimizu K (2006) Effect of *lpdA* gene knockout on the metabolism in *Escherichia coli* based on enzyme activities, intracellular metabolite concentrations and metabolic flux analysis by ^{13}C-labeling experiments. J Biotechnol 122:254–266
45. Li M, Ho PY, Yao SJ, Shimixu K (2006) Effect of *sucA* or *sucC* gene knockout on the metabolism in *Escherichia coli* based on gene expressions, enzyme activities, intracellular metabolite concentrations and metabolic fluxes by ^{13}C-labeling experiments. Biochem Eng J 30:286–296
46. Abdel-Hamid AM, Attwood MM, Guest JR (2001) Pyruvate oxidase contributes to the aerobic growth efficiency of *Escherichia coli*. Microbiology 147:1483–1498
47. Dietrich J, Henning U (1970) Regulation of pyruvate dehydrogenase complex synthesis in *Escherichia coli* K12. Eur J Biochem 14: 258–269

Chapter 17

^{13}C-Based Metabolic Flux Analysis of Recombinant *Pichia pastoris*

Pau Ferrer and Joan Albiol

Abstract

Overexpression of a foreign protein may negatively affect several cell growth parameters, as well as cause cellular stress. Central (or core) metabolism plays a crucial role since it supplies energy, reduction equivalents, and precursor molecules for the recombinant product, cell's maintenance, and growth needs. However, the number of quantitative physiology studies of the impact of recombinant protein production on the central metabolic pathways of yeast cell factories has been traditionally rather limited, thereby hampering the application of rational strain engineering strategies targeting central metabolism.

The development and application of quantitative physiology and modelling tools and methodologies is allowing for a systems-level understanding of the effect of bioprocess parameters such as growth rate, temperature, oxygen availability, and substrate(s) choice on metabolism, and its subsequent interactions with recombinant protein synthesis, folding, and secretion.

Here, we review the recent developments and applications of ^{13}C-based metabolic flux analysis (^{13}C-MFA) of *Pichia pastoris* and the gained understanding of the metabolic behavior of this yeast in recombinant protein production bioprocesses. We also discuss the potential of multilevel studies integrating ^{13}C-MFA with other omics analyses, as well as future perspectives on the metabolic modelling approaches to study and design metabolic engineering strategies for improved protein production.

Key words *Pichia pastoris*, Recombinant protein production, Metabolic burden, Metabolic flux ratio (METAFoR) analysis, Instationary ^{13}C-metabolic flux analysis (INST-^{13}C-MFA), NMR, MS, Metabolomics

1 Introduction

Since the establishment of recombinant DNA technology for protein production, it is well recognized and documented that overexpression of foreign proteins has an impact on host metabolism (also known as metabolic burden). In yeast, overexpression of foreign proteins has been reported to negatively affect growth parameter such as growth rate, biomass yield, and specific substrate consumption rate [1–5]. Also, it has been shown in several cases that increasing synthesis of recombinant secretory proteins (e.g., by increasing their gene dosage) over a certain limit may overload

Jens O. Krömer et al. (eds.), *Metabolic Flux Analysis: Methods and Protocols*, Methods in Molecular Biology, vol. 1191, DOI 10.1007/978-1-4939-1170-7_17, © Springer Science+Business Media New York 2014

the secretory pathway, triggering the unfolded protein stress response (reviewed in refs. 6–9).

Central (or core) metabolism plays a crucial role in protein synthesis, since it supplies energy, reduction equivalents, and precursor molecules for the recombinant product, as well as cell maintenance and growth needs. The expression of a foreign protein at a high level may either directly limit other cellular processes by competing for their precursors or indirectly interfere with metabolism, if their synthesis is blocked, thus inducing a cellular stress reaction.

There are several successful engineering strategies reported for improved protein production in *Pichia pastoris.* Most of them target protein folding and secretion as a major bottleneck (reviewed in refs. 8, 9), but not the central metabolism, despite the increasing evidence pointing at it as a target [5, 10–12]. This is mainly due to the fact that metabolic engineering requires a detailed knowledge of the cellular metabolic response to protein overexpression for the identification of promising targets. The impact of recombinant protein production on the central metabolic pathways is still little studied, thereby limiting the application of metabolic engineering strategies. In this regard, systems biotechnology provides an increasing range of analytical and modelling tools allowing for the quantitative analysis of *P. pastoris* physiology (reviewed in ref. 13), expanding the knowledge base for the design of new strategies for rational strain improvement. This review summarizes recent developments in ^{13}C-based metabolic flux analysis (MFA) methodologies for quantitative analysis of *P. pastoris* physiology, as well as highlights the major outcomes of the MFA studies reported so far.

2 Methods of ^{13}C-Based Metabolic Flux Analysis Used in *P. pastoris*

Determination of metabolic fluxes is one of the key steps in the analysis of the cell response to any implemented metabolic engineering strategy. This field has been thoroughly reviewed in several surveys (e.g., [14, 15]), so in the following only the methods that have been applied to *P. pastoris* will be mentioned. The most extended approaches for metabolic flux determination rely on the assumption that metabolic networks reach a (pseudo/quasi)steady state, that is, transport and metabolic reaction rates and metabolite concentrations are constant over time. This assumption appears to be valid in most of the cases where culture conditions are chosen to guarantee a constant growth rate. An exception to this general assumption is the oscillatory metabolic behavior that some microorganisms such as *S. cerevisiae* can show under certain growth conditions in chemostat cultures [16, 17]. Hence, since metabolite concentrations and fluxes are assumed to remain constant under steady-state conditions, the metabolite mass balances can be represented by a linear system of equations for which there are well-known available solution methods. As with any linear system of equations, a unique solution can only be found if the system is determined or in certain cases overdetermined. Unfortunately, all genome-scale metabolic models are underdetermined due to the intricacy of the networks including, among other traits, parallel and redundant pathways. Reduction of genome-scale metabolic networks to key relevant parts, such as central carbon metabolism, and calculating in advance the amount of central carbon intermediates consumed for biomass synthesis, decreases the size of the problem. Nevertheless, even in those cases, useful networks remain underdetermined.

Initial approaches to metabolic flux analysis, sometimes referred to as "classical" or "stoichiometric" metabolic flux analysis, were based on a limited number of metabolic reactions, which were typically extracted from available literature on central carbon metabolism. These systems of reaction equations were formulated in such a way that the remaining degrees of freedom allowed to find a solution using only information derived from measuring extracellular (cell's input-output) fluxes. This means that parallel pathways and metabolic cycles cannot be resolved. To partially alleviate the lack of information, the mass balances of common reduction equivalent carriers (NADH, NADPH) were initially included. However, this strategy has strong limitations, as it is known that some metabolic steps can be catalyzed using both type of carriers. Furthermore, there are several reported examples where generated NADPH does not match the consumed amounts for biosynthesis [12, 18, 19], pointing at the existence of relevant missing electron exchange steps that are not usually included in reduced networks. In some cases, if additional data is not available, it can be considered that a global electron balance has to be fulfilled, obtaining one further balance equation to solve a minimal metabolic model. Examples of this type of approach can also be found in the *P. pastoris* literature [20].

With the advances in the field, particularly in genome sequencing, the used metabolic networks have been increasing in size. Undoubtedly, the most useful and interesting information derives from networks of higher dimensions. If no additional information is available, useful insights can also be obtained by studying the network structure characteristics by elementary mode analysis [21], by applying a range of assumptions and optimization criteria such as minimization of intracellular protein accumulation [22], by determining the minimum cost solution (maximum possibility) in a constraint-based model [23, 24], or by determining ranges for the non-measured fluxes [25]. Nevertheless, to be able to solve networks of increased complexity it is necessary to obtain additional information than the input-output fluxes. The current most useful approach for gathering additional information (constraints) is ^{13}C-labelling techniques.

Feeding biomass with a convenient mixture of ^{13}C-labelled and non-labelled carbon source is the starting point of a number of techniques for metabolic flux determination. It relies on the precise and nonrandom carbon redistribution along different metabolic pathways. The labelled metabolites obtained in those experiments can be analyzed either by using nuclear magnetic resonance (NMR) or mass spectrometry (MS) combined with a separation technique such as gas chromatography (GC) or liquid chromatography (LC). Those analytical techniques, combined with different calculatory approaches, allow obtaining different levels of information on the operation of metabolic pathways.

As for NMR, the most frequently reported method in early ^{13}C-based MFA studies was the measurement of fractional enrichments of specific carbon atoms. Nevertheless, one of the most successfully applied techniques uses two-dimensional NMR (2D ^{13}C-^{1}H-COSY spectra)-derived information from labelled protein amino acids, an approach known as biosynthetically directed fractional (BDF) labelling [26]. In this strategy, the incorporation of intact two-carbon- and three-carbon-labelled fragments from a single carbon source molecule into proteinogenic amino acids is monitored. Probabilistic equations allow relating the observed multiplet intensities of the ^{13}C fine structures to the relative abundance of intact carbon fragments—thus originating from the same source molecule—in the cell's protein amino acids. These equations are based on the observation that the probability of finding two adjacent carbons labelled by random incorporation, from a mixture of 10 % uniformly labelled substrate and 90 % naturally labelled substrate, is of about 1 %. However, the probability that both labelled positions derive from the same labelled molecule increases up to about 10 %. Combining the information of the relative intact carbon fragments in amino acids with the known metabolic network underlying amino acid biosynthesis, highly valuable information is derived about the relative activity of the contributing pathways. This approach allowed among others the determination of active biochemical pathways, the semiquantitative assessment of exchange fluxes, and, particularly important, the determination of metabolic flux ratios [27]. This approach is known as metabolic flux ratio (METAFoR) analysis [28] and is essentially independent of the absolute global flux distribution of the entire metabolic network, as it only determines ratios among fluxes and, therefore, pathways. Flux ratios can be calculated even without information on external fluxes or biomass composition of the cell, but it is not possible to assign absolute flux values to the metabolic network with only this information. This approach was the first experimentally applied to *P. pastoris* [29, 30]. If additional information on external fluxes and biomass composition is available it is possible to set up a metabolic network where this information combined with the metabolic flux ratios, either derived from NMR or MS data, is combined to calculate a metabolic flux distribution [31]. In *P. pastoris*, this approach has been applied to the metabolic analysis of metabolism under protein production conditions, using glucose as a sole carbon source [5, 13], or in glucose/methanol mixtures [12].

The previous approaches consider only the net contribution of each metabolic reaction to the flux distribution of the network. However, it can be considered that most biological reactions operate significantly in both directions at the same time due to their intrinsic reversible nature. That is, the observed net flux is only the overall result of both activities. Nevertheless, simultaneous operation of reversible reactions in both directions often has a significant impact on the labelling distribution. In these cases, it is necessary to consider the activity in both directions to properly fit experimental labelling data. Furthermore, the information contained in the labelling patterns can also be used to calculate the degree of reversibility existing in some metabolic reactions. This fact is closely related to the reaction proximity to equilibrium and may provide clues on the level of control exerted by different metabolic steps. For example, those steps close to their equilibrium will show a high degree of reversibility and will probably exert a low level of control on the network metabolic fluxes. Thus, information on the level of reversibility can be useful. Nevertheless, its determination has to be carefully examined, as the reversibility of many reactions is not observable from the available labelling patterns. Due to its interest and impact on the labelling patterns, different calculatory approaches include bidirectional fluxes in their approximations. These methods are based on the calculation of the complete labelling distribution within the network and the calculation of the labelling pattern of the measured metabolites. The calculated labelling patterns are fitted to the experimental ones usually following a weighted least squares approach. The main difference among the different methods relies on the approach used to calculate the labelling distribution; this step requires intensive calculations. These different approaches use a variety of isotopomer combinations (cumomers [32], bondomers [33], elementary metabolite units [34]) so as to reach the most efficient calculation routines. The most established of these methodologies uses steady-state labelling patterns and has also been applied to the *P. pastoris* metabolic studies [19].

In addition to those well-established steady-state approaches to calculate metabolic fluxes from ^{13}C-labelling patterns, new methodologies are being developed. Strategies based on the detection of ^{13}C-labelling patterns of proteinogenic amino acids require long labelling phases to reach isotopic steady state, so that enrichment of carbon backbones of key precursor metabolites is also present in the amino acids of the cell protein. Long labelling times result in expensive experiments, mainly due to the cost of labelled material. Also, in some systems it is difficult to maintain a steady state for long periods of time. It is therefore highly desirable to shorten as much as possible the labelling periods. In this case, one can consider mainly two scenarios. In the first one, the metabolic system is maintained in steady state but the labelling process is followed during a relatively short period of time after a shift from unlabelled to labelled feed medium. That is, the metabolite labelling status is in non-steady state, as it is continuously changing over time. This approach is often referred to as "instationary." In the second scenario, nor the metabolite nor the labelling is in steady state. This is the more complex case requiring following the labelling procedure as well as the time evolution of the metabolite pools. Due to its complexity, it is experimentally highly demanding and, to our knowledge, there are so far no examples of applications in *P. pastoris*. Therefore it will not be further discussed in this review.

Recently, an "instationary" ^{13}C-MFA approach has been established for *P. pastoris* [35]. This technique can be directly applied to the metabolites of the central carbon pathways, which potentially results in shorter labelling times, additional information, and, consequently, more detailed estimation of fluxes—including fluxes beyond central carbon metabolism [14, 36]. The implementation of this approach requires the availability of a fast sampling procedure to allow following the labelling process dynamics and measure metabolite pool sizes. This in turn requires a well-established quenching procedure allowing freezing the metabolic state in (sub-)seconds, thereby preserving the information on metabolite levels and labelling patterns. Different quenching protocols can be found in the literature which must be tested and adapted to each biological species. From a general point of view, these protocols mostly rely on a fast (in the sub-second time frame) freezing step, usually below −40 °C, in the presence of an organic solvent such as methanol or ethanol. This step could damage the cell membrane, resulting in the release (leakage) of intracellular metabolites into the solvent media, resulting in an underestimation of the metabolite levels. This effect seems to be species specific and, consequently, it is of key importance to initially test and adapt any metabolic quenching protocol to the used organism in order to minimize metabolite leakage. To this purpose, a quenching and metabolite extraction method has been recently developed for *P. pastoris* cells growing on glucose [37], opening the door to quantitative metabolomics and metabolic flux analysis studies based on direct measurements of isotopic enrichments of the intracellular metabolites. In this case, to determine the metabolic fluxes, the calculation procedure includes the solution of a system of ordinary differential equations describing the labelling transients of the measured isotopomers. The simulated transient labelling can be iteratively fitted to the measured isotopic enrichments, by current weighted quadratic minimization approaches. The quenching and extraction method, together with the iterative calculation approach, has been successfully applied for metabolic flux determination in cells growing on a glucose/methanol mixture [38]. A shortcoming of methods based on free intracellular metabolites labelling patterns is that for some metabolites it may take longer times than expected before they completely reach isotopic steady state as a result of their exchange with large intracellular pools or storage compounds (e.g., trehalose, glycogen) and turnover of cell components such as RNA and proteins. During the short time labelling experiments, these pools act as a source of unlabelled material, thus resulting in a dilution of the provided label. Consequently, proper labelling and metabolic flux identification are not possible without taking them into account. Moreover, additional advances are required to fully estimate all cellular fluxes including compartmentalization. The labelling data of metabolites present in more than one compartment are in most techniques lumped information, without the possibility to discriminate between, e.g., cytosol and mitochondrion. With the most advanced metabolic data in hand, the genome-scale metabolic model of *P. pastoris* is still highly underdetermined and thus does not allow the estimation of a unique flux distribution. Therefore, most metabolic flux calculations are still based on reduced metabolic networks derived from genome-scale reconstructed metabolic networks, following the application of a number of simplifying rules, as well as taking into account a range of additional physiological information on the network topology (e.g., lack of certain enzyme activities under a given growth condition). Nevertheless, the comparison of the metabolic flux distributions obtained using such simplified models provides highly valuable insights into the cell's adaptation to a changed environment or genetic modification.

In the following section, we summarize the insights derived from the application of these methodologies to *P. pastoris.*

3 ^{13}C-Based Metabolic Flux Analysis of Recombinant *P. pastoris*

3.1 Impact of Bioprocess Parameters on Central Metabolism

It is well known that the protein secretion rate, i.e., the specific productivity (q_P), is related to the specific growth rate [39–41]. Also, several studies have shown that the use of mixed carbon sources, that is, methanol plus a multicarbon source such as glycerol, sorbitol, or glucose, can boost process productivities in PAOX1-based recombinant protein production in *P. pastoris* (i.e., where foreign protein expression is under the methanol-inducible alcohol oxidase 1 promoter) [22, 42–45]. Moreover, it has been suggested that growth under mixed feed conditions may alleviate the metabolic burden related to stress responses [43] and cell viability [46], both triggered by protein overproduction and secretion.

In this context, the pioneering applications of ^{13}C-based metabolic analyses in *P. pastoris* were focused on the analysis of carbon source(s) and growth rate on the central carbon metabolism of this yeast [29, 30]. In these studies, the application of the classic METAFoR analysis provided a first quantitative snapshot of *P. pastoris* metabolic behavior when growing aerobically in chemostat cultures at two different dilution rates (0.05 and 0.16 h^{-1}, corresponding to near maximum specific growth rates, μ_{max}, as determined in batch cultures in minimal medium with methanol and glycerol, respectively), with glucose, glycerol, or different glycerol/methanol mixtures (80/20, 60/40, and 40/60, w/w) as carbon source(s). These series of experiments were performed with a *P. pastoris* Mut^+ strain secreting a recombinant *Rhizopus oryzae* lipase (Rol), but no reference strain was used for comparison. Although no metabolic fluxes were calculated using these metabolic flux ratio datasets, the METAFoR results allowed for the comparative profiling of *P. pastoris* aerobically grown on glycerol and glucose [30], pointing at some significant differences in fluxes through the TCA cycle. For instance, the fraction of mitochondrial oxaloacetate (OAA) arising from phosphoenolpyruvate (PEP), that is, the major anaplerotic supply of the TCA cycle, was 1.26-fold higher in glycerol-grown cells than on glucose. Unfortunately, the use of the C3 source glycerol for ^{13}C labelling of proteinogenic amino acids enabled the determination of the flux ratios for reactions associated with the TCA cycle, but not for those related to glycolysis and the pentose phosphate pathway (PPP), as uniformly labelled glycerol being metabolized through gluconeogenesis and oxidative PPP does not produce labelling patterns that are sufficiently distinct from those generated when glycerol is channelled through the non-oxidative PPP [30]. Further METAFoR analyses of *P. pastoris* growing in chemostat cultures with varying mixtures of glycerol and methanol at two dilution rates allowed to study the impact of these parameters on the metabolic pathways associated with TCA cycle intermediates, as well as inferring some insights into methanol co-metabolism. Remarkably, distinct flux ratios were registered at each of the two growth rates, regardless of the methanol fraction in the feed: The relative TCA activity was reduced at the higher dilution rate, in concomitance to a largely unidirectional flux of OAA from the cytosol to the mitochondria when the cells were growing close to their maximal growth rate, whereas a very fast exchange of cytosolic and mitochondrial OAA pools was observed at the lower growth rate. This finding is consistent with the observation that TCA cycle activity in *S. cerevisiae* is strongly correlated with the environmentally determined growth rate [47]. Conversely, rather little flux changes were observed when the fraction of methanol in the medium was varied. Notably, during growth at the lower dilution rate, when the residual concentration in the bioreactor of both substrates was neglligible, the incorporation of methanol into the biomass increased as the methanol fraction in the feed was increased. This was reflected in the impact of methanol co-assimilation on the labelling patterns of PPP intermediates. Such impact on the PPP was not observed at the higher dilution rate, suggesting that the distribution of methanol carbon into assimilatory (involving C5 intermediates) and dissimilatory (direct oxidation to CO_2) pathways (Fig. 1) is different at the two dilution rates.

Recently, new experimental data for validation of the *P. pastoris* chemostat culture series reported by Solà and co-workers [29]

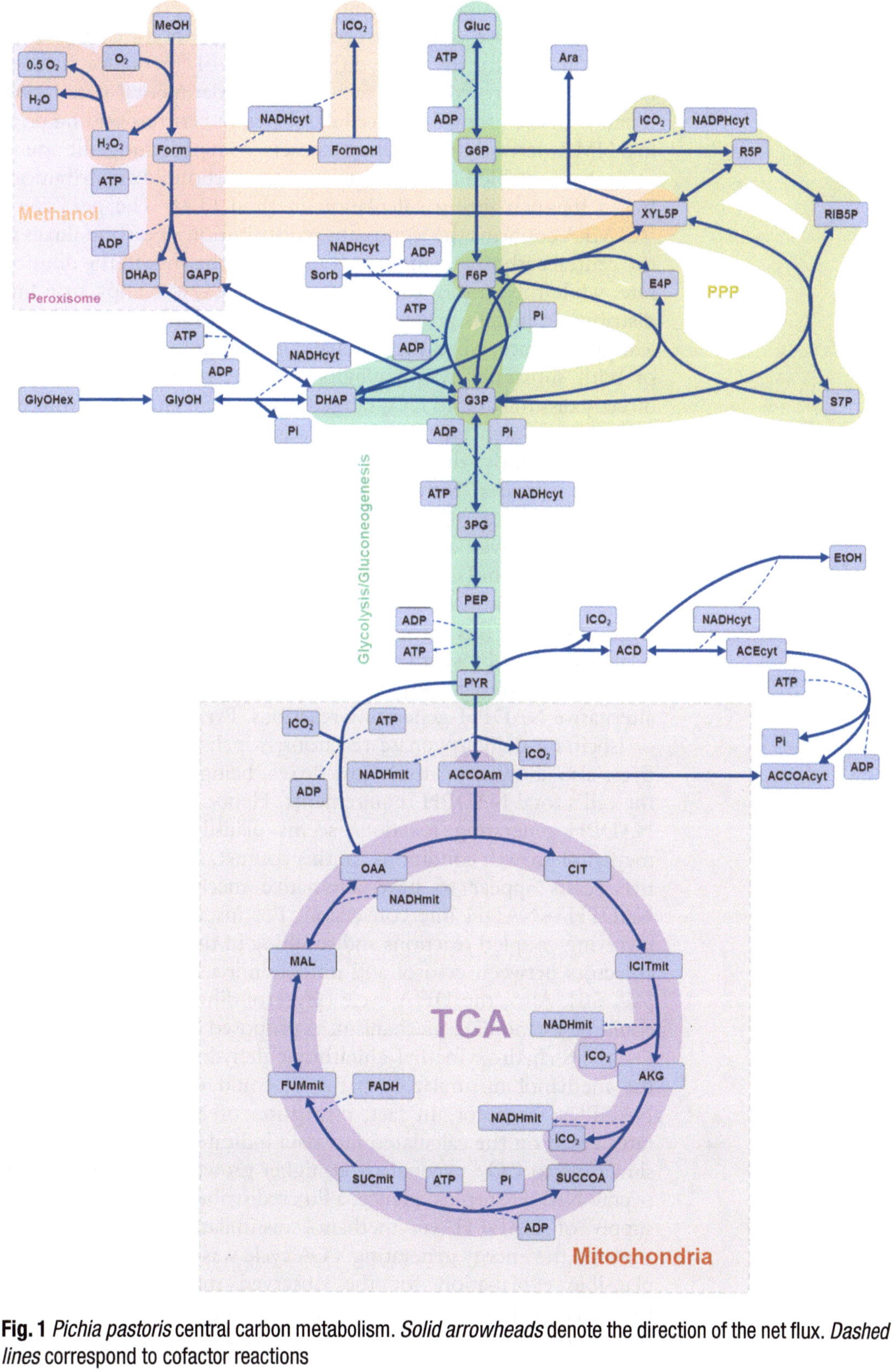

Fig. 1 *Pichia pastoris* central carbon metabolism. *Solid arrowheads* denote the direction of the net flux. *Dashed lines* correspond to cofactor reactions

have been generated, including a validation of extracellular fluxes and determination of biomass compositional data for each growth condition, i.e., for the three different glycerol/methanol ratios at two different growth rates each. These experimental data, combined with the original extracellular fluxes and corresponding original NMR-derived ^{13}C-labelling datasets from proteinogenic amino acids, have enabled the calculation of the carbon flux distributions by an iterative fitting calculation method [19]. The performed ^{13}C-MFA confirms the significant redistribution of carbon fluxes in the central carbon metabolism as a result of the shift in the dilution rate, while the ratio of carbon sources has a lower impact on flux distribution in cells growing at the same dilution rate. Notably, at low growth rate, when the residual concentration in the bioreactor of both substrates was neglligible, the percentage of methanol directly dissimilated to CO_2 ranged between 50 and 70 %, whereas at higher growth rate, methanol was completely dissimilated to CO_2. Coherently, the pentose flux to the methanol assimilation pathway increased with the methanol fraction in the feed at the lower growth rate, whereas in the higher growth rate this flux was substantially lower, almost zero for the 60/40 and 40/60 (w/w) glycerol/methanol cultures. Strikingly, the flux through the oxidative branch of the PPP was almost negligible in virtually all tested conditions. The flux through this pathway is generally directly correlated with the biosynthetic demand for NADPH when glucose is used as a C-source [48]. Hence, this points at the existence of alternative NADPH-generating reactions. Pyruvate dehydrogenase or isocitrate dehydrogenase reactions, which are possible alternatives, also showed relatively low fluxes, being insufficient to fulfil the cell's total NADPH requirements. Hence, the activity of other NADPH-generating reactions seems plausible under glycerol/methanol growth conditions. In this context, some Crabtree negative yeasts appear to have alternative mechanisms involved in NADPH$\leftrightarrow$NADH interconversion. For instance, several shuttles involving coupled reactions and resulting in the exchange of redox cofactors between cytosol and mitochondria have been described [49, 50]. Also, the DHA cycle (glycerol-dihydroxyacetone cycle) could be a candidate mechanism, as proposed for *S. cerevisiae* [51]. Or, the S-(hydroxymethyl)glutathione dehydrogenase, involved in the methanol dissimilatory pathway, could use either NAD^+ or $NADP^+$ as acceptor. In fact, estimation of NADH regeneration rates based on the calculated flux data indicates that methanol dissimilation to CO_2, particularly at higher growth rates, results in an overall NADH surplus. Hence, a flux redistribution enabling a high supply of NADPH via methanol dissimilation, while the flux through the energy generating TCA cycle was reduced, might be a plausible explanation for the observed metabolic phenotype. Oxygen availability is another key parameter in *P. pastoris* fermentation processes, as growth on methanol is stoichiometrically linked to heat generation and oxygen uptake, showing significant heat and mass requirements that often become the limiting factor in scale-up. Several oxygen-limited cultivation strategies have been evaluated to limit the oxygen uptake rate (and concomitant required heat removal rate) in PAOX-based systems, particularly for Mut$^+$ *P. pastoris* strains [52, 53], i.e., strains with wild-type *M*ethanol *ut*ilization phenotype (both alcohol oxidase genes, *AOX1* and *AOX2*, are functional).

Also, Baumann and co-workers [54] have shown that hypoxia enhances specific and volumetric productivities of recombinant proteins expressed constitutively under the control of the glycolytic *GAP* promoter when glucose is used as a sole C source. Recently, the impact of reduced oxygen supply on *P. pastoris* growing in glucose-limited chemostats has been characterized systematically in a multi-level genome-scale analysis involving transcriptomics, proteomics, and ^{13}C-based MFA using metabolic flux ratios determined by METAFoR as constraints [13]. In general terms, the metabolic adaptation from oxidative to respiro-fermentative growth, reflected in a reduction of the biomass yield and production of ethanol and arabitol as major by-products, was accompanied by complex carbon flux redistribution throughout the whole central carbon metabolism. Under normoxic conditions there was an important net contribution of the PPP to glucose catabolism (up to 40–50 % of PEP was originated from the pentose phosphate pool), but, as oxygen availability was decreased and, particularly, when arabitol was produced from the pentose phosphate pool (Fig. 1), this contribution was clearly reduced. Such decrease in the relative PPP flux was the result of both an increased glycolytic flux and a decrease in the specific flux through the oxidative branch of the PPP. As previously observed in *S. cerevisiae* [55], the glycolytic flux increased progressively as the oxygen availability decreased. However, whereas the correlation between transcript levels of glycolytic genes and the corresponding in vivo fluxes in *S. cerevisiae* is very weak, *P. pastoris* glycolytic pathway is strongly regulated at the transcriptional level upon oxygen deprivation [20]. As expected, limitation in oxygen availability reduced the respiratory net carbon flux through the TCA cycle and increased the flux through the pyruvate decarboxylase pathway, reflecting the production of ethanol. Moreover, these fluxes were also positively correlated with the levels of the respective transcripts. As observed in the pioneering studies in *P. pastoris* [29, 30], and other yeast (e.g., *S. cerevisiae* and *P. stipitis*) [55, 56], cells growing aerobically in glucose-limited chemostats show a bidirectional transport of OAA and/or other TCA cycle intermediates across the mitochondrial membrane, pointing at the participation of redox shuttles for translocation of NADH [49]. Such interconversion of cytosolic and mitochondrial TCA cycle intermediate pools was decreased under reduced oxygen availability conditions.

More recently, new flux ratios for methanol assimilation pathways have been defined for *P. pastoris* cells growing on an 80/20 (w/w) glucose/methanol mixture, thereby extending the METAFoR formalism [12]. This study has allowed a direct comparison of metabolic flux distributions in cells growing aerobically in carbon-limited chemostats with glucose or glucose/methanol. Specifically, the split ratio between the glycolytic and the oxidative branch of the PPP was clearly shifted to the latter pathway in glucose/methanol-grown cells, probably reflecting the demand of pentose phosphates for methanol assimilation. Also, cells growing on the mixed substrate presented a reduced flux through the TCA cycle. However, most methanol (about 80 %) was directly dissimilated to CO_2, yielding two NADH mol per methanol mol. Overall, this resulted in increased NADH regeneration when methanol was used as a co-substrate. That is, the reduction equivalents produced by direct dissimilation of methanol to CO_2 appear to compensate the decrease in TCA cycle. This pattern is similar to that described above for glycerol/methanol chemostat cultures at the lower dilution rate, where the TCA cycle flux also decreased—in parallel to an increase in the methanol dissimilatory flux to CO_2—when the methanol fraction in the feed was increased [19]. Methanol co-assimilation has an important effect on the bioenergetics of methylotrophic yeasts. Early studies on methanol metabolism showed that the enzymes for primary oxidation and assimilation of methanol or formaldehyde are under inhibitory control of energy equivalents, e.g., NADH and ATP (formaldehyde and formate dehydrogenases), ADP (dihydroxyacetone kinase), and AMP (fructose-1,6-bisphosphatase) [57]. In methylotrophic yeasts, these metabolic pathways operate in accordance with the balance of consumed and produced energy equivalents, thus providing energetic regulation of formaldehyde oxidation and assimilation. Overall, it seems plausible that the glucose-only cultivation is NADH limited and the co-assimilation of methanol as auxiliary substrate may provide an extra direct source of NADH [12, 58].

The metabolic flux profile for glucose/methanol-grown cells obtained from the classic NMR-derived stationary state labelling data of proteinogenic amino acids (METAFoR analysis) has been validated by means of instationary ^{13}C-based MFA using GC-MS and LC-MS [59]. Moreover, this approach revealed additional insights such as the extent of the trehalose-recycling futile cycle (about 1.5 % of the glucose uptake rate) and the reversibility of the glycolytic/gluconeogenesis, pointing at a significant gluconeogenic flux from dihydroxyacetone phosphate/glyceraldehyde pool to glucose-6P. This could be related to carbon recycling from the methanol assimilatory pathway to the pentose phosphate pool. In addition, high exchange fluxes of oxaloacetate with aspartate and malate indicated the activity of the malate/Asp shuttle. Conversely, the ratio of methanol oxidation vs. utilization appeared to be lower than that estimated by the METAFoR-based MFA methodology (54 % vs. 79 % assimilated methanol directly oxidized to CO_2, respectively).

3.2 Effect of Recombinant Protein Production on Central Carbon Metabolism

Omics analytical platforms and systems biology (e.g., genome-scale metabolic modelling) are gaining importance for the quantitative study of *P. pastoris* physiology at a systems level, as well as a source of information for strain engineering (reviewed in ref. 9). Typically, these have been comparative analyses of recombinant producing strains vs. their reference counterparts under well-defined and controlled environmental conditions. In some cases, these comparative omics analyses have identified limited but significant physiological differences in the central carbon metabolism between producing and their corresponding control strains, even in the case that foreign protein amounts were very small relative to the total cell protein. For instance, combined proteomic and classical MFA of *P. pastoris* growing on glucose as C-source indicated that several enzymes involved in the TCA cycle, amino acid biosynthesis, and pentose phosphate pathway show different abundance between a recombinant protein (Fab fragment) producing strain and its wild-type control when cultivated at a lower temperature (20 °C), where q_p (specific productivity) was threefold higher compared to the higher temperature (30 °C) condition [60]. Such increased productivity at lower cultivation temperature was accompanied by a reduced flux through the TCA cycle, as well as reduced levels of proteins involved in energy metabolism (particularly TCA cycle), oxidative stress response, and chaperones. The authors speculated that at decreased temperature, reduced stress due to reduced demand for protein (re)folding and degradation might also lead to lower energy demand, thereby minimizing metabolic burden. In contrast, energy production at 30 °C was increased, but this increase was bigger in the production strain than in the control strain, as reflected in the higher impact of growth temperature on proteins involved in the TCA cycle and energy production, i.e., making metabolic burden evident. Notably, such metabolic burden was not reflected in the flux distribution through the central carbon metabolism when the same strains were compared under hypoxic conditions, where q_p was 2.5-fold higher than in normoxic conditions [13]. Such increase in specific productivity was—at least partially—the result of increased transcriptional levels of glycolytic genes under hypoxic conditions and, therefore, of the product gene under the control of the glycolytic GAP promoter. Nevertheless, these cultivations were performed at an intermediate temperature (25 °C).

The effect of different recombinant protein expression levels was recently compared with ^{13}C-MFA using METAFoR datasets as constraints, using a series of *P. pastoris* strains producing intracellularly a model protein (a bacterial β-aminopeptidase) in glucose batch and fed-batch cultures [5, 10]. The aminopeptidase gene was expressed under the constitutive GAP promoter. These studies enabled to discard a potential metabolic burden related to bottlenecks in the secretory pathway. By using ^{13}C-based MFA in glucose-based fed-batch experiments at a controlled growth rate of 0.12 h^{-1}, where a final β-aminopeptidase concentration of 2.0 g/L was achieved, the recombinant strain showed an increased relative TCA cycle flux compared with the reference strain (from 55 to 75 %, normalized to the glucose uptake rate). This resulted in significantly increased NADH and ATP regeneration rates in the recombinant strain compared to the reference strain. In addition, the recombinant strain had a slight decrease in the biomass yield compared to the reference strain, which correlated with PPP activity. Overall, this suggested an increased demand for energy to compensate the metabolic burden of recombinant protein production. Notably, overexpression of β-aminopeptidase resulted in reduced growth rates in batch cultures (from 0.28 to 0.16 h^{-1}). Such effect could be partially restored by providing all 20 proteinogenic amino acids in the fermentation medium, increasing recombinant product synthesis twofold [10].

The studies by Heyland and co-workers [5, 10] provided direct evidences for a direct response of *P. pastoris*' metabolic network to recombinant protein production, which can not only be explained by the direct resources necessary to produce it. This becomes even more evident in the light of recent ^{13}C-based MFA and metabolomic studies on the metabolic burden caused by secretion of a recombinant lipase from *R. oryzae* (Rol), expressed under the control of the methanol-inducible *AOX1* promoter [12, 38]. This protein has been previously shown to trigger the unfolded protein response in *P. pastoris*, even though it is secreted at relatively moderate levels [61]. In a first experimental approach, quantification of the impact of Rol on the central carbon metabolism was performed by means of a comparative MFA analysis using ^{13}C-NMR-derived constraints (METAFoR) of two recombinant strains secreting different amounts of lipase vs. the corresponding reference strain. These were grown aerobically in C-limited chemostat cultures with an 80/20 (w/w) glucose/methanol mix at a dilution rate of 0.09 h^{-1}. The metabolic impact of Rol secretion could already be inferred from the slightly reduced biomass yield, as well as higher CER and OUR values in the Rol-producing strain, compared to the reference strain. MFA allowed identification of limited but statistically significant changes in the fluxes through the central carbon metabolism. Specifically, the glycolytic flux was increased in the Rol-producing strains compared to the reference strain. Also, the flux of pentose phosphates to the methanol assimilation pathway was significantly decreased in the Rol-producing strains, consistent with lower fluxes of methanol being assimilated to glyceraldehyde 3-phosphate. Remarkably, although the biomass yield of the Rol-producing strains was somewhat lower compared to the reference strain, the flux through the PPP appeared to be constant in all strains. Considering that a correlation between biomass yield and the activity of the oxidative branch of the PPP is usually observed in yeast [48], such effect could be the result of an increased demand of reduction equivalents (NADPH) to regenerate reduced glutathione (GSH) in the endoplasmatic reticulum, the electron donor in the oxidative protein-folding process [62]. This observation would support that oversecretion of a recombinant protein difficult to fold such as Rol may result in an increased oxidation of glutathione [63], thereby increasing NADPH demand to maintain redox balance. Interestingly, secretion of a recombinant enzyme in *Aspergillus niger* also leads to a flux redistribution enabling an elevated supply of NADPH via activation of the PPP and malic enzyme [64]. It is therefore tempting to speculate that the metabolic burden exerted by secreted recombinant proteins may be at least partially related to redox processes involved in protein folding. Further evidence for a metabolic burden derived from Rol expression was indicated by a significantly increased flux through the TCA cycle (normalized to the glucose uptake rate) in the Rol-producing strains, as well as a tendency to increase the flux through the dissimilatory pathway to CO_2. This was reflected in the respiratory rate (CER, OUR values), and leads to a significant increase in NADH regeneration rates in the producing strains compared to the reference. Again, we cannot exclude the possibility that some mechanism involved in NADH → NADPH conversion takes place, as discussed in the previous section.

In a more recent study [38], combined quantitative metabolomic and instationary ^{13}C-MFA has been performed on the same experimental setup (i.e., comparative analyses of a Rol-producing and control strains growing in 80/20 (w/w) glucose/methanol chemostats). Importantly, the INST-^{13}C-MFA methodology yielded coherent results with previous METAFoR-based MFA studies, confirming the limited but significant impact of Rol overexpression on the methanol metabolism and TCA cycle, as well as in energetic parameters. Remarkably, the flux distribution in the trehalose cycle was clearly altered by Rol production. The absolute flux through the trehalose cycle was significantly higher (twofold) in the Rol-expressing strain. Furthermore, this was consistent with the twofold higher trehalose intracellular levels observed in this strain, further supporting increased recycling of this molecule via an ATP-futile cycle [65]. Higher trehalose levels could confer increased protection under stress conditions derived from the UPR, e.g., contributing to avoid death under UPR stress. Based on the ^{13}C-flux estimations (which only balance carbon and labelling), the cofactor balances were recalculated. The energy required for maintenance in the Rol-producing strain was slightly higher compared with the one reported for the control strain.

Nevertheless, this difference was not statistically significant due to the high relative standard deviation of this calculation.

In summary, although the precise reasons behind metabolic burden caused by recombinant protein production, and particularly secretion, remain largely unknown, recent fluxomic and metabolomic studies are bringing some light, pointing at increased energy requirements, e.g., to meet protein folding and ER stress demands.

The major outcomes of the ^{13}C-MFA studies so far reported on *P. pastoris* are summarized in Table 1.

4 Conclusions and Outlook

Systems biology and particularly ^{13}C-based MFA have been proven to be a valuable tool to analyze quantitatively the metabolic burden caused by protein production and secretion. It has provided direct evidences for a direct response of *P. pastoris* metabolic network to recombinant protein production, which can not only be explained by the direct resources necessary to produce it, but rather points at a cross talk with other cellular processes such as protein folding and related stress responses. Yet, ^{13}C-MFA is still limited to rather reduced metabolic networks. Quantitative metabolomics and instationary ^{13}C-MFA offers the potential to characterize metabolic fluxes over extended metabolic networks. Nevertheless, subcellular compartmentalization of metabolites remains challenging [66]. Quantitative metabolomics of cofactors (glutathione, $NAD(P)^+/NAD(P)H$, ATP/ADP/AMP) to allow for deeper understanding of the correlation between the flux distribution through the central carbon metabolism, energy status, and redox balance is needed. In this respect, important developments are being made available for *P. pastoris* metabolomics [67]. Three genome-scale metabolic models of *P. pastoris* have been recently published [68–70], and compared in a recent review [71]. These models have been used to predict recombinant protein productivity under different environmental conditions such as oxygen availability, or cell growth under different substrate mixtures. They provide a valuable tool for exploring and understanding the metabolic potential of *P. pastoris*. Moreover, they can be a scaffold for multilevel data (i.e., multi-omics studies) integration to facilitate data interpretation and analysis, and ultimately for simulation and in silico strain design improvement strategies. Although no ^{13}C-MFA-based metabolic engineering strategies for improved protein production have been so far reported for *P. pastoris*, the reported studies clearly point at novel targets in the central carbon metabolism/energy metabolism, so a significant impact of systems metabolic engineering on *P. pastoris* research and strain improvement programs can be expected.

Table 1
Summary of reported ^{13}C-based MFA studies of *Pichia pastoris*. Experimental setup, methodology, and major outcomes are outlined

Aim and overall experimental design	Cultivation conditions and ^{13}C-MFA methodology	Description	Reference
Impact of specific growth rate and glycerol/methanol ratio	Glycerol and glycerol/methanol (80/20, 60/40, and 40/60, w/w) chemostat cultures at 0.05 and 0.16 h^{-1}. Mut^+ strain METAFoR analysis based on NMR-based labelling data from proteinogenic amino acids	Correlation of relative TCA cycle activity and growth rate: Increased relative anaplerotic supply of the TCA cycle (i.e., reduced relative TCA cycle activity) and largely unidirectional flux of cytosolic oxaloacetate (OAA) to the mitochondria at higher dilution rates compared to fast exchange of OAA cytosolic and mitochondrial pools at lower growth rates. The dependency of the flux ratios involving TCA cycle intermediates on the varying fraction of methanol in the medium is less pronounced.	[29, 30]
Impact of specific growth rate and glycerol/methanol ratio	Glycerol and glycerol/methanol (80/20, 60/40, and 40/60, w/w) chemostat cultures at 0.05 and 0.16 h^{-1}. Mut^+ strain MFA based on iterative fitting of NMR-derived ^{13}C-labelling data from proteinogenic amino acids	Split ratio between methanol assimilatory and dissimilatory pathways correlated with growth rate: 50–70 % of methanol is dissimilated directly to CO_2 at the lower growth rate, whereas almost 100 % of methanol is dissimilated at the higher growth rate. Flux of pentose phosphate to the methanol assimilation pathway correlated with the methanol fraction in the feed. Negligible flux through the oxidative branch of the PPP under virtually all conditions points at other NADPH-generating reactions.	[19]
Impact of methanol and glucose co-assimilation on central carbon metabolism	Glucose and glucose/methanol (80/20, w/w) chemostat cultures at 0.09 h^{-1}. Mut^+ strain ^{13}C-constrained MFA using METAFoR data derived from labelling patterns of proteinogenic amino acids analyzed by NMR (glucose/methanol fluxes validated by instationary ^{13}C-MFA using GC-MS and LC-MS data)	Split ratio between glycolytic and the oxidative branch of the PPP shifted to the latter pathway in glucose/methanol-grown cells (about 55 % of the glucose-6-P enters the oxidative branch of PPP). Reduced flux through the TCA cycle in cells growing on mixed substrates, but most of the methanol (80 % according to the NMR-based methodology or 54 % according to the MS-based methodology). This results in increased NADH regeneration when methanol is used as co-substrate. Methanol co-assimilation results in a significant gluconeogenic flux to recycle carbon from the methanol assimilatory pathway to the pentose phosphate pool.	[12, 59]

(continued)

Table 1 (continued)

Aim and overall experimental design	Cultivation conditions and ^{13}C-MFA methodology	Description	Reference
Impact of oxygen availability	Glucose chemostat cultures at 0.1 h^{-1}. ^{13}C-constrained MFA using METAFoR data derived from labelling patterns of proteinogenic amino acids analyzed by NMR Multi-level transcriptomic and proteomic analysis	Reduced oxygen availability (hypoxia) leads to increased glycolytic and TCA cycle flux, decreased flux through the oxidative branch PPP, as well as shift to respiro-fermentative metabolism (ethanol and arabitol formation). Strong correlation of glycolytic and TCA cycle fluxes with corresponding transcript levels. Bidirectional transport of oxaloacetate and/or other TCA cycle intermediates across mitochondrial membrane, but anaplerotic flux through the pyruvate carboxylase pathway decreases when oxygenation is reduced.	[13]
Impact of intracellular recombinant protein production on metabolism	Glucose fed-batch cultures at a controlled growth rate (0.12 h^{-1}). ^{13}C-constrained MFA using METAFoR data derived from labelling patterns of proteinogenic amino acids analyzed by GC-MS	Increased TCA cycle flux and ATP regeneration rate in the recombinant strain compared to the reference strain. Correlation of biomass yield and PPP (slightly decreased in the producing strain compared to the reference strain).	[5]
Impact of intracellular recombinant protein production on metabolism	Glucose batch cultures (shake flasks), with ^{13}C labelling under pseudo-steady state (exponential growth-phase) conditions Comparison of three strains producing different amounts of recombinant protein plus reference strain. METAFoR analysis and ^{13}C-constrained MFA using METAFoR data derived from labelling patterns of proteinogenic amino acids analyzed by GC-MS	Increasing recombinant protein production results in reduced growth rate but increased biomass yield. Additional resources required for protein synthesis are apparently compensated by reducing by-product formation and by increasing energy generation via the TCA cycle (reflected in an increased respiration rate). However, the absolute net flux through TCA cycle was constant irrespective of the strain, suggesting an upper limit of TCA cycle activity. Reduced growth rate partially restored by amino acid supplementation in the growth medium.	[10]

Impact of recombinant protein secretion on metabolism	Glucose/methanol (80/20, w/w) chemostat cultures at 0.09^{-1}. Mut^+ lipase-producing and reference strains ^{13}C-constrained MFA using METAFoR data derived from labelling patterns of proteinogenic amino acids analyzed by NMR	Increased glycolytic and TCA cycle fluxes in the producing strains compared to the reference strain. Decreased flux of pentose phosphate to the methanol assimilation pathway in the producing strains, consistent with higher fluxes through the methanol dissimilatory pathway to CO_2. Flux through the oxidative branch of the PPP was constant in all strains, but the biomass yield of the producing strains was slightly lower, thereby pointing at an increased NADPH demand in the producing strains. Overall increased NADH regeneration rates in the producing strains compared to the reference strain.	[12]
Impact of recombinant protein secretion on metabolism	Glucose/methanol (80/20, w/w) chemostat cultures at 0.09^{-1}. Mut^+ lipase-producing and reference strains Instationary ^{13}C-MFA based on MS data Multi-level metabolomics study	Carbon flux distributions coherent with previous ^{13}C-constrained MFA using ^{13}C-NMR data. Twofold higher absolute flux through the trehalose cycle (i.e., an ATP-futile cycle) in the producing strain. Results consistent with twofold higher trehalose levels in this strain. Energy required for maintenance slightly higher in the producing strain in relation to the reference strain.	[38]

Acknowledgement

We appreciate the support by the Spanish Ministry of Science and Innovation (CICYT project CTQ2010-15131) and the Catalan Government (contract grant 2009-SGR-281, Xarxa de Referència en Biotecnologia and AGAUR Project grant 2010 CONE3-00063).

References

1. Gorgens JF, van Zyl WH, Knoetze JH, Hahn-Hagerdal B (2001) The metabolic burden of the PGK1 and ADH2 promoter systems for heterologous xylanase production by *Saccharomyces cerevisiae* in defined medium. Biotechnol Bioeng 73:238–245
2. Vigentini I, Brambilla L, Branduardi P, Merico A, Porro D, Compagno C (2005) Heterologous protein production in *Zygosaccharomyces bailii*: physiological effects and fermentative strategies. FEMS Yeast Res 5:647–652
3. Cos O, Serrano A, Montesinos JL, Ferrer P, Cregg JM, Valero F (2005) Combined effect of the methanol utilization (Mut) phenotype and gene dosage on recombinant protein production in *Pichia pastoris* fed-batch cultures. J Biotechnol 116:321–335
4. Krogh AM, Beck V, Christensen LH, Henriksen CM, Møller K, Olsson L (2008) Adaptation of *Saccharomyces cerevisiae* expressing a heterologous protein. J Biotechnol 137:28–33
5. Heyland J, Fu J, Blank LM, Schmid A (2010) Quantitative physiology of *Pichia pastoris* during glucose-limited high-cell density fed-batch cultivation for recombinant protein production. Biotechnol Bioeng 107:357–368
6. Gasser B, Saloheimo M, Rinas U, Dragosits M, Rodriguez-Carmona E, Baumann K, Giuliani M, Parrilli E, Branduardi P, Lang C et al (2008) Protein folding and conformational stress in microbial cells producing recombinant proteins: a host comparative overview. Microb Cell Fact 7:11
7. Idiris A, Tohda H, Kumagai H, Takegawa K (2010) Engineering of protein secretion in yeast: strategies and impact on protein production. Appl Microbiol Biotechnol 86:403–417
8. Damasceno LM, Huang CJ, Batt CA (2012) Protein secretion in *Pichia pastoris* and advances in protein production. Appl Microbiol Biotechnol 93:31–39
9. Gasser B, Prielhofer R, Marx H, Maurer M, Nocon J, Steiger M, Puxbaum V, Sauer M, Mattanovich D (2013) *Pichia pastoris*: protein production host and model organism for biomedical research. Future Microbiol 8:191–208
10. Heyland J, Fu J, Blank LM, Schmid A (2011) Carbon metabolism limits recombinant protein production in *Pichia pastoris*. Biotechnol Bioeng 108:1942–1953
11. Carnicer M, ten Pierick A, van Dam J, Heijnen JJ, Albiol J, van Gulik W, Ferrer P (2012) Quantitative metabolomics analysis of amino acid metabolism in recombinant *Pichia pastoris* under different oxygen availability conditions. Microb Cell Fact 11:83
12. Jordà J, Jouhten P, Cámara E, Maaheimo H, Albiol J, Ferrer P (2012) Metabolic flux profiling of recombinant protein secreting *Pichia pastoris* growing on glucose:methanol mixtures. Microb Cell Fact 11:57
13. Baumann K, Carnicer M, Dragosits M, Graf AB, Stadlmann J, Jouhten P, Maaheimo H, Gasser B, Albiol J, Mattanovich D et al (2010) A multi-level study of recombinant *Pichia pastoris* in different oxygen conditions. BMC Syst Biol 4:141
14. Sauer U (2006) Metabolic networks in motion: ^{13}C-based flux analysis. Mol Syst Biol 2:62
15. Kohlstedt M, Becker J, Wittmann C (2010) Metabolic fluxes and beyond-systems biology understanding and engineering of microbial metabolism. Appl Microbiol Biotechnol 88: 1065–1075
16. Xu Z, Tsurugi K (2006) A potential mechanism of energy-metabolism oscillation in an aerobic chemostat culture of the yeast *Saccharomyces cerevisiae*. FEBS J 273:1696–1709
17. Hans MA, Heinzle E, Wittmann C (2003) Free intracellular amino acid pools during autonomous oscillations in *Saccharomyces cerevisiae*. Biotechnol Bioeng 82:143–151
18. Marx A, de Graaf AA, Wiechert W, Eggeling L, Sahm H (1996) Determination of the fluxes in the central metabolism of *Corynebacterium glutamicum* by nuclear magnetic resonance spectroscopy combined with metabolite balancing. Biotechnol Bioeng 49:111–129
19. Jordà J, Santos de Jesus S, Peltier S, Ferrer P, Albiol J (2014) Metabolic flux analysis of recombinant *Pichia pastoris* growing on different glycerol/methanol mixtures by iterative fitting of

NMR-derived ^{13}C-labelling data from proteinogenic amino acids. New Biotechnol 31(1): 120–132

20. Baumann K, Dato L, Graf AB, Frascotti G, Dragosits M, Porro D, Mattanovich D, Ferrer P, Branduardi P (2011) The impact of oxygen on the transcriptome of recombinant *S. cerevisiae* and *P. pastoris*—a comparative analysis. BMC Genomics 12:218
21. Unrean P, Nguyen NH (2012) Metabolic pathway analysis of *Scheffersomyces (Pichia) stipitis*: effect of oxygen availability on ethanol synthesis and flux distributions. Appl Microbiol Biotechnol 94:1387–1398
22. Celik E, Calik P, Oliver SG (2010) Metabolic flux analysis for recombinant protein production by *Pichia pastoris* using dual carbon sources: effects of methanol feeding rate. Biotechnol Bioeng 105:317–329
23. Tortajada M, Llaneras F, Picó J (2010) Validation of a constraint-based model of *Pichia pastoris* metabolism under data scarcity. BMC Syst Biol 4:115
24. Tortajada M, Llaneras F, Ramón D, Picó J (2012) Estimation of recombinant protein production in *Pichia pastoris* based on a constraint-based model. J Process Control 22: 1139–1151
25. Llaneras F, Picó J (2007) A procedure for the estimation over time of metabolic fluxes in scenarios where measurements are uncertain and/or insufficient. BMC Bioinformatics 8:421
26. Szyperski T (1995) Biosynthetically directed fractional ^{13}C-labeling of proteinogenic amino acids. An efficient analytical tool to investigate intermediary metabolism. Eur J Biochem 232: 433–448
27. Szyperski T, Glaser RW, Hochuli M, Fiaux J, Sauer U, Bailey JE, Wuthrich K (1999) Bioreaction network topology and metabolic flux ratio analysis by biosynthetic fractional ^{13}C labeling and two-dimensional NMR spectroscopy. Metab Eng 1:189–197
28. Sauer U, Lasko DR, Fiaux J, Hochuli M, Glaser R, Szyperski T, Wuthrich K, Bailey JE (1999) Metabolic flux ratio analysis of genetic and environmental modulations of *Escherichia coli* central carbon metabolism. J Bacteriol 181:6679–6688
29. Solà A, Jouhten P, Maaheimo H, Sanchez-Ferrando F, Szyperski T, Ferrer P (2007) Metabolic flux profiling of *Pichia pastoris* grown on glycerol/methanol mixtures in chemostat cultures at low and high dilution rates. Microbiology 153:281–290
30. Solà A, Maaheimo H, Ylonen K, Ferrer P, Szyperski T (2004) Amino acid biosynthesis and metabolic flux profiling of *Pichia pastoris*. Eur J Biochem 271:2462–2470
31. Fischer E, Zamboni N, Sauer U (2004) High-throughput metabolic flux analysis based on gas chromatography-mass spectrometry derived ^{13}C constraints. Anal Biochem 325:308–316
32. Wiechert W, Mollney M, Isermann N, Wurzel M, de Graaf AA (1999) Bidirectional reaction steps in metabolic networks: III. Explicit solution and analysis of isotopomer labeling systems. Biotechnol Bioeng 66:69–85
33. van Winden WA, Heijnen JJ, Verheijen PJ (2002) Cumulative bondomers: a new concept in flux analysis from 2D [^{13}C,^{1}H] COSY NMR data. Biotechnol Bioeng 80:731–745
34. Antoniewicz MR, Kelleher JK, Stephanopoulos G (2007) Elementary metabolite units (EMU): a novel framework for modeling isotopic distributions. Metab Eng 9:68–86
35. Nöh K, Wiechert W (2006) Experimental design principles for isotopically instationary ^{13}C labeling experiments. Biotechnol Bioeng 94:234–251
36. Wiechert W, Wurzel M (2001) Metabolic isotopomer labeling systems. Part I: global dynamic behavior. Math Biosci 169:173–205
37. Carnicer M, Canelas AB, ten Pierick A, Zeng Z, van Dam J, Albiol J, Ferrer P, Heijnen JJ, van Gulik W (2012) Development of quantitative metabolomics for *Pichia pastoris*. Metabolomics 8:284–298
38. Jordà J, Suarez C, Carnicer M, ten Pierick A, Heijnen JJ, van Gulik W, Wahl A, Ferrer P, Albiol J (2014) Quantitative metabolomics and instationary ^{13}C-metabolic flux analysis reveals impact of recombinant protein production on trehalose and energy metabolism in *Pichia pastoris*. Metabolites 4(2):281–299
39. Hensing MC, Rouwenhorst RJ, Heijnen JJ, van Dijken JP, Pronk JT (1995) Physiological and technological aspects of large-scale heterologous-protein production with yeasts. Antonie Van Leeuwenhoek 67:261–279
40. Resina D, Cos O, Ferrer P, Valero F (2005) Developing high cell density fed-batch cultivation strategies for heterologous protein production in *Pichia pastoris* using the nitrogen source-regulated FLD1 Promoter. Biotechnol Bioeng 91:760–767
41. Maurer M, Kuhleitner M, Gasser B, Mattanovich D (2006) Versatile modeling and optimization of fed batch processes for the production of secreted heterologous proteins with *Pichia pastoris*. Microb Cell Fact 5:37
42. Cos O, Ramon R, Montesinos JL, Valero F (2006) Operational strategies, monitoring and control of heterologous protein production in

the methylotrophic yeast *Pichia pastoris* under different promoters: a review. Microb Cell Fact 5:17
43. Ramon R, Ferrer P, Valero F (2007) Sorbitol co-feeding reduces metabolic burden caused by the overexpression of a *Rhizopus oryzae* lipase in *Pichia pastoris*. J Biotechnol 130:39–46
44. Jungo C, Marison I, von Stockar U (2007) Mixed feeds of glycerol and methanol can improve the performance of *Pichia pastoris* cultures: a quantitative study based on concentration gradients in transient continuous cultures. J Biotechnol 128:824–837
45. Arnau C, Casas C, Valero F (2011) The effect of glycerol mixed substrate on the heterologous production of a *Rhizopus oryzae* lipase in *Pichia pastoris* system. Biochem Eng J 57:30–37
46. Wang Z, Wang Y, Zhang D, Li J, Hua Z, Du G, Chen J (2010) Enhancement of cell viability and alkaline polygalacturonate lyase production by sorbitol co-feeding with methanol in *Pichia pastoris* fermentation. Bioresour Technol 101:1318–1323
47. Blank LM, Sauer U (2004) TCA cycle activity in *Saccharomyces cerevisiae* is a function of the environmentally determined specific growth and glucose uptake rates. Microbiology 150: 1085–1093
48. Blank LM, Lehmbeck F, Sauer U (2005) Metabolic-flux and network analysis in fourteen hemiascomycetous yeasts. FEMS Yeast Res 5:545–558
49. Bakker BM, Overkamp KM, van Maris AJ, Kotter P, Luttik MA, van Dijken JP, Pronk JT (2001) Stoichiometry and compartmentation of NADH metabolism in *Saccharomyces cerevisiae*. FEMS Microbiol Rev 25:15–37
50. Valadi A, Granath K, Gustafsson L, Adler L (2004) Distinct intracellular localization of Gpd1p and Gpd2p, the two yeast isoforms of NAD^+-dependent glycerol-3-phosphate dehydrogenase, explains their different contributions to redox-driven glycerol production. J Biol Chem 279:39677–39685
51. Celton M, Goelzer A, Camarasa C, Fromion V, Dequin S (2012) A constraint-based model analysis of the metabolic consequences of increased NADPH oxidation in *Saccharomyces cerevisiae*. Metab Eng 14:366–379
52. Charoenrat T, Ketudat-Cairns M, Stendahl-Andersen H, Jahic M, Enfors SO (2005) Oxygen-limited fed-batch process: an alternative control for *Pichia pastoris* recombinant protein processes. Bioprocess Biosyst Eng 27: 399–406
53. Berdichevsky M, d'Anjou M, Mallem MR, Shaikh SS, Potgieter TI (2011) Improved production of monoclonal antibodies through oxygen-limited cultivation of glycoengineered yeast. J Biotechnol 155:217–224
54. Baumann K, Maurer M, Dragosits M, Cos O, Ferrer P, Mattanovich D (2008) Hypoxic fed-batch cultivation of *Pichia pastoris* increases specific and volumetric productivity of recombinant proteins. Biotechnol Bioeng 100:177–183
55. Jouhten P, Rintala E, Huuskonen A, Tamminen A, Toivari M, Wiebe M, Ruohonen L, Penttilä M, Maaheimo H (2008) Oxygen dependence of metabolic fluxes and energy generation of *Saccharomyces cerevisiae* CEN.PK113-1A. BMC Syst Biol 2:60
56. Fiaux J, Cakar ZP, Sonderegger M, Wuthrich K, Szyperski T, Sauer U (2003) Metabolic-flux profiling of the yeasts *Saccharomyces cerevisiae* and *Pichia stipitis*. Eukaryot Cell 2:170–180
57. Trotsenko YA, Bystrykh LV, Ubiyvovk VM (1984) Regulatory aspects of methanol metabolism in yeasts. In: Crawford RL, Hanson RS (eds) Proceedings of the 4th international symposium on microbial growth on one carbon compounds. ASM Press, Washington, DC, pp 118–122
58. Babel W, Muller RH (1985) Mixed substrate utilization in microorganisms—biochemical aspects and energetics. J Gen Microbiol 131:39–45
59. Jordà J, Suarez CA, Carnicer M, ten Pierick A, Heijnen JJ, van Gulik W, Ferrer P, Albiol J, Wahl A (2013) Glucose-methanol co-utilization in *Pichia pastoris* studied by metabolomics and instationary ^{13}C flux analysis. BMC Syst Biol 7:17
60. Dragosits M, Stadlmann J, Albiol J, Baumann K, Maurer M, Gasser B, Sauer M, Altmann F, Ferrer P, Mattanovich D (2009) The effect of temperature on the proteome of recombinant *Pichia pastoris*. J Proteome Res 8: 1380–1392
61. Resina D, Bollok M, Khatri NK, Valero F, Neubauer P, Ferrer P (2007) Transcriptional response of *P. pastoris* in fed-batch cultivations to *Rhizopus oryzae* lipase production reveals UPR induction. Microb Cell Fact 6:21
62. Delic M, Rebnegger C, Wanka F, Puxbaum V, Haberhauer-Troyer C, Hann S, Kollensperger G, Mattanovich D, Gasser B (2012) Oxidative protein folding and unfolded protein response elicit differing redox regulation in endoplasmic reticulum and cytosol of yeast. Free Radic Biol Med 52:2000–2012
63. Zhu T, Guo M, Zhuang Y, Chu J, Zhang S (2011) Understanding the effect of foreign gene dosage on the physiology of *Pichia pastoris* by transcriptional analysis of key genes. Appl Microbiol Biotechnol 89:1127–1135
64. Driouch H, Melzer G, Wittmann C (2012) Integration of in vivo and in silico metabolic

fluxes for improvement of recombinant protein production. Metab Eng 14:47–58

65. Parrou JL, Teste MA, Francois J (1997) Effects of various types of stress on the metabolism of reserve carbohydrates in *Saccharomyces cerevisiae*: genetic evidence for a stress-induced recycling of glycogen and trehalose. Microbiology 143:1891–1900
66. Niklas J, Schneider K, Heinzle E (2010) Metabolic flux analysis in eukaryotes. Curr Opin Biotechnol 21:63–69
67. Haberhauer-Troyer C, Delic M, Gasser B, Mattanovich D, Hann S, Koellensperger G (2013) Accurate quantification of the redox-sensitive GSH/GSSG ratios in the yeast *Pichia pastoris* by HILIC-MS/MS. Anal Bioanal Chem 405:2031–2039
68. Sohn SB, Graf AB, Kim TY, Gasser B, Maurer M, Ferrer P, Mattanovich D, Lee SY (2010) Genome-scale metabolic model of methylotrophic yeast *Pichia pastoris* and its use for in silico analysis of heterologous protein production. Biotechnol J 5:705–715
69. Chung BK, Selvarasu S, Andrea C, Ryu J, Lee H, Ahn J, Lee H, Lee DY (2010) Genome-scale metabolic reconstruction and in silico analysis of methylotrophic yeast *Pichia pastoris* for strain improvement. Microb Cell Fact 9:50
70. Caspeta L, Shoaie S, Agren R, Nookaew I, Nielsen J (2012) Genome-scale metabolic reconstructions of *Pichia stipitis* and *Pichia pastoris* and in silico evaluation of their potentials. BMC Syst Biol 6:24
71. Chung BK, Lakshmanan M, Klement M, Ching CB, Lee DY (2013) Metabolic reconstruction and flux analysis of industrial *Pichia* yeasts. Appl Microbiol Biotechnol 97:1865–1873

INDEX

Jens O. Krömer et al. (eds.), *Metabolic Flux Analysis: Methods and Protocols*, Methods in Molecular Biology, vol. 1191, DOI 10.1007/978-1-4939-1170-7, © Springer Science+Business Media New York 2014

MIX
Papier aus verantwortungsvollen Quellen
Paper from responsible sources
FSC® C105338

If you have any concerns about our products,
you can contact us on
ProductSafety@springernature.com

In case Publisher is established outside the EU,
the EU authorized representative is:
Springer Nature Customer Service Center GmbH
Europaplatz 3, 69115 Heidelberg, Germany

Printed by Libri Plureos GmbH
in Hamburg, Germany